Computational Statistical Physics

Springer
Berlin
Heidelberg
New York
Barcelona
Hong Kong
London
Milan
Paris
Tokyo

Physics and Astronomy | **ONLINE LIBRARY**

http://www.springer.de/phys/

Karl Heinz Hoffmann   Michael Schreiber

Computational Statistical Physics

From Billiards to Monte Carlo

With 141 Figures

Springer

Professor Karl Heinz Hoffmann
Professor Michael Schreiber
Institut für Physik
Technische Universität Chemnitz
09107 Chemnitz

hoffmann@physik.tu-chemnitz.de
schreiber@physik.tu-chemnitz.de

Cover picture see page 20

Library of Congress Cataloging in Publication Data applied for.

Die Deutsche Bibliothek - CIP-Einheitsaufnahme
Hoffmann, Karl Heinz: Computational statistical physics : from billiards to Monte Carlo / Karl Heinz Hoffmann ; Michael Schreiber. - Berlin ; Heidelberg ; New York ; Barcelona ; Hong Kong ; London ; Milan ; Paris ; Tokyo : Springer, 2002 (Physics and astronomy online library) ISBN 3-540-42160-2

ISBN 3-540-42160-2 Springer-Verlag Berlin Heidelberg New York

Springer-Verlag Berlin Heidelberg New York
a member of BertelsmannSpringer Science+Business Media GmbH

http://www.springer.de

© Springer-Verlag Berlin Heidelberg 2002
Printed in Germany

Typesetting by the authors using a Springer TeX macro package.
Cover design: *design & production* GmbH, Heidelberg

Printed on acid-free paper   SPIN 10832556   57/3141/mf   5 4 3 2 1 0

Preface

Physics deals with systems ranging from one or a few elementary particles and elementary interactions to large scale systems like the climate or even stars. Although in principle each of the macroscopic systems can be described on a microscopic level, that would lead to a huge number of variables and make a detailed treatment hopeless.

However, in many cases, if not in most cases, a very small number of variables suffice to describe the system reasonably well. The reason is that many degrees of freedom are not individually relevant, but only their overall behavior is of significance at the macroscopic level. Statistical physics provides a conceptual answer to the problem of handling the majority of the degrees of freedom, namely by treating them in a statistical way. This reduction in the number of degrees of freedom has made it possible to gain a deep understanding of many different physical systems such as radiation, gases, or granular matter.

The power of statistical physics has increased significantly with the advent of computers. The possibility of simulating physical systems, rather than just describing them, has opened up a whole new range of methods and allows us nowadays to analyze the properties of large systems quantitatively.

Hence with the advance of computational resources, the area of application of today's statistical physics has grown wider than ever. New methods have been developed enabling us to treat systems which were previously not open to analysis owing to the limited power of available analytical methods. Nowadays systems can be analyzed which allow no solution in closed form, i.e., over-complicated terms which are interesting and important need no longer be omitted from the defining equations of the model. In addition to the available solutions, the advent of computers and their graphics capabilities has allowed the visualization of many phenomena. This in turn has improved insight into the physics and demonstrated its beauty in many cases.

While there are certainly several good textbooks for the field of statistical physics as a whole, and for parts of it, numerical methods and topical developments which are impossible without the massive use of computers are usually inadequately presented in textbooks, if they are presented at all. On the other hand, available review articles usually concentrate on one particular field of research. In this book, we try to fill this gap by covering a large number of topics in computational statistical physics on an introductory level, as well as leading the reader to related research problems currently under investigation.

The book is aimed at graduate students and researchers and tries to reveal broad areas in which methods of statistical physics are now applied. It is especially devoted to computational approaches. In addition, the book demonstrates that the powerful methods developed in computational statistical physics are now successfully used in neighbouring areas and even in several rather different fields. Subjects range from the application of statistical methods in solid state physics to their application in financial modelling, from neural networks to billiards and from traffic modelling to game theory.

The easy-to-read articles emphasize topical computational methods and give readers ready access to the most advanced methods used today in these fields. Contributions by the various authors, experts in the relevant fields, are arranged in such a way that this percolation of important physical methods into other areas becomes apparent, showing how interdisciplinary science is progressing today. One example is provided by methods developed to analyse chaotic motion in billiards, which are then used in the optimization of combinatorial problems. Another example is the application of Monte Carlo methods such as the Metropolis algorithm, which is then used in simulated annealing to solve important industrial optimization problems. The analysis of time series is a further example which shows how methods developed in statistical physics, e.g., to analyse temperature profiles, can be used in the description of financial time series. Likewise, random matrix theory and the statistical analysis of eigenvalue distributions has become a powerful tool, not only in statistical physics, but also in semiconductor physics and metal physics, describing metal–insulator transitions, electrons in quantum dots and quasicrystals.

Several authors have provided simple programs for training purposes. These programs are not included in this book, but we have archived them at

http://www.tu-chemnitz.de/physik/HERAEUS/2000/Springer.html

so that they can be downloaded. In this way it is also possible to update the programs and data, and offer new links to topical developments for the various subjects.

The articles are based on topics which were presented to about 50 students who took part in a Heraeus Physics School, organized at the Chemnitz University of Technology in the fall of 2000. We thank all the lecturers for their contributions, which made it possible for us to provide our students with this wide overview of the field of computational statistical physics. Knowing the shortage of time and the many obligations our coauthors have, we would like to express our gratitude to our colleagues who have found time among many administrative, research, and teaching duties to prepare the high quality articles presented here. Finally, we would like to thank Peter Blaudeck, Angelique Gaida, Karin Nerger and Steffen Seeger for their technical help, and Springer-Verlag for making this volume a reality. Last, but not least, we thank the Wilhelm und Else Heraeus-Stiftung for the generous financial support which has made the school possible and which has also helped to make this book available.

Chemnitz,   *Karl Heinz Hoffmann*
June 2001   *Michael Schreiber*

Contents

List of Contributors

Marcel Ausloos
GRASP[1] and SUPRAS[2]
B5 Sart Tilman Campus
B-4000, Liège, Belgium
marcel.ausloos@ulg.ac.be

Robert Barlović
Physik von Transport und Verkehr
Gerhard-Mercator-Universität
Duisburg
Lotharstr. 1
D-47048 Duisburg, Germany
barlovic@uni-duisburg.de

Barbara Drossel
Physics Department
Sackler Faculty of Exact Sciences
Tel Aviv University
Postbox 39040
Tel Aviv 69978, Israel
barbara@gina.tau.ac.il

Werner Ebeling
Institute of Physics
Humboldt–University Berlin
Invalidenstr. 110
D-10115 Berlin, Germany
ebeling@physik.hu-berlin.de

Andreas Engel
Institut für Theoretische Physik
Otto-von-Guericke-Universität
Postfach 4120
D-39016 Magdeburg, Germany
andreas.engel@physik.
uni-magdeburg.de

Peter Grassberger
John von Neumann-Institut für
Computing
Forschungszentrum Jülich
D-52425 Jülich, Germany
p.grassberger@fz-juelich.de

Uwe Grimm
Applied Mathematics Department
Faculty of Mathematics
and Computing
The Open University
Milton Keynes MK7 6AA, U.K.
u.g.grimm@open.ac.uk

Karl Heinz Hoffmann
Institut für Physik
Technische Universität Chemnitz
D-09107 Chemnitz, Germany
hoffmann@physik.tu-chemnitz.de

---

[1] Group for Research in Applied Statistical Physics
[2] Services Universitaires Pour la Recherche et les Applications en Supraconductivité

Wolfgang Kinzel
Institut für Theoretische Physik
Universität Würzburg
Am Hubland
D-97074 Würzburg, Germany
kinzel@physik.uni-wuerzburg.de

Hubert Klüpfel
Physik von Transport und Verkehr
Gerhard-Mercator-Universität
Duisburg
Lotharstr. 1
D-47048 Duisburg, Germany
kluepfel@uni-duisburg.de

Wolfgang Knospe
Physik von Transport und Verkehr
Gerhard-Mercator-Universität
Duisburg
Lotharstr. 1
D-47048 Duisburg, Germany
knospe@uni-duisburg.de

Hans Jürgen Korsch
Fachbereich Physik
Universität Kaiserslautern
D-67653 Kaiserslautern, Germany
korsch@physik.uni-kl.de

Bernhard Kramer
I. Institut für Theoretische Physik
Universität Hamburg
Jungiusstraße 9
D-20355 Hamburg, Germany
kramer@physnet.uni-hamburg.de

Frank Milde
Robert Bosch GmbH
K5/ESM5, Building Fe053/1
P.O. Box 300220
D-70442 Stuttgart, Germany
frank.milde@de.bosch.com

Ingo Morgenstern
Institut für Theoretische Physik
Universität Regensburg
Universitätsstr. 31
D-93053 Regensburg, Germany
ingo.morgenstern@physik.
uni-regensburg.de

Walter Nadler
John von Neumann-Institut für
Computing
Forschungszentrum Jülich
D-52425 Germany
w.nadler@fz-juelich.de

Rudolf A. Römer
Institut für Physik
Technische Universität
D-09107 Chemnitz, Germany
rar@physik.tu-chemnitz.de

Pál Ruján
SER Systems AG
Knowledge Processing Team
Sandweg 236
D-26135 Oldenburg, Germany
pal.rujan@ser.de

Erich Runge
Institut für Physik
Humboldt-Universität zu Berlin
Hausvogteiplatz 5-7
D-10117 Berlin, Germany
runge@physik.hu-berlin.de

Klaus Schenk
Lehrstuhl 5 für Theoretische Physik
Physik Department
TU München
James-Franck-Str.
D-85747 Garching, Germany
klaus.schenk@epost.de

Johannes Schneider
Physik-Institut
Universität Zürich-Irchel
Winterthurerstrasse 190
CH-8057 Zürich, Switzerland
jsch@physik.unizh.ch

Michael Schreckenberg
Physik von Transport und Verkehr
Gerhard-Mercator-Universität
Duisburg
Lotharstr. 1
D-47048 Duisburg, Germany
schreck@uni-duisburg.de

Michael Schreiber
Institut für Physik
Technische Universität
D-09107 Chemnitz, Germany
schreiber@physik.tu-chemnitz.de

Franz Schwabl
Institut für theoretische Physik
Physik Department T34
TU München
James-Franck-Str.
D-85747 Garching, Germany
schwabl@ph.tum.de

Thomas Vojta
Theoretical Physics
University of Oxford
Oxford OX3 1NP, UK
vojta@thphys.ox.ac.uk

Frank Zimmer
Fachbereich Physik
Universität Kaiserslautern
D-67653 Kaiserslautern, Germany
zimmer@physik.uni-kl.de

1 Game Theory and Statistical Mechanics

Andreas Engel

Summary. Methods from the statistical mechanics of disordered systems are beginning to permeate game theory. If the game theoretical problem involves a large number of strategies and random payoff matrices, these methods may be efficiently used to determine the typical values of specific parameters characterizing the game and the corresponding optimal strategies. The present article gives a short pedagogical introduction to some of the relevant concepts of game theory and their relation to statistical mechanics, with a special emphasis on matrix and bimatrix games.

1.1 Introduction

Many problems in economics, sociology and politics involve decision-making in complex situations in which the result of an action depends not only on the action itself but also on the simultaneous actions of others. The aim of game theory is to study simple but generic situations of this kind, which can be analyzed in mathematical terms. Although first attempts of this kind apparently occurred rather long ago, game theory only became an established field of mathematics in the last century, following the seminal work by John von Neumann, collected in the influential book [1]. Meanwhile game theory has developed into a flourishing field of mathematics. It has interesting connections with evolutionary biology and population dynamics and a growing influence on economic research.[1]

A standard game theoretic problem is given by a set of *players* $\{X, Y, \dots\}$ choosing between different *strategies* $\{X_i\}, \{Y_j\}, \dots$. The outcome of the game is specified by the *payoffs* $P_X(X_i, Y_j, \dots), P_Y(X_i, Y_j, \dots), \dots$ every player stands to receive. The game is repeated many times and every player strives to maximize his cumulative payoff, which depends also on the strategies chosen by the other players.

At present, fruitful interactions between game theoretical investigations and theoretical physics exist mainly via two directions, using methods either from the theory of dynamical systems or from statistical mechanics. In both cases, also the corresponding numerical techniques are relevant.

Connections with the theory of dynamical systems arise because, for the qualitative analysis of the dynamics of an iterated game, concepts such as fixed points, attractors and bifurcations are often appropriate. In the framework of

---

[1] The 1994 Nobel prize for economics was awarded to J.C. Harsanyi, J.S. Nash and R. Selten for their work in game theory.

evolutionary game theory, dealing with spatially distributed situations, the notions of phase equilibria and nucleation may be used to some advantage.

On the other hand, the methods of statistical mechanics come into play if there is emphasis on *large* game theoretic setups, involving many different strategies. In these situations it is often hard or even impossible to pin down the specific values of the entries in payoff matrices. As in many other cases of disordered physical or non-physical complex systems (such as electrons in random potentials, glasses, or polymer solutions and artificial neural networks, error correcting codes, or hard optimization problems, respectively), it may then be sensible to investigate situations with *random* parameters and to look for *self-averaging* quantities describing the *typical* case. Since problems in game theory are often characterized by conflicting interests among the various players, the concept of *frustration* is likely to play a role, so that the methods of spin-glass theory [2] prove very valuable.

The present chapter aims at a pedagogical introduction to some of the basic notions of game theory and their relation to statistical mechanics. The emphasis is on intuitive understanding rather than mathematical rigour. In the next section, matrix games will be discussed as one of the basic setups considered in game theory. Section 1.3 contains a very simple example of a more complex situation and introduces the concept of Nash equilibria, of central importance in game theory. In Sect. 1.4, bimatrix games will be studied, including a description of the famous prisoner's dilemma problem and the possibility of calculating the typical number of Nash equilibria for large random payoff matrices. Finally, a miscellaneous section brings together some other interesting relations between game theory and statistical mechanics.

1.2   Matrix Games

The simplest scenario in game theory, which was also the starting point of von Neumann's classical analysis, involves just two players X and Y who may choose between N strategies X_i and M strategies Y_j, respectively. Additionally, it is assumed that the payoffs $P_X(X_i, Y_j) =: c_{ij}$ and $P_Y(X_i, Y_j) := -c_{ij}$ always sum to zero so that they may be coded in a single payoff matrix c_{ij}. Hence, in the situation of a *zero-sum* game, what one player gains is always the other player's loss, and the goals of the players are totally conflicting.

To get some impression of what an advantageous strategy in a matrix game might look like, it is useful to consider some simple examples. Hence, consider the payoff matrix given by

$$c_{ij} = \begin{pmatrix} -8 & 2 & -4 & -11 \\ 6 & 3 & 2 & 17 \\ 12 & 7 & 1 & 3 \end{pmatrix}. \tag{1.1}$$

A possible 'worst case' kind of reasoning for player X is as follows: playing strategy i, the player will receive at least the payoff $\min_j c_{ij}$. Accordingly, in the example, choosing strategy $1, 2$ or 3 would guarantee payoffs $-11, 2$ and 1, respectively. It then seems best to choose the *maximum* of these and to play strategy i^* satisfying $\min_j c_{i^*j} = \max_i \min_j c_{ij}$. In the example, this is strategy 2. Likewise player Y, attempting to maximize his payoff, which is equivalent to minimizing c_{ij}, first looks for the *largest* losses he may encounter when playing a given strategy, i.e., this player determines $\max_i c_{ij}$ for all j. In the example, he may lose $12, 7, 2$ and 17 when playing strategies $1, 2, 3$, and 4, respectively. He therefore chooses the strategy j^* satisfying $\max_i c_{ij^*} = \min_j \max_i c_{ij}$ and hence *minimizing* the maximal possible loss. In the example, the best choice for player Y would therefore be to play strategy 3. The matrix game specified by (1.1) is therefore characterized by

$$\max_i \min_j c_{ij} = \min_j \max_i c_{ij} , \qquad (1.2)$$

and is said to have a *saddle point*. In this case, the determination of the optimal strategies for both players is rather simple. Player X should always play his second strategy and player Y always his third. If one player deviates from this prescription and the other one stays with it, the payoff received by the deviating player would decrease.

The situation is less obvious if the payoff matrix has no saddle point. Let us consider the slightly changed payoff matrix

$$c_{ij} = \begin{pmatrix} -8 & 2 & -4 & -11 \\ 6 & 3 & 6 & 17 \\ 12 & 7 & 1 & 3 \end{pmatrix} , \qquad (1.3)$$

which differs from (1.1) just by the value of c_{23}. Now we get

$$\max_i \min_j c_{ij} = 3 < 6 = \min_j \max_i c_{ij} , \qquad (1.4)$$

and the gain guaranteed for player X is *less* than the loss player Y is expecting. This is in fact the generic case since one may easily prove that, for every matrix c_{ij},

$$\max_i \min_j c_{ij} \leq \min_j \max_i c_{ij} , \qquad (1.5)$$

with equality holding only in the case of a saddle point. Intuitively, it is clear now that the players should try to move into the gap left by inequality (1.4). In other words, they should replace their pessimistic 'worst case' reasonings described above by more sophisticated strategies. An interesting observation in this respect is that for any given strategy of player Y, player X may on average obtain a larger payoff than 3 by *switching at random* between strategies 2 and 3. The randomness in the choice between the different strategies is crucial since the choice must not be anticipated by the opponent. The central idea of von Neumann's classical analysis of matrix games was hence that the players may improve their payoffs if they *mix* their strategies *at random*. This allows them, in a manner unpredictable for the other player, to use all their strategies including those promising high gain at the risk of equally high losses.

Accordingly a *mixed strategy* for player X is a vector of probabilities $\boldsymbol{x} = \{x_1, \ldots, x_N\}$ for playing the different *pure* strategies $X_1, \ldots, X_N$. If equivalently $\boldsymbol{y} = \{y_1, \ldots, y_M\}$ denotes the mixed strategy of player Y the *average* payoff for player X is given by

$$\nu = \sum_{i,j} c_{ij} \, x_i \, y_j \, , \tag{1.6}$$

whereas the payoff for Y is just $-\nu$, as before.

It turns out that, using such mixed strategies, it is possible for the two players to utilize all the remaining possibilities specified by the inequality (1.4) and to increase their respective payoffs up to the theoretical limit. This is guaranteed by the famous *minimax theorem* of von Neumann which states that, for all payoff matrices c_{ij}, there exists a *saddle point of mixed strategies*, i.e., there are *optimal mixed strategies* $\boldsymbol{x}^*$ and $\boldsymbol{y}^*$ such that [compare with (1.2)]

$$\max_{\boldsymbol{x}} \min_{\boldsymbol{y}} \sum_{i,j} c_{ij} \, x_i \, y_j = \sum_{i,j} c_{ij} \, x_i^* \, y_j^* = \min_{\boldsymbol{y}} \max_{\boldsymbol{x}} \sum_{i,j} c_{ij} \, x_i \, y_j \, . \tag{1.7}$$

The expected payoff for the optimal mixed strategies

$$\nu_c := \sum_{i,j} c_{ij} \, x_i^* \, y_j^* \tag{1.8}$$

is called the *value of the game*. For games with saddle points, the optimal strategies are realized as *pure strategies* of the form $x_i = \delta_{im}$, $y_j = \delta_{jn}$ and the value of the game is just the respective element c_{mn} of the payoff matrix, as in the first example (1.1) discussed above.

How difficult is it to actually *find* the optimal strategies $\boldsymbol{x}^*$ and $\boldsymbol{y}^*$, given the explicit form of the payoff matrix c_{ij}? To answer this question the following simple but important condition for $\boldsymbol{x}^*$ to be an optimal strategy is very helpful.

A mixed strategy $\boldsymbol{x}^*$ is optimal if and only if

$$\sum_{i} c_{ij} x_i^* \geq \nu_c \qquad \text{for all } j \, . \tag{1.9}$$

The condition is necessary since, if violated for some j, player Y could increase his payoff by always playing strategy j. It is sufficient since multiplying equations (1.9) by y_j of some mixed strategy $\boldsymbol{y}$ and summing over j, one finds that the payoff for X playing $\boldsymbol{x}^*$ can never be smaller than ν_c. Note also that the conditions (1.9) no longer explicitly involve the strategy of player Y.

Being probabilities, the components of vector $\boldsymbol{x}$ must be positive and satisfy the normalization condition

$$\sum_{i} x_i = 1 \, . \tag{1.10}$$

Equations (1.9) and (1.10) define a problem of *linear optimization*. For the mapping on the standard form of linear programming, a good reference is [3], which also gives a nice intuitive discussion and provides some sample programs on the attached diskette. The most common procedure for solving problems of linear optimization is the simplex algorithm (see [4], for example). The explicit determination of optimal strategies for a matrix game may thus be mapped on a standard problem of numerical analysis.

A very simple example is given by the well-known children's game of rock–scissors–paper. In this case the payoff matrix is given by

$$c_{ij} = \begin{pmatrix} 0 & 1 & -1 \\ -1 & 0 & 1 \\ 1 & -1 & 0 \end{pmatrix}, \tag{1.11}$$

which obviously has no saddle point. For symmetry reasons the value of the game is $\nu_c = 0$ and, from (1.9), we find $x_2 \geq x_3$, $x_3 \geq x_1$, $x_1 \geq x_2$ giving rise to $x_1 = x_2 = x_3$. From (1.10), this results in $x^* = (1/3, 1/3, 1/3)$. The best way to play this game is hence to choose rock, scissors and paper with equal probabilities at random, which ensures an average payoff of nothing. All other strategies result asymptotically in losses, if the opponent reacts in an optimal manner.

In order to elucidate the role statistical mechanics may play in the analysis of matrix games, let us consider the 'thermodynamic limit' $N \to \infty$, $M \to \infty$ with $\alpha := M/N$ remaining of order 1 corresponding to a very large game in which each player may choose between a large variety of different strategies. As discussed above, the explicit form of the optimal strategies will depend on the specific values of the $O(N^2)$ entries of the payoff matrix c_{ij}, which will hardly ever be available in practice. To get an impression of the general features of optimal strategies in such complex situations, even without knowing all the details of the relevant payoff matrix, it is sensible to generate the c_{ij} at *random* according to some probability distribution and to look for characteristics of the game depending only on the parameters of this distribution and not on the individual realizations of the payoff matrix. This kind of reasoning is completely analogous to the approach taken in the determination of the density of states of a disordered solid, for example. There the detailed potential the electrons are moving in is not known and differs from sample to sample. Nevertheless, the density of states is known to be reproducible in experiments. One may therefore try to calculate the density of states for electrons in a random potential and will find that it is *self-averaging*, i.e., in the thermodynamic limit, it is the same for almost all realizations of the random potential.

Let us therefore assume that the payoffs c_{ij} are independent Gaussian random variables with zero mean and variance $1/N$. This particular scaling of the second moment with the size of the system is chosen to yield a non-trivial thermodynamic limit. The connection with quantities typical for statistical mechanics can now be established, exploiting once more the condition (1.9) for a mixed strategy being optimal. Using (1.9), we may introduce an *indicator function* for an optimal mixed strategy of the form

$$\chi(\boldsymbol{x}) = \prod_{j=1}^{M} \Theta\Big(\sum_i c_{ij}x_i - \nu_c\Big), \tag{1.12}$$

where $\Theta(x)$ denotes the Heaviside function taking value 1 for $x > 0$ and 0 otherwise. Accordingly, $\chi(\boldsymbol{x})$ is zero if (1.9) is violated for at least one j and 1 otherwise. Placing this function under an integral over all possible mixed strategies $\{x_i\}$ satisfying the normalization condition (1.10) would therefore project out the optimal strategy $\boldsymbol{x}^*$. Unfortunately, the function (1.12) involves the unknown value of the game ν_c, a complicated function of the payoff matrix which has to be determined self-consistently with the optimal strategy. This can be done by replacing ν_c by a so far unspecified parameter ν and determining the phase space volume

$$\Omega(\nu, c_{ij}) := \frac{\int_0^1 \prod_i \mathrm{d}x_i\, \delta\left(\sum_i x_i - 1\right) \prod_{j=1}^{M} \Theta\left(\sum_i c_{ij}x_i - \nu\right)}{\int_0^1 \prod_i \mathrm{d}x_i\, \delta\left(\sum_i x_i - 1\right)} \tag{1.13}$$

describing the fraction of the complete phase space of all possible mixed strategies corresponding to those which fulfill (1.9) with threshold ν for the realization c_{ij} of the elements of the payoff matrix. The central quantity is then given by

$$s(\nu) = \lim_{N \to \infty} \frac{1}{N} \langle\langle \ln \Omega(\nu, c_{ij}) \rangle\rangle \tag{1.14}$$

which is the equivalent of an *average microcanonical entropy* with ν playing the role of the energy. The double brackets $\langle\langle \ldots \rangle\rangle$ denote the average over the distribution of the payoffs c_{ij}.

This entropy was calculated explicitly in [5] using the standard repertoire of mean field spin-glass theory [2,6]. The behaviour of $s(\nu)$ is shown qualitatively in Fig. 1.1. By definition, it is always negative since $\Omega(\nu)$ is a fraction [see (1.13)]. Moreover, it decreases with increasing ν since fewer and fewer strategies are able to realize a payoff larger than the threshold. Finally, for $\nu \to \nu_c$, only a single strategy survives, namely $\boldsymbol{x}^*$, and therefore $s \to -\infty$. It is thus possible, by calculating the average entropy defined in (1.14) and considering the limit $s \to -\infty$, to get some information about the statistical properties of the optimal strategies of large random matrix games. The general procedure is very similar to elucidating the ground state properties of a disordered magnetic system.

Some of the results obtained in this way in [5] are displayed in Fig. 1.2. They include both the typical value of the game and the fraction of strategies $\theta(0)$ entering the optimal mixed strategy of player X as functions of the parameter $\alpha = M/N$ characterizing the a priori variability of the two players. For $\alpha < 1$, player X has more strategies at his disposal than player Y, whereas for $\alpha > 1$, he has fewer. Accordingly, the average payoff is positive for $\alpha < 1$, meaning that X is superior, whereas it is negative if $\alpha > 1$. For $\alpha = 1$, we find $\nu_c = 0$, consistent with the symmetry of the situation. The analytical results agree well

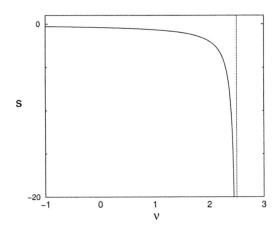

Fig. 1.1. Qualitative behaviour of the average entropy defined in (1.14) as a function of the payoff threshold ν. If ν approaches the value of the game ν_c from below, the entropy diverges to $-\infty$

with numerical simulations performed using the simplex algorithm to determine the optimal strategies for different realizations of the payoffs c_{ij}.

In order to characterize the structure of the optimal strategy, it is instructive to calculate the function

$$\theta(a) := \left\langle\!\left\langle \sum_i \Theta(x_i^* - a) \right\rangle\!\right\rangle \tag{1.15}$$

determining the fraction of strategies which are played with a probability larger than a within the optimal mixed strategy. In particular, $\theta(0)$ gives that fraction of strategies which are actually used in the optimal strategy. As can be seen from Fig. 1.2, this fraction is always smaller than one. In fact, for $\alpha = 1$, half of all the strategies available to either player are such that they should *never* be played in order to behave optimally! With increasing α, $\theta(0)$ also increases, meaning that player X has to use a larger fraction of his smaller set of available strategies in order to perform optimally. In a complementary manner, for $\alpha > 1$, Y may choose from a larger pool of strategies and, due to their random nature, there is an appreciable probability of having pure strategies that perform well. It is then possible for player Y to compose his optimal strategy from only a few of these high performance strategies. Accordingly, his $\theta(0)$ decreases with increasing α, whereas his payoff increases.

1.3   Nash Equilibria

The assumption of zero overall payoff for matrix games is rather restrictive since it makes the goals of the players totally conflicting. In order to highlight some of the interesting new features which may arise if this condition is relaxed, let us

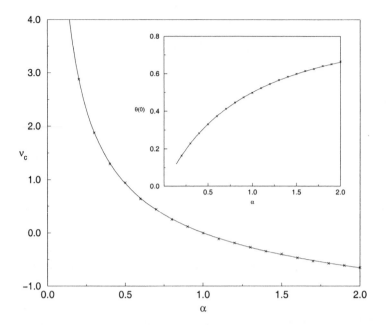

Fig. 1.2. Results of a statistical mechanical analysis of a large matrix game with random payoff matrix of size $N \times \alpha N$. The value of the game ν_c is shown and, in the *inset*, the fraction $\theta(0)$ of pure strategies entering the optimal mixed strategy, both as a function of α. The *full lines* are analytical results using the limit $N \to \infty$, and *crosses* denote results of numerical simulations for $N = 200$, averaged over 200 realizations of the random payoffs. The statistical error in the simulations is smaller than the symbol size

consider a simple example from an economic context. Three players, companies X, Y, Z, compete on a market by producing amounts x, y, z assumed to be integer numbers of the same product. The price p of the product is determined by supply and demand and is assumed to be given by

$$p = \begin{cases} 20 - (x + y + z) & \text{if } 20 - (x + y + z) > 0 , \\ 0 & \text{otherwise .} \end{cases} \tag{1.16}$$

How much of the product should the companies produce in order to maximize their respective profits $P_X = px, P_Y = py, P_Z = pz$? In order to investigate this question, some instructive examples have been compiled in Table 1.1. In the first case company Y makes the highest profit. However, this does not imply that it has played optimally. As shown in the second row, reducing its production, it may increase its profit (and that of the other players) due to the increasing price. Looking for optimal situations, it is not difficult to show that under the condition $x = y = z$, the third case gives the maximum profit for all companies. However, this situation is *unstable*, since it is advantageous for an individual player to deviate from this strategy if the others stick to it. This is exemplified

Table 1.1. Production numbers and prices for a very simple model of an economic market (see text)

Case	Production numbers			Price	Profits		
	x	y	z	p	P_X	P_Y	P_Z
1	4	8	6	2	8	16	12
2	4	5	6	5	20	25	30
3	3	3	3	11	33	33	33
4	3	5	3	9	27	45	27
5	5	5	5	5	25	25	25
6	5	6	5	4	20	24	20
7	5	4	5	6	30	24	30

by the fourth case. Hence situations like the third case, although theoretically advantageous, are not viable in real situations. Things are different in the fifth case. Again all companies produce the same amount and consequently receive the same profit, which, however, is less than the one in case 3. But now the situation is *stable* since, as shown by the last two rows of the table, a player cannot increase his payoff by deviating from the prescription with all the others sticking to their strategy. If arrangements and coordinated action of the players are not allowed, it is therefore the best for all to produce 5 products as in case 5. This is a simple example of a *Nash equilibrium* [7], defined as a situation in which the players cannot improve their payoffs by deviating from their current strategy, under the condition that all the other players do not change their strategy.[2]

Matrix games have just *one* Nash equilibrium, defined by the optimal mixed strategies of both players. In contrast, for more general game theoretic settings without the zero-sum constraint, there may be a great many different Nash equilibria. In fact the determination of the number and structure of Nash equilibria and methods for accessing them in a game are among the most interesting current problems in game theory. As will be shown in the next section, statistical mechanics has recently contributed to answering this kind of question.

1.4 Bimatrix Games

An interesting realization of games without the zero-sum constraint, but which is still simple enough to be amenable to mathematical analysis, is given by a game with two players X and Y in which each player has his *own* payoff matrix a_{ij} and b_{ij}, respectively. Games of this kind are called *bimatrix games*. Denoting the mixed strategies of X and Y by x and y as before, the average payoffs of

---

[2] Another nice example is provided by the so-called 'stable marriage problem' [8], the importance of which is, of course, beyond all question.

the players are given by $\nu_X = \sum_{ij} a_{ij} x_i y_j$ and $\nu_Y = \sum_{ij} b_{ij} x_i y_j$. Due to the absence of the constraint $a_{ij} = -b_{ij}$ which was characteristic for matrix games, it is now possible as in the last section to find strategies which are advantageous for both players.

The most popular bimatrix game is the so-called *prisoner's dilemma*, describing the following situation. Two criminals were arrested after having committed a crime together and wait for their trial in different cells. The body of evidence is such that if they were both claiming to be innocent, it would be possible to punish them for a minor part of the crime only. If they blame each other for the main part in the crime, they both receive a higher punishment. Finally, if one claims innocence for both and the other one defects, the latter will go free, while the former gets a maximal punishment. Hence, each player may choose between two strategies, namely cooperating and defecting, and the payoff matrices for such a situation may look like

$$a_{ij} = \begin{pmatrix} 3 & 5 \\ 0 & 1 \end{pmatrix} \quad \text{and} \quad b_{ij} = \begin{pmatrix} 3 & 0 \\ 5 & 1 \end{pmatrix} . \tag{1.17}$$

For either choice of the opponent, it is thus advantageous to defect. On the other hand if both defect, the payoff remains less than in the case where both cooperate. This is the origin of the dilemma.

If the situation occurs just once for the two players, it is indeed best for both to defect. In a more interesting variant, the so-called *iterated* prisoner's dilemma, however, there is a constant probability w for another round. Then the choice of strategy becomes non-trivial. In fact, one may show that under rather general conditions for w as a function of the payoffs, there is no 'best' strategy at all [9]. The main problem is to find a way to encourage the other player to cooperate without being exploited too often. Whilst in zero-sum games it was crucial to select strategies *independently* and at random in order to avoid the opponent being able to anticipate one's choice, the opposite is now true. By making strongly *correlated* moves, one tries to demonstrate reliability and to encourage cooperation. Note that the parameters in (1.17) were chosen such that cooperation is superior to alternating exploitation.

In both theoretical biology and sociology, the iterated prisoner's dilemma has become a paradigmatic example for the emergence of altruism and the evolution of cooperation in competing societies. It gives the nutshell situation in which individuals may benefit from cooperation but do even better by exploiting the cooperative efforts of others. An original and thought-provoking discussion of these issues is given by Axelrod [10], who provides striking biological and historical examples. He also organized two round-robin tournaments in which computer strategies for the iterated prisoners dilemma were competing with each other. The most remarkable among the surprising results of these contests was probably the unexpected success of the simple *tit-for-tat* strategy which starts with cooperation and then simply does what the opponent did the last time.

With just two possible strategies for each player the prisoner's dilemma is a small bimatrix game. It has just a single Nash equilibrium, given by the situation in which both players defect. If the number of available strategies increases, so does the number of Nash equilibria. In the asymptotic regime of $N \to \infty$, methods of statistical mechanics may again be used to determine the typical number and some properties of the Nash equilibria. Let us therefore reconsider random payoff matrices a_{ij} and b_{ij}. As in Sect. 1.2, we may choose the distribution such that $\langle\langle a_{ij} \rangle\rangle = \langle\langle b_{ij} \rangle\rangle = 0$ and $\langle\langle a_{ij}^2 \rangle\rangle = \langle\langle b_{ij}^2 \rangle\rangle = 1/N$. Moreover, it is useful to introduce a *correlation* between elements of the two matrices with the same indices, of the form $\langle\langle a_{ij} b_{kl} \rangle\rangle = (\kappa/N)\,\delta_{ik}\delta_{jl}$. The parameter κ characterizes the degree of competition in the game. For $\kappa = 1$, both players receive identical payoffs $a_{ij} = b_{ij}$, whilst the case $\kappa = -1$ brings us back to zero-sum games. For $\kappa = 0$, the payoffs are uncorrelated and equally many conflicting and cooperating situations are possible.

The condition for two mixed strategies $\boldsymbol{x}$ and $\boldsymbol{y}$ forming a Nash equilibrium with respective payoffs ν_X and ν_Y can be found by generalizing (1.9) and takes the form [11]

$$\sum_j a_{ij} y_j \leq \nu_X \quad \text{and} \quad x_i \left(\sum_j a_{ij} y_j - \nu_X \right) = 0 \quad \forall i\,, \tag{1.18}$$

$$\sum_i b_{ij} x_i \leq \nu_Y \quad \text{and} \quad y_j \left(\sum_i b_{ij} x_i - \nu_Y \right) = 0 \quad \forall j\,. \tag{1.19}$$

These conditions are less obvious than their zero-sum counterpart (1.9). In particular, it is no longer possible to separate the condition for $\boldsymbol{x}$ from the one for $\boldsymbol{y}$. Roughly speaking, the first of the conditions ensures that, for a given strategy of the opponent, players cannot increase their payoffs beyond the value characterizing the Nash equilibrium, whereas the second guarantees that the respective payoffs are indeed realized at the Nash equilibrium.

Using conditions (1.18) and (1.19), one may now introduce an indicator function for Nash equilibria with the help of which the number of Nash equilibria $\mathcal{N}(\nu_X, \nu_Y; a_{ij}, b_{ij})$ with prescribed values for the payoffs of the two players for a given realization of the payoff matrices may be determined [12]. Again, the appropriate quantity to average over the distribution of the payoff matrices is the logarithm of this number, giving rise to the expression

$$s(\nu_X, \nu_Y) := \lim_{N \to \infty} \frac{1}{N} \left\langle\!\left\langle \ln \mathcal{N}(\nu_X, \nu_Y; a_{ij}, b_{ij}) \right\rangle\!\right\rangle \tag{1.20}$$

for the *entropy of Nash equilibria*. The explicit calculation of $s(\nu_X, \nu_Y)$ using the machinery of spin-glass theory is quite involved. We therefore only display some results in Fig. 1.3. For a detailed exposition see [12].

For all $\kappa > -1$, there are values of ν_X and ν_Y such that $s(\nu_X, \nu_Y) > 0$. This implies that there is typically an *exponential number* of Nash equilibria in large bimatrix games. On the other hand this total number is (again exponentially) dominated by Nash equilibria with payoffs $\nu_X^{(\mathrm{max})}$ and $\nu_Y^{(\mathrm{max})}$ which realize the maximum of $s(\nu_X, \nu_Y)$. This implies that choosing one of the many Nash equilibria of a large bimatrix game at random will yield one with payoffs $\nu_X^{(\mathrm{max})}$

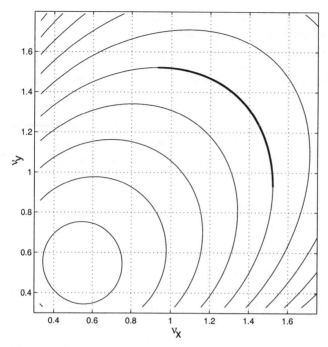

Fig. 1.3. Contour plot of the average entropy of Nash equilibria for a large bimatrix game with random payoff matrices and correlation coefficient $\kappa = -0.8$ as a function of the expected payoffs ν_X and ν_Y of the two players. *Curves* correspond to $S = 0.8, 0.6, 0.4, 0.2, 0, -0.2$ (from *bottom left* to *top right*). The *thick part* of the $S = 0$ contour denotes those combinations of payoffs ν_X and ν_Y for which, with probability 1, there is no Nash equilibrium with higher payoff for one player without decreasing the payoff of the other player

and $\nu_Y^{(\max)}$ with probability 1. Also interesting is the curve in the ν_X–ν_Y plane at which $s(\nu_X, \nu_Y)$ becomes zero (see Fig. 1.3) since it separates regions with exponentially many Nash equilibria from those with at most a few, i.e., sub-exponentially many. Since with increasing κ the goals of the two players become less and less conflicting, when $\kappa \to 1$, one may expect this curve to be shifted to larger and larger values of ν_X and ν_Y, in such a way that Nash equilibria with high payoffs for both players become possible. It is a rather surprising result of the statistical mechanical analysis that this transition occurs quite early on. Already for $\kappa > \kappa_c \cong -0.58$, one finds $s(\nu, \nu) > 0$ for arbitrarily large ν. This result implies that, in the given setup of a large random bimatrix game, even in the regime where the interests of the players are on average highly conflicting, there is an exponential number of Nash equilibria offering both players equal and arbitrarily high payoffs.

1.5   What Else?

There are several other interesting and potentially important connections between game theory and physics. One is related to *evolutionary game theory* [13]
in which the game models the competition between different species and the
payoff determines the *fitness* of the species. In the simplest case, the latter is
just the number of offspring in the next generation. In this setting, the components x_i of a mixed strategy determine the fraction of individuals in a population
playing the pure strategy i. The players of the game are repeatedly selected at
random from the competing populations and, by Darwin's principle, those with
lower payoffs disappear with time. From the point of view of modeling biological
evolution, the framework of evolutionary game theory is appealing since it uses
a fitness function that depends on the behaviour of other species and since it
allows us to investigate the role of cooperation in biological evolution which,
though undoubtedly relevant, has not been much elucidated.

Mathematically, the dynamics is described by the *replicator model* [14], given
by the set of differential equations

$$\dot{x}_i = x_i f_i - x_i \sum_j f_j x_j \ .$$

(1.21)

The fitness f_i is determined by the expected payoff of strategy i

$$f_i = \sum_j a_{ij} x_j \ .$$

(1.22)

The first term in (1.21) describes self-reproduction with a rate proportional to
the fitness, whereas the second ensures normalization $\sum_i x_i = 1$, thereby exerting
selection pressure. Many aspects of the replicator model have been analyzed in
mathematical biology. In the present context, the analysis of the properties of
large random payoff matrices using methods of spin-glass theory, as performed
in [15], are of particular interest.

The central new concept in evolutionary game theory is that of an *evolutionarily stable strategy*. By definition, this is a strategy which, if all members of a
population adopt it, cannot be invaded by a mutant strategy under the sole influence of natural selection. In the evolutionary variant of the prisoner's dilemma,
for example, the strategies of always defecting and 'tit-for-tat' are evolutionarily
stable. The invasion of a superior strategy may occur in different ways, which
become even more diverse if *spatially distributed* systems are considered. In this
case also the neighborhood becomes important. Most results in this field are
from numerical simulations using the machinery of cellular automata.

Finally, there is growing interest in models of simple economic situations,
including the dynamics of financial markets which are amenable to methods
from theoretical physics. For this field of research, notions of 'econophysics' and
'phynance' have been coined, which may be taken as useful starting points for
a Web search on recent activities. A simple example that has generated an
avalanche of activity is the so-called *minority game*. In this game, N agents have

to decide between two options (buy or sell, leave or stay, etc.) and those on the *minority* side receive a reward. The problem was known for some time as the 'El Farol bar problem' in sociology [16], but activity in the physics community only started after a phase transition was found in numerical simulations signalling the emergence of cooperation. A well documented Web site concerning recent work on this problem is [17], including an introduction and a selection of key papers.

In summary there are several promising connections between game theory and statistical mechanics and it is very likely that the next few years will bring increased research activity at the boundary of these two disciplines, hopefully with stimulating results for both fields.

Acknowledgements. I would like to thank Johannes Berg for pleasant cooperation on the issues discussed in this paper.

References

1. J. von Neumann, O. Morgenstern: *Theory of Games and Economic Behaviour* (Princeton Press, Princeton 1953)
2. M. Mezard, G. Parisi, M.A. Virasoro: *Spin Glass Theory and Beyond* (World Scientific, Singapore 1987)
3. W. Kinzel, G. Reents: *Physik per Computer* (Spektrum, Heidelberg 1996)
4. W.H. Press, B.P. Flannery, S.A. Teukolsky, W.T. Vetterling: *Numerical Recipes in C* (Cambridge University Press, Cambridge 1990)
5. J. Berg, A. Engel: Phys. Rev. Lett. **81**, 4999 (1998)
6. A. Engel, C. van den Broeck: *Statistical Mechanics of Learning* (Cambridge University Press, Cambridge 2000)
7. J.F. Nash: Annals of Mathematics **54**, 268 (1951)
8. M. Szierzawa, M.-J. Oméro: http://xxx.lanl.gov/cond-mat/0007321
9. R. Axelrod, W.D. Hamilton: Science **211**, 1390 (1981)
10. R. Axelrod: *The Evolution of Cooperation* (Basic Books, New York 1984)
11. Wang Jianhua: *The Theory of Games* (Oxford University Press, Oxford 1988)
12. J. Berg: Phys. Rev. **E61**, 2327 (2000), Statistical Mechanics of Random Games, PhD Thesis, Otto-von-Guericke-University, Magdeburg (1999)
13. J. Maynard Smith: *Evolution and the Theory of Games* (Cambridge University Press, Cambridge 1993)
14. J. Hofbauer, K. Sigmund: *The Theory of Evolution and Dynamical Systems* (Cambrigde University Press, Cambridge 1988)
15. S. Diederich, M. Opper: Phys. Rev. **A39**, 4333 (1989); M. Opper, S. Diederich: Phys. Rev. Lett. **69**, 1616 (1992)
16. W.B. Arthur: Am. Econ. Assoc. Papers and Proc. **84**, 406 (1994)
17. http://www.unifr.ch/econophysics/minority/

2 Chaotic Billiards

Hans Jürgen Korsch and Frank Zimmer

Summary. The frictionless motion of a particle on a plane billiard table bounded by a closed curve provides a very simple example of a conservative classical system with non-trivial, chaotic dynamics. The limiting cases of strictly regular ('integrable') and strictly irregular ('ergodic') systems can be illustrated, as well as the typical case, which displays an intricate mixture of regular and irregular behavior. Irregular orbits are characterized by an extreme sensitivity to initial conditions. Such billiard systems are extremely well suited for educational purposes as models of simple systems with complicated dynamics, as well as for far-reaching fundamental investigations.

2.1 Introduction

In the past decades, classical physics has witnessed unexpected and impetuous developments, which have led to an entirely new understanding of the classical dynamics of simple systems, an area of physics that has usually been presumed to be generally understood and concluded. It became evident, however, that – contrary to the concepts conveyed by most physics textbooks – even the simplest, completely deterministic systems may show irregular, chaotic behavior, that is as unpredictable as the tossing of a coin. Commonly, one accepted a random, stochastic behavior only for a system with a large number of degrees of freedom ($\approx 10^{23}$), e.g., a gas. It has now been established that such a behavior can be exhibited by systems with merely two degrees of freedom, i.e., a very small number indeed. The motion of a particle in a 2-dimensional conservative force field, i.e., in a potential, typically shows chaotic behavior. Since the first indication of chaos in strictly deterministic systems, a torrent of scientific studies have been carried out, in which the existence of such 'deterministic chaos' has been analyzed and further verified.

Chaotic systems in physics can be divided into two groups. There are so-called dissipative systems, where friction is present, and then there are conservative systems, where energy is a constant of the motion. We shall deal exclusively with the latter case here. As an introduction to the dynamics of conservative systems, the textbooks by Lichtenberg and Liebermann [1] and Schuster [2], as well as the excellent review article by Berry [3], are recommended.

The description of a system with N degrees of freedom requires $2N$ coordinates, namely N position coordinates $q = (q_1, q_2, \ldots, q_N)$ and N canonical momenta $p = (p_1, p_2, \ldots, p_N)$. The time-evolution of the system is therefore given by $(q(t), p(t))$, a curve in a $2N$-dimensional space, the so-called phase space.

This curve is determined by the Hamiltonian differential equations of motion

$$\dot{p}_i = -\frac{\partial H}{\partial q_i} , \quad \dot{q}_i = \frac{\partial H}{\partial p_i} , \quad i = 1, \dots, N . \tag{2.1}$$

The textbooks in classical mechanics, with some rare exceptions, deal with so-called *integrable* systems, i.e., systems in which there exist N independent constants of motion F_j, $j = 1, \dots, N$. These are functions on the phase space whose values do not change along the trajectory. In addition, one requires these functions $F_j(q, p)$ to be 'in involution', i.e., their Poisson brackets vanish:

$$\{F_j, F_k\} = \sum_i \left(\frac{\partial F_j}{\partial p_i} \frac{\partial F_k}{\partial q_i} - \frac{\partial F_k}{\partial p_i} \frac{\partial F_j}{\partial q_i} \right) = 0 . \tag{2.2}$$

Because of the N conditions $F_j(q, p) = f_j = \text{const.}$, the motion is restricted to an N-dimensional manifold in $2N$-dimensional phase space whose topology is that of an N-torus [3–5]. Moreover, this must be true for any trajectory and therefore the phase space is densely filled with these nested tori. Such dynamics is called regular.

In a so-called conservative Hamiltonian system, the energy is one of the constants of the motion and consequently a 1-dimensional conservative system is always integrable. In the following we shall discuss the simplest non-trivial case, namely 2-dimensional conservative systems, which are thus integrable if there exists yet another independent conserved quantity F, i.e., with $\{H, F\} = 0$. Contrary to popular belief, integrable systems are extremely rare. The probability for the integrability of a randomly chosen system with more than one degree of freedom is equal to zero. However, integrability can often be related to symmetry: the motion of a particle in a central force field belongs to the few examples of integrable systems. The opposite of an integrable system is an *ergodic* system, for which almost every orbit fills the available phase space (for this energy!) densely. Such an orbit is called *irregular* or *chaotic*. There exist only a few systems for which ergodicity has been rigorously proven. One of these systems is the stadium billiard, i.e., a rectangle with semicircular ends [6]. The typical case is a system which is neither fully regular nor chaotic and contains both regular and irregular orbits. Well-known examples are the double pendulum and the three-body problem of celestial mechanics.

In order to visualize the complicated dynamics and to simplify its handling, one uses a reduction of information by introducing a surface of section in phase space: instead of studying the entire orbit, one keeps track only of the sequence (q_n, p_n), $n = 0, 1, 2, \dots$ of its intersection points with this surface of section. In this manner we obtain a discrete mapping

$$(q_n, p_n) \xrightarrow{T} (q_{n+1}, p_{n+1}) , \tag{2.3}$$

which associates each intersection point with its successor. For a Hamiltonian system, such a Poincaré map T is area preserving. The set of all intersection points is called the *Poincaré section* of the orbit. In such a section one can easily distinguish the different types of orbits.

Billiard systems are perfectly suited as models for educational purposes (as well as for far-reaching fundamental investigations!) since the billiard motion is easy to comprehend and, unlike many other systems, the numerical treatment does not require numerical integration of differential equations. This is an important advantage, because such a computation is comparatively time-consuming, even using modern computers, especially since chaotic phenomena are exhibited in the long-time behavior of an orbit. Furthermore, numerical methods for solving differential equations are not exact, and show instabilities, which cannot always be clearly distinguished from true chaotic behavior.

2.2   Billiard Systems

The 2-dimensional billiard problem [6,7] describes a point particle moving without friction on a plane billiard table, bounded by a closed curve. Between the impacts at the boundary, the particle moves on straight lines with constant velocity. It is reflected at the boundary according to the reflection law: the angle of incidence is equal to the angle of reflection. We shall deal with convex billiards here, i.e., a straight line has at most two intersection points with the boundary curve, which reads in polar coordinates

$$r = r(\varphi) \ . \tag{2.4}$$

For a sufficiently smooth boundary curve, such billiards are nonergodic [6,8].

In the billiard system, the Poincaré section evolves quite naturally from the boundary curve. This means that the intersection points are represented by the data at impact with the boundary, namely, the angle φ at this point and the direction of the trajectory after the impact, which can be measured by the angle α with respect to the tangent (see Fig. 2.1). It is more convenient, however, to

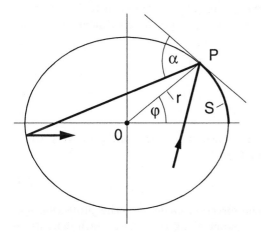

Fig. 2.1. Boundary curve $r(\varphi)$ of a billiard system, an initial part of a billiard trajectory and the coordinates: the arc length $S(\varphi)$ and $p = \cos\alpha$ (α = angle between trajectory and tangent of the boundary curve)

use the projection onto the tangent

$$p = \cos \alpha \,, \tag{2.5}$$

and the arc length divided by the total length L of the boundary curve

$$S(\varphi) = \frac{1}{L} \int_0^\varphi \sqrt{r^2(\varphi') + (dr/d\varphi')^2} \, d\varphi' \,. \tag{2.6}$$

The Poincaré map T relates the data of the nth impact $P_n = (S_n, p_n)$ to the next one $P_{n+1} = (S_{n+1}, p_{n+1})$. It can be shown that this mapping is area preserving using these variables:

$$\det \frac{\partial(S_{n+1}, p_{n+1})}{\partial(S_n, p_n)} = 1 \,. \tag{2.7}$$

More details about this mapping are given in the Appendix. Numerically, the problem is quite simple: one has to compute the intersection of the trajectory (a straight line) with the boundary curve, which yields the next impact angle φ_{n+1}. Then the arc length S_{n+1} is evaluated from (2.6) and the angle α_{n+1} between the trajectory and the tangent of the boundary curve at this point is determined, as well as $p_{n+1} = \cos \alpha_{n+1}$ and the reflected trajectory. This process is continued to obtain the next impact data.

2.2.1   The Billiard Computer Program

For a comfortable study of billiard dynamics, an interactive computer program BILLIARD has been developed which allows efficient computations without any prior knowledge of computing. This program is contained in a collection of PC programs illustrating chaotic dynamics for a selection of systems with applications in physics [9]. In the billiard program, some preset billiards can be chosen or an arbitrary boundary curve $r(\varphi)$ can be inserted via the keyboard. The program shows position and phase space presentations of an orbit for variable initial conditions. Comments concerning the numerical algorithm can be found in the Appendix. In the following, we describe some numerical experiments which can be carried out using this program.

2.3   Integrable Systems

2.3.1   Circular Billiards

For a circular billiard $r(\varphi) = r_0$, the reflections of the orbit are very easily evaluated. One finds the explicit equation

$$\varphi_n = \varphi_0 + 2\alpha n \,, \tag{2.8}$$

for the angle of the nth impact. The direction angle α or the projection $p = \cos \alpha$ onto the tangent direction is constant along the orbit, which constitutes conservation of angular momentum L. We have a second constant of motion L, and the circular billiard is therefore integrable. In phase space, each orbit lies on a 2-dimensional surface, whose intersection with the (S, p) plane is a curve. Since the coordinate p represents a conserved quantity, these invariant curves appear as horizontal lines. For orbits with a rational angle ratio

$$\alpha_n = \frac{m}{k}\pi , \quad m, k \in \mathbb{N} , \quad m < k , \tag{2.9}$$

we have

$$\varphi_n = \varphi_0 + \frac{m}{k} 2\pi n , \tag{2.10}$$

i.e., the orbits are periodic and close after k cycles. Such an orbit is called k-periodic. In this case, the Poincaré section consists of a series of k discrete points (S_n, p_n), $n = 1, \ldots, k$, which are traversed periodically:

$$T^k(S_n, p_n) = (S_n, p_n) . \tag{2.11}$$

For this reason the k points (S_n, p_n) are called *fixed points* of the mapping

$$T^k = \underbrace{T \circ T \circ \cdots \circ T}_{k \text{ times}} . \tag{2.12}$$

For angles that are not rational multiples of π, the iterated points (S_n, p_n) fill a horizontal line densely, with growing n. This is a so-called *invariant curve*, because it is invariant under the Poincaré map T. Figure 2.2 shows such an

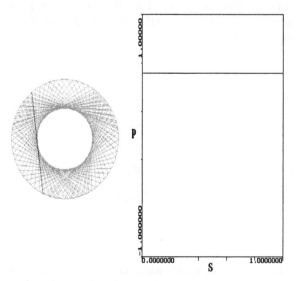

Fig. 2.2. Circular billiard. A non-periodic orbit in position space (*left*) appears as a straight line in the phase space diagram (S, p) (*right*)

irrational orbit. Radial distances $< p$ are forbidden due to conservation of angular momentum.

The motion along the orbit can be divided into two components: an oscillation between the inner envelope and the outer boundary curve and a rotation about the center. For a k-periodic orbit with $\alpha = m\pi/k$, the frequencies related to these partial motions have the ratio

$$\frac{\omega_1}{\omega_2} = \frac{m}{k} \,, \tag{2.13}$$

i.e., for $\pi/5$, we find five oscillations between the inner and outer boundary curves and a single rotation of 2π, whilst for $2\pi/5$, we have two such rotations.

2.3.2 Elliptical Billiards

We shall now consider an elliptical billiard with eccentricity ϵ, i.e., the boundary curve is given by

$$r(\varphi) = \frac{b}{\sqrt{1 - \epsilon^2 \cos^2 \varphi}} \,, \tag{2.14}$$

with semi-minor axis b and semi-major axis $a = b(1 - \epsilon^2)^{-1/2}$. Figure 2.3 depicts such an ellipse with $\epsilon = 0.3$. For large values of p, the orbits are similar to those for a circular billiard. Such an orbit fills an annular area in position space and possesses an enveloping curve that separates a forbidden inner region. It can be shown that this enveloping curve – called a caustic – is again an ellipse with the same foci as the boundary ellipse. The orbit always intersects the large diameter of the ellipse outside the line connecting both foci. For small values of p, a different type of motion appears: the position space orbit fills an area which is bounded by two confocal hyperbolic curves.

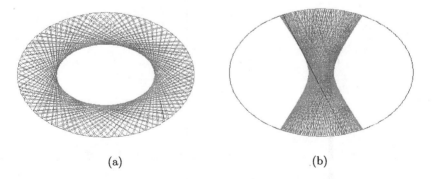

(a)   (b)

Fig. 2.3. Elliptical billiard (eccentricity $\epsilon = 0.3$). Orbits which intersect the main axis in the sections outside the two foci are bounded by an elliptical envelope (**a**). Orbits passing between the two foci have a hyperbolic envelope (**b**)

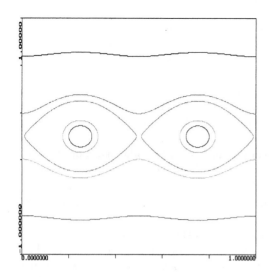

Fig. 2.4. Elliptical billiard (eccentricity $\epsilon = 0.3$). A Poincaré section is shown in the (S, p) plane. Different types of orbits can be distinguished (compare Fig. 2.3)

Figure 2.4 shows a Poincaré section in the (S, p) plane for a number of trajectories with different initial conditions. Two different kinds of orbits can be distinguished, as described above (compare also Fig. 2.3). For large p, the iterated phase space points trace out a more or less undulating curve. For small values of p, the phase space points alternate between two islands. When iterated, the orbit fills two disconnected closed curves. The two types of motion are separated in phase space by a separation curve (*separatrix*), that approximately satisfies the equation

$$p(S) \approx \pm \epsilon \sin(2\pi S) \tag{2.15}$$

(see below). In the center of the separatrix, we find a 2-periodic orbit along the minor axis (with length $l = 2b$). In other words, the mapping T^2 possesses a fixed point there. This orbit is stable, i.e., sufficiently close orbits always remain in its neighborhood [6]. This numerically observed stability can also be verified analytically using (2.38) and (2.41) in the Appendix. For the stability matrix of the diametrical two-bounce orbit, these yield

$$\mathbf{M}_2 = \begin{pmatrix} 2\left(\dfrac{l}{\rho}-1\right)^2 - 1 & 2l\left(1-\dfrac{l}{\rho}\right) \\ \dfrac{2}{\rho}\left(\dfrac{l}{\rho}-1\right)\left(2-\dfrac{l}{\rho}\right) & 2\left(\dfrac{l}{\rho}-1\right)^2 - 1 \end{pmatrix} , \tag{2.16}$$

where ρ is the radius of curvature at the points of impact. The stability condition $|\mathrm{Tr}\,\mathbf{M}_2| < 2$ [see (2.43)] is in this case simply

$$l < 2\rho , \tag{2.17}$$

which immediately implies stability of the orbit along the minor axis ($l = 2b < 2\rho$) and instability of the orbit along the major axis (here we have $l = 2a > 2\rho$).[1]

Orbits on the separatrix pass through both foci. These orbits clearly demonstrate the focal properties of an ellipse. The separatrix orbit with $p = 0$ is a 2-periodic orbit along the large diameter, which is unstable, in contrast to the small diameter orbit. This means that even the smallest deviations from the initial values yield orbits that do not remain in its vicinity. As with the circle, either the impact positions of an orbit constitute fixed points of the Poincaré presentation (rational circular frequency), implying a periodic orbit, or else they fill a curve densely. In this case we have an irrational circular frequency and the orbit densely fills a 2-dimensional surface in phase space, whose section is then an invariant curve in Fig. 2.4. Since this holds true for all orbits – invariant curves fill the phase space section completely – the elliptical billiard and the circular billiard represent integrable systems. Aside from the energy, there exists yet another conserved quantity, F. A simple consideration [7,9] shows that the product of the angular momenta with respect to the two foci remains unchanged in a collision with an elliptical boundary. After an elementary calculation described in the Appendix, one obtains for the invariant

$$F(\varphi, \alpha) = \frac{1}{2}\left[r^2 - \epsilon^2 a^2 + \left(a^2 - \epsilon^2 r^2 \cos^2\varphi\right)\cos(2\alpha)\right] . \tag{2.18}$$

Minima of F lie at $\varphi = \pi/2$ or $3\pi/2$ with $\alpha = \pi/2$, where F has the value $-\epsilon^2 a^2$. On the separatrix, we have $F = 0$ (the orbit passes through both foci). For small values of ϵ, one obtains the expression (2.15) for the separatrix. The circular billiard with its constant of motion, the angular momentum L, appears as a special case of the elliptical billiard.

2.4   'Typical' Billiards

The elliptical billiard is the only convex billiard with a smooth boundary curve that leads to integrable dynamics. This conjecture by Poritsky (1950) [10] was proven in 1991 by Amiran (see [11], p. 120). It is therefore a very atypical system. As an example of a 'typical' billiard, we shall study the boundary curve

$$r(\varphi) = 1 + \epsilon \cos\varphi . \tag{2.19}$$

(A similar one has been investigated by Robnik [8].) With increasing deformation ϵ, this curve deforms from a circle for $\epsilon = 0$, into a cardioid-like curve for $\epsilon = 1$. In the region between $0 \leq \epsilon \leq 0.5$, the boundary curve is convex. We shall limit ourselves to this case in the following discussion. Phase space presentations for $\epsilon = 0.1$ to $\epsilon = 0.5$ are displayed in Figs. 2.5a to d and 2.6. One encounters a qualitatively different behavior from the elliptical billiard. The differences become

---

[1] From (2.39), we easily deduce $\rho = a^2/b$ for the minor-axis orbit and $\rho = b^2/a$ for the major axis.

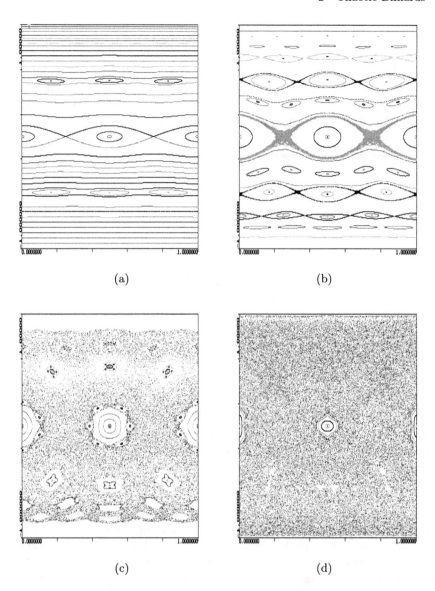

(a)   (b)

(c)   (d)

Fig. 2.5. Phase space diagram (S, p) for the typical billiard (2.19) with various values of the deformation parameter: (**a**) $\epsilon = 0.1$, (**b**) 0.2, (**c**) 0.3, and (**d**) 0.4

more distinct with growing ϵ. For small deformations $\epsilon = 0.1$ or $\epsilon = 0.2$, structurally speaking, the picture closely resembles that of an elliptical deformation of the circular billiard: the 2-periodic orbits for $p = 0$ are broken apart into a stable and an unstable orbit, which bounce back and forth between $(S, p) = (0, 0)$ and $(1/2, 0)$ or $(1/4, 0)$ and $(3/4, 0)$, respectively. One should note, however, that for the curve (2.19), the diameter of the billiard in the horizontal direction is independent of ϵ, while the diameter in the vertical direction increases with ϵ. For small values of ϵ, we find a maximum of the diameter at $\cos\varphi \approx \epsilon$ with magnitude $a \approx 1 + \epsilon^2/2$. An ellipse with the same diameters therefore has an eccentricity $[1 - (b/a)^2]^{1/2} \approx \epsilon$. If one approximates the billiard by an ellipse, the latter appears to be turned through $90°$ compared with the one studied in Sect. 2.3. Apart from this, the basic structures of the phase space diagrams for $\epsilon = 0.2$, for example, are very similar (compare Fig. 2.5b and Fig. 2.4).

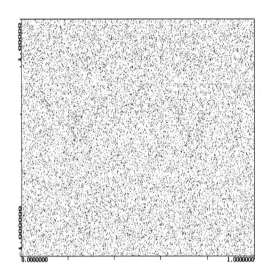

Fig. 2.6. Phase space diagram (S, p) for the typical billiard (2.19) with a deformation parameter $\epsilon = 0.5$

The resemblance with an elliptical billiard for small values of ϵ is merely superficial. Even for $\epsilon = 0.1$ some island chain structures can be clearly discerned, a first indication of the non-integrability of our billiard. If one studies the island structures more closely, one finds in the center of these islands an n-periodic orbit, for example the 3-periodic orbit in Fig. 2.5a for $\alpha = \pi/3$ ($p = \cos\alpha \approx 0.5$), the 4-periodic orbit for $\alpha \approx \pi/4$ ($p \approx 0.707$) or both 5-periodic orbits at $\alpha \approx \pi/5$ ($p \approx 0.809$) and $\alpha \approx 2\pi/5$ ($p \approx 0.309$). The centers of the island chains in Fig. 2.5a consist of stable fixed points of T^n for $n = 2, 3, \ldots$. Between these stable fixed points one finds unstable fixed points of T^n. This appears to be quite similar to the metamorphosis of the 2-periodic orbits in the transition circle $\rightarrow$ ellipse (Fig. 2.4). The second major difference in the phase space presentation of the ellipse is the appearance of irregular orbits. One can recognize orbits in the vicinity of the unstable fixed points whose Poincaré sections no longer trace out a curve but fill an area in phase space. This necessitates the non-existence of a constant of motion, and it is therefore quite astonishing that the majority of orbits behave as though one existed. Upon iteration, most orbits again fill an invariant curve.

We shall formulate the behavior of our circular billiard with a small perturbation in a somewhat different manner. The so-called KAM theorem is important here, named after the mathematical physicists A.N. Kolmogorov, V.I. Arnold, and J. Moser, [1–4]. In the proof of this theorem, a sufficiently differentiable potential is a prerequisite, and this is not fulfilled by our hard-bounded billiard. The KAM theorem can therefore only be applied with some caution. The KAM theorem states that, for a perturbed integrable system, those invariant curves for which the frequency ratio between radial and angular oscillations is sufficiently irrational remain unchanged. More precisely, all invariant orbits with

$$\left| \frac{\omega_1}{\omega_2} - \frac{m}{k} \right| > \frac{C(\epsilon)}{k^{5/2}} \tag{2.20}$$

for arbitrary coprime natural numbers m and k remain unchanged (with the assumption $\omega_1 < \omega_2$).

$C(\epsilon)$ is a constant depending on the perturbation ϵ of the integrable system and approaches 0 for $\epsilon = 0$. Around each rational frequency ratio $\omega_1/\omega_2 = m/k$, there exists a narrow region of size $C(\epsilon)k^{-5/2}$, in which (2.20) is not satisfied. For a more irrational frequency ratio, i.e., a larger denominator k in m/k, the region excluded by (2.20) appears to be narrower. Since the set of rational numbers ω_1/ω_2 is dense in the interval $[0, 1]$ and, for every rational frequency ratio m/k, an entire interval

$$\left| \frac{\omega_1}{\omega_2} - \frac{m}{k} \right| \leq \frac{C(\epsilon)}{k^{5/2}} \tag{2.21}$$

is excluded by (2.20), one could assume that (2.20) is practically never satisfied. However, this is not the case. A simple estimate yields, for the union of all intervals violating (2.20),

$$\sideset{}{'}\sum_{m<k} \left| \frac{\omega_1}{\omega_2} - \frac{m}{k} \right| < \sum_{k=1}^{\infty} \sum_{m=1}^{k} \left| \frac{\omega_1}{\omega_2} - \frac{m}{k} \right| \leq \sum_{k=1}^{\infty} k\, C(\epsilon) k^{-5/2} = C(\epsilon) \sum_{k=1}^{\infty} k^{-3/2} \, , \tag{2.22}$$

where the primed sum denotes that m and k have no common integer divisor. Since the sum $\sum_k k^{-3/2}$ converges, the interval sum in (2.22) goes with ϵ to zero [as $C(\epsilon)$ does], i.e., for a sufficiently small perturbation ϵ, the area which is not filled with invariant curves can be made arbitrarily small. The majority of invariant curves remain unchanged even in the perturbed system. A closer inspection of the phase space diagrams Figs. 2.5a, b confirms these statements.

The invariant curves in the zones excluded by the KAM condition are typically destroyed. Rational orbits with $\omega_1/\omega_2 = m/k$ decay into ℓk stable and ℓk unstable fixed points, where the natural number ℓ is often equal to one (compare with the theorem of Poincaré and Birkhoff [3]). This is also confirmed by Figs. 2.5a, b.

These statements are valid for small deformations ϵ. With increasing ϵ, the destroyed zones grow, and the chaotic area-filling orbits in phase space increase. The 'chaotic sea' is, at the beginning, still enclosed by intact invariant curves (compare Fig. 2.5b). With increasing ϵ, a growing number of invariant curves is destroyed. For $\epsilon = 0.3$, we find an extended chaotic region. All points in this wide chaotic band $|p| \approx 0.77$ are created by a single orbit. Only small islands with invariant curves remain. For $\epsilon = 0.4$, these regions are further diminished and for $\epsilon = 0.5$, we find only very small visible islands (for example at $\alpha = 90°$, $\varphi = 143.13°$). All points in Fig. 2.6 originate from a single orbit.

2.5   Further Computer Experiments

2.5.1   Uncertainty and Predictability

In the previous section we have seen that regular and irregular or 'chaotic' orbits can be distinguished by their phase space behavior. There exists yet another characteristic of chaotic orbits: in the regular case, initially neighboring orbits remain quite close. For irregular, chaotic orbits this does not hold true and trajectories separate extremely fast. Here very important questions arise concerning the long-time predictability of strictly deterministic processes.

For the simple case of a circular billiard, the effects of an initial uncertainty $(\delta\varphi_0, \delta\alpha_0)$ can easily be derived. We find

$$\delta\varphi_n = \delta\varphi_0 + 2\delta\alpha_0\, n\,, \tag{2.23}$$

i.e., the errors grow linearly. Similar relations are valid for other integrable cases and for regular orbits in general. The irregular case is completely different: here the orbits are very sensitive to small deviations in the initial conditions and the initial uncertainties grow exponentially:

$$|\delta\varphi_n| = |\delta\varphi_0|\, e^{\lambda n}\,, \tag{2.24}$$

where the coefficient λ is known as the *Lyapunov exponent*.

Figure 2.7 shows the angular separation for two orbits of the chaotic billiards (2.19), for $\epsilon = 0.5$ with $\varphi_0 = 0$ and $\alpha_0 = 70°$ and $70.1°$, as a function of n on a logarithmic scale. The exponential law (2.24) is approximately satisfied with

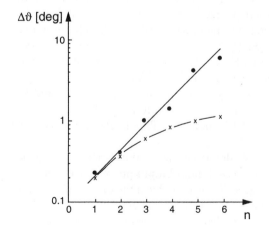

Fig. 2.7. Angle difference $\delta\varphi_n$ of two initially close orbits as a function of the number of impacts n for a chaotic orbit ($\bullet$) of the billiard (2.15) with $\epsilon = 0.5$ and for a regular orbit ($\times$) of the circle billiard

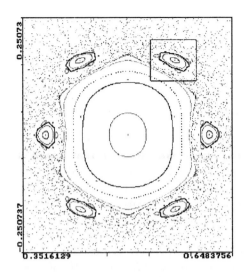

Fig. 2.8. Magnification of the neighborhood of the central fixed point in Fig. 2.5c

$\lambda = 0.7$. For comparison, the separation of two regular orbits of a circular billiard is also plotted, where the errors grow linearly.

One consequence of the depicted behavior is that an orbit is completely unpredictable if, for example, the angular precision reaches the value $\delta\varphi_{max} = 2\pi$. Under the conditions of Fig. 2.7, the destiny of the regular orbit is predictable up to about 1800 impacts with the boundary, whereas in the irregular case the number of predictable impacts is merely

$$n_{max} = \frac{1}{\lambda} \ln \frac{\delta\varphi_{max}}{\delta\varphi_0} \approx \frac{1}{0.7} \ln \frac{360}{0.1} \approx 12 \ . \tag{2.25}$$

Doubling the initial precision doubles the predictability of the regular case to 3600 impacts, whilst in the irregular case it only increases by $(\ln 2)/\lambda$ to $n_{max} \approx 12 + (\ln 2)/0.7 \approx 13$. This exponential increase in the uncertainty and the merely logarithmic increase in the predictability is characteristic of chaotic

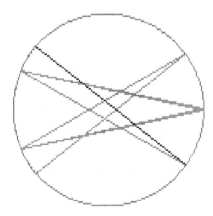

Fig. 2.9. Periodic orbit in the center of the six islands of Fig. 2.8

orbits. A predictability of about 100 impacts with the conditions of Fig. 2.7 yields a required initial precision in the angle of $2\pi e^{-70} \approx 2 \cdot 10^{-30}$. It is therefore impossible, or only possible with an unrealistic effort, to make long term predictions, even for strictly deterministic dynamics.

2.5.2 Fine Structure in Phase Space

The last example showing the complexity of non-integrable dynamics draws our attention to the fine structure of the phase space. We have seen that invariant curves with a rational frequency ratio break up into chains of stable and unstable fixed points. Now each of these stable fixed points is itself the center of a system of invariant curves, which can be even further broken apart. Figure 2.8 shows a magnification of the vicinity of the central 2-periodic fixed point of Fig. 2.5c. The neighborhood of this fixed point shows a system of invariant curves with narrow, broken-up rational orbits. This becomes especially evident for the six outer islands: they belong to a periodic orbit with period 12 in the center of these islands. Figure 2.9 shows these orbits in position space. Magnifying once more, for example, the island at the upper right corner of Fig. 2.8, one finds a similar island structure again, as displayed in Fig. 2.10. A stable fixed point of T^{12} is located in the center, which is again surrounded by a system of invariant curves and by regions broken apart into stable and unstable fixed points. Figure 2.11 shows a further magnification of an island in Fig. 2.10. In principle, we can continue this magnification into deepest depths. Each fixed point "is a microcosmos of the whole, down to arbitrarily small scales" (M.V. Berry [3]).

In the present study, we can only give a glimpse of the myriad of fascinating phenomena in chaotic dynamics, which can be illustrated in a simple fashion by

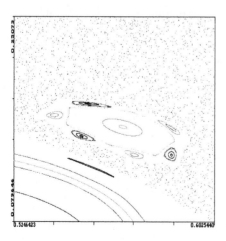

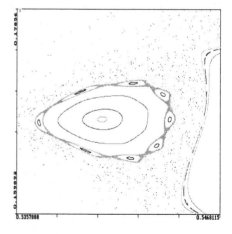

Fig. 2.10. Magnification of the island in the upper right corner of Fig. 2.8

Fig. 2.11. Magnification of the island in the upper left corner of Fig. 2.10

a billiard system. Many features cannot be mentioned here, such as the meta-morphosis of stable periodic orbits with increasing perturbation parameter by a sequence of period-doublings. We refer to the literature [1–4,6], where these and other phenomena are discussed.

2.6 Gravitational Billiards

Another class of frequently studied billiard systems are *gravitational billiards*. A mass point moves freely in a homogeneous gravitational field which is reflected elastically from a hard convex surface (in three space dimensions) or a hard convex boundary curve (in two space dimensions). Here we only consider the latter case. The most prominent example of such a system is the wedge billiard [12–15]. Figure 2.12 illustrates the motion in such a symmetric wedge, which consists of a sequence of reflections at the boundary (two straight lines for the wedge billiard) connected by parabolic trajectories.

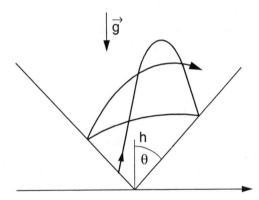

Fig. 2.12. Hopping of a mass point in a wedge under the action of a gravitational field

The dynamics can be worked out analytically in this case, leading to a simple 2-dimensional mapping, which can be most conveniently formulated in velocity space. We use the components v_x and v_y parallel and orthogonal to the wedge, respectively, or even better,

$$X = \frac{v_x}{\cos\theta}\,, \qquad Y = \frac{v_y}{\sin\theta}\,, \tag{2.26}$$

where θ is half the opening angle of the wedge. We can then derive a simple mapping equation

$$X_{n+1} = F(X_n, Y_n)\,, \quad Y_{n+1} = G(X_n, Y_n)\,, \tag{2.27}$$

where F and G are simple elementary functions (see [12–15] for details). Graph-ically, the iteration can be conveniently displayed in terms of the variables X_n and $Z_n = Y_n^2$, where the mapping

$$(X_n, Z_n) \longrightarrow (X_{n+1}, Z_{n+1})\,, \tag{2.28}$$

is again area preserving. These equations are used for the numerical iterations and the theoretical analysis. In the displayed velocity space sections, $x = v_x$ and $z = v_y^2$ are plotted. First of all, conservation of the energy

$$E = \frac{m}{2}\left(v_x^2 + v_y^2\right) + mgh \,, \tag{2.29}$$

where h is the height with respect to the vertex of the wedge or, using scaled units,

$$1 = x^2 + z + h \,, \tag{2.30}$$

restricts the dynamics in velocity space to the parabolic region

$$0 \le z = 1 - x^2 - h \le 1 - x^2 \,. \tag{2.31}$$

Trajectories directly hitting the vertex ($h = 0$) map to the parabolic boundary $z = 1 - x^2$. The base line $z = 0$ describes a sliding motion along the wedge and points on the line $x = 0$ are trajectories orthogonal to the wedge. Contrary to the billiard systems discussed in the preceding sections, the dynamics of the wedge billiard depends on a single parameter, the wedge angle θ.

The program WEDGE – again chosen from the collection of programs for chaotic systems [15] – can be used to explore the interesting dynamical features of the wedge billiard.

As an example, Fig. 2.13 shows the velocity space Poincaré sections for $\theta = 44°$ and $\theta = 17°$. One again observes an interesting island and sub-island structure (the islands are generated by stable periodic trajectories) embedded in a more or less extended chaotic sea. With varying wedge angle θ, the pattern undergoes interesting structural changes, related to bifurcations in the stability properties of the underlying skeleton of periodic orbits.

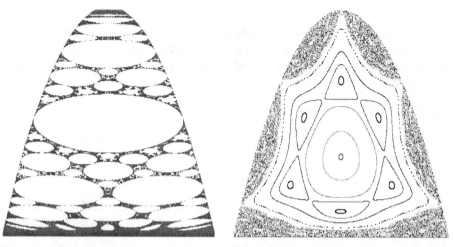

Fig. 2.13. Poincaré section in velocity space (x, z) for a wedge billiard with wedge half angle $\theta = 44°$ (*left*) and $\theta = 17°$ (*right*)

More numerical experiments exploring the wedge billiard can be found in H.J. Korsch, H.-J. Jodl: *Chaos – A Program Collection for the PC* [15]. Let us finally note that gravitational billiards with a smooth boundary curve have also been explored in the context of atom trapping in gravitational cavities (see [16,17], for example). In particular, it has been shown that the parabolic gravitational billiard is integrable.

2.7 Quantum Billiards

As demonstrated above, billiard systems help us to investigate and illustrate the fascinating features of chaotic dynamics. However, these systems are classical and, as we all know, on small scales, we enter the world of quantum mechanics. Immediately a seemingly simple question arises: does chaos also exist for quantum systems? Up to now, this question has not been fully answered. (As a simple exercise, the inexperienced reader should note that quantum dynamics is governed by linear equations, whereas classical chaos originates from nonlinearity.) In a milder formulation, one could pose the question: are there signatures of classical chaotic dynamics in quantum systems? A discussion of this problem can be found in recent textbooks by F. Haake [18] and M. Gutzwiller [19] (see also [20–22]).

Most of our knowledge in this field is again based on computational (and, more recently, experimental) studies of some model systems. Here we will confine ourselves to quantum studies of billiard systems, or closely related studies of wave dynamics in cavities.

Figure 2.14 shows an example of such a study of 'postmodern quantum mechanics' [21]. The figure shows the motion of a wave packet in the stadium billiard [21,23], which is classically ergodic. After a short time the quantum wave function is completely delocalized.

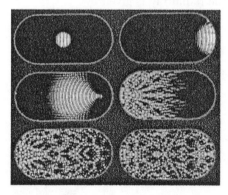

Fig. 2.14. Motion of an initially localized wavepacket in a stadium billiard [21]

Fig. 2.15. Quantum corral on a metal surface [24]

Related experimental studies investigate waves on a metal surface and quantum corrals [24], see Fig. 2.15, the transport of electrons through billiard shaped quantum dots (see, e.g., [25]) or wave propagation in various macroscopic systems, which serve as substitute for quantum dynamics because of the similarity between the Helmholtz and the Schrödinger equations. Such systems are microwave cavities (see [26–28], for example), light propagation in optical cavities [29], water surface waves in water tanks [30], or even vibrating soap films [31].

In such studies of quantum (or wave) dynamics, the classical chaos is manifested in different ways. A prominent example is the nearest-neighbour spacing distribution [18,19], which is Poisson distributed for classically integrable and Wigner distributed for classically ergodic systems. In many cases, one also observes so-called 'scars', i.e., states whose wave function localizes on (unstable) classically periodic orbits. It is a topic of contemporary semiclassical analysis to investigate the connection between the classical periodic orbits (note that these orbits form the skeleton of the classical phase space structure) and the quantum eigenstates [19].

Finally, we would like to mention the recent experiments by M. Raizen investigating the (quantum) dynamics of atoms in strong laser fields, where a gravitational wedge billiard (compare the end of the preceding section) is created by (blue-detuned) laser light which reflects the atoms. The chaotic motion of the atoms is then used in a cooling scheme, aiming to produce a Bose–Einstein condensate [32].

2.8 Appendixes

2.8.1 Billiard Mapping

Following [9], we give a brief description of the evaluation of the billiard mapping. Let us assume that the angles φ_n and α_n are given. Then the angle ϑ_n between the positive direction of the tangent and the radial ray is given by

$$\tan \vartheta_n = \left. \frac{r(\varphi)}{dr/d\varphi} \right|_n , \tag{2.32}$$

and the direction β_n of the trajectory (i.e., the angle it forms with the $\varphi = 0$ direction) is

$$\beta_n = \pi + \varphi_n + \alpha_n - \vartheta_n . \tag{2.33}$$

Hence, the straight line trajectory after impact n is given in polar form by

$$R(\varphi) = r(\varphi_n) \frac{\sin (\beta_n - \varphi_n)}{\sin (\beta_n - \varphi)} . \tag{2.34}$$

The next impact coordinate φ_{n+1} is determined by the intersection of the line (2.34) with the boundary $r(\varphi)$, i.e., the solution of the equation

$$R(\varphi) - r(\varphi) = 0 . \tag{2.35}$$

When $r(\varphi)$ is convex, there exists only one further solution, φ_{n+1}, in addition to φ_n, which is numerically extracted by the Newton iteration scheme. The angle between the trajectory and the tangent at φ_{n+1} is

$$\alpha_{n+1} = \varphi_{n+1} - \varphi_n + \vartheta_n - \vartheta_{n+1} - \alpha_n , \tag{2.36}$$

as illustrated in Fig. 2.16. Further boundary reflections are computed by repeating these steps.

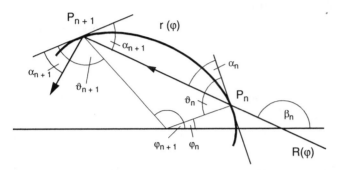

Fig. 2.16. Billiard mapping $(\alpha_n, \varphi_n) \to (\alpha_{n+1}, \varphi_{n+1})$

2.8.2 Stability Map

For an investigation of the stability properties of an orbit, a useful linearization of the billiard Poincaré mapping $(S_{n+1}, p_{n+1}) \xrightarrow{T} (S_n, p_n)$ and its linearized matrix form

$$\begin{pmatrix} dS_{n+1} \\ dp_{n+1} \end{pmatrix} = \frac{\partial(S_{n+1}, p_{n+1})}{\partial(S_n, p_n)} \begin{pmatrix} dS_n \\ dp_n \end{pmatrix} = \mathbf{M}_{n+1,n} \begin{pmatrix} dS_n \\ dp_n \end{pmatrix} \tag{2.37}$$

is known [6]. We formulate it here for $n = 0$:

$$\mathbf{M}_{1,0} = \begin{pmatrix} -\dfrac{q_0}{q_1} + \dfrac{l_{10}}{q_1 \rho_0} & -\dfrac{l_{10}}{q_0 q_1} \\[2ex] -\dfrac{l_{10}}{\rho_0 \rho_1} + \dfrac{q_1}{\rho_0} + \dfrac{q_0}{\rho_1} & -\dfrac{q_1}{q_0} + \dfrac{l_{10}}{q_0 \rho_1} \end{pmatrix} , \tag{2.38}$$

where $q_i = \sin \varphi_i$. The length of the straight line segment from P_0 to P_1 is denoted by l_{10}, and ρ_i is the radius of curvature at φ_i given by

$$\rho(\varphi) = \frac{(r^2 + r'^2)^{3/2}}{r^2 + 2r'^2 - rr''} , \tag{2.39}$$

with $r = r(\varphi)$, $r' = dr/d\varphi$ and $r'' = d^2r/d\varphi^2$. The determinant of $\mathbf{M}_{1,0}$ is equal to unity, i.e., the mapping T is area preserving.

The linearization of the iterated map $T^n(S_0, p_0)$ is

$$\mathbf{M}_{n,0} = \prod_{i=1}^{n} \mathbf{M}_{i,i-1} , \qquad (2.40)$$

where the $\mathbf{M}_{i,i-1}$ have the form (2.38). For the special case of a periodic n-bounce orbit, we have $(S_n, p_n) = (S_0, p_0)$, and the deviation map

$$\mathbf{M}_n = \mathbf{M}_{0,n-1} \mathbf{M}_{n-1,n-2} \ldots \mathbf{M}_{1,0} \qquad (2.41)$$

determines its stability: the eigenvalues of the stability matrix $\mathbf{M}$ are given by

$$\lambda_\pm = \frac{1}{2} \left[\mathrm{Tr}\,\mathbf{M} \pm \sqrt{(\mathrm{Tr}\,\mathbf{M})^2 - 4} \right] , \qquad (2.42)$$

($\det \mathbf{M} = 1$) and therefore the condition for stability is (see [15], p. 40, for example)

$$|\mathrm{Tr}\,\mathbf{M}| < 2 . \qquad (2.43)$$

In this case the eigenvalues are complex conjugate with modulus unity and small deviations from the fixed point remain small, whereas in the opposite case we have a pair of real-valued eigenvalues, where one of them has modulus bigger than one, i.e., a typical deviation from the fixed point will blow up.

2.8.3 Elliptical Billiard: Constant of Motion

It is easy to construct the invariant for the elliptical billiard. Let r_1 be the vector from the focal point F_1 to the point of impact. Before the collision with the boundary, the angular momentum with respect to F_1 is

$$L_1 = p\,r_1 \sin \gamma_1 , \qquad (2.44)$$

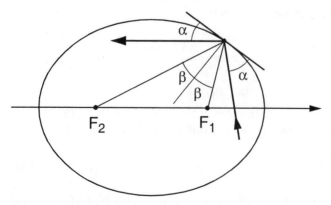

Fig. 2.17. Billiard mapping $(\alpha_n, \varphi_n) \to (\alpha_{n+1}, \varphi_{n+1})$

where p is the (constant) momentum and γ_1 is the angle between r_1 and the trajectory. This angle is determined by the angle α between the trajectory and the tangent, and the angle β between the focal ray and the normal at the point of impact by $\gamma_1 = \pi/2 - \alpha - \beta$ (compare Fig. 2.17). Similarly we have

$$L_2 = p\,r_2 \sin\gamma_2 \,, \tag{2.45}$$

with $\gamma_2 = \pi/2 - \alpha + \beta$. (Note that the normal bisects the angle between the focal rays.) After the collision with the boundary, we have angular momenta $L_1 = p\,r_1 \sin\gamma_1'$ and $L_2 = p\,r_2 \sin\gamma_2'$ with $\gamma_1' = \gamma_2$ and $\gamma_2' = \gamma_1$, as can be seen from Fig. 2.17. The product is thus conserved, i.e.,

$$L_1'L_2' = L_1L_2 \,, \tag{2.46}$$

and elementary algebra yields (2.18) for the constant of motion $F = L_1L_2/p^2$.

References

1. A.J. Lichtenberg, M.A. Lieberman: *Regular and Stochastic Motion* (Springer, New York 1991)
2. H.G. Schuster: *Deterministic Chaos* (VCH, Weinheim 1988)
3. M.V. Berry: 'Regular and irregular motion'. In: *Topics in Nonlinear Dynamics*, ed. by S. Jorna (Am. Inst. Phys. Conf. Proc. Vol. 46, 1978) p. 16. Reprinted in R.S. MacKay, J.D. Meiss: *Hamiltonian Dynamical Systems* (Adam Hilger, Bristol 1987)
4. M.V. Berry: 'Semi-classical mechanics of regular and irregular motion' In: *Les–Houches Summer School 1981 on Chaotic Behaviour of Deterministic Systems*, ed. by G. Iooss, H.G. Helleman, R. Stora, (North-Holland, Amsterdam 1983) p. 171
5. V.I. Arnold: *Mathematical Methods of Classical Mechanics* (Springer, New York 1978)
6. M.V. Berry: Eur. J. Phys. **2**, 91 (1981)
7. H.J. Korsch, B. Mirbach, H.-J. Jodl: Praxis d. Naturwiss. (Phys.) **36/7**, 2 (1987)
8. M. Robnik: J. Phys. A **16**, 3971 (1983)
9. H.J. Korsch, H.-J. Jodl: *Chaos – A Program Collection for the PC* (Springer, Berlin 1998)
10. H. Poritsky: Ann. Math. **51**, 446 (1950)
11. V.F. Lazutkin: *KAM Theory and Semiclassical Approximations to Eigenfunctions* (Springer, New York 1991)
12. H.E. Lehtihet, B.N. Miller: Physica D **21**, 93 (1986)
13. B.N. Miller, H. Lehtihet: Chaotic Dynamics: An Instructive Model. In: *Computers in Physics Instruction*. ed. by E.F. Redish, J.S. Risley, (Addison–Wesley, New York 1990)
14. P.H. Richter, H.-J. Scholz, A. Wittek: Nonlinearity **3**, 45 (1990)
15. H.J. Korsch, H.-J. Jodl: *Chaos – A Program Collection for the PC*, 2nd edn. (Springer, Heidelberg, New York 1998)
16. H.J. Korsch, J. Lang: J. Phys. A **24**, 45 (1990)
17. H. Wallis, J. Dalibard, C. Cohen-Tanoudji: Appl. Phys. B **54**, 407 (1992)
18. F. Haake: *Quantum Signatures of Chaos* (Springer, Berlin, Heidelberg, New York 1992)

36   Hans Jürgen Korsch and Frank Zimmer

19. M.C. Gutzwiller: *Chaos in Classical and Quantum Mechanics* (Springer, New York 1990)
20. M.C. Gutzwiller: Scientific American (Jan. 1992) p.26. See also www.scitec.auckland.ac.nz/~king/Preprints/book/quantcos/qchao/quantc.htm
21. E.J. Heller, S. Tomsovic: Physics Today (July 1993) p.38. See also http://www.physics.wsu.edu/Research/tomsovic/chaospage.htm
22. R. Blümel, W.P. Reinhardt: *Chaos in atomic physics* (Cambridge University Press, Cambridge 1997)
23. S. Tomsovic, E.J. Heller: Phys. Rev. E **47**, 282 (1993)
24. M.F. Crommie, C.P. Lut, D.M. Eigler, E.J. Heller: Surface Review and Letters **2**, 127 (1995). See also http://www.almaden.ibm.com/vis/stm/corral.html
25. T. Dittrich, P. Hänggi, G.-L. Ingold, B. Kramer, G. Schön, W. Zwerger: *Quantum Transport and Dissipation* (Wiley-VCH, Weinheim 1998)
26. H.-D. Gräf, H.L. Harney, H. Lengeler, C.H. Lewenkopf, C. Rangacharyulu, A. Richter, P. Schardt, H.A. Weidenmüller: Phys. Rev. Lett. **69**, 1296 (1992)
27. J. Stein, H.-J. Stöckmann: Phys. Rev. Lett. **68**, 2867 (1992)
28. A. Richter: 'Playing Billiards with Microwaves – Quantum Manifestations of Classical Chaos'. In: *Emerging Applications of Number Theory*, ed. by D.A. Hejhal, J. Friedman, M.C. Gutzwiller, A.M. Odlyzk, The IMA Volumes in Mathematics and its Applications, Vol. 109 (Springer, New York 1999) p. 479
29. J.U. Nöckel, A. D. Stone: Nature **385**, 45 (1997)
30. R. Blümel, I.H. Davidson, W.P. Reinhardt, H. Lin, M. Sharnoff: Phys. Rev. Lett. **64**, 241 (1990)
31. E. Arcos, G. Báez, P.A. Cuatláyol, M.L.H. Prian, R.A. Méndez-Saldaña: Am. J. Phys. **66**, 601 (1998)
32. M. Raizen, N. Davidson: private communication (2000)

3 Combinatorial Optimization and High Dimensional Billiards

Pál Ruján

Summary. Combinatorial optimization deals with algorithms for finding extrema of functions subject to a (possibly large) number of constraints. Bayesian inference also requires averages over such extrema. In this chapter we show how simple dynamic systems like billiards can be used to find solutions for such problems. The topics covered are linear and quadratic programming, classification, and Bayesian mixture problems.

3.1 Introduction

Everyone who has studied physics remembers with somewhat mixed feelings the d'Alembert principle and the Lagrange equations of motion of the first kind. On the one hand, one could argue that it was exactly the cumbersome mathematics of taking into account boundary conditions which led to the development of Hamiltonian mechanics. On the other hand, like it or not, most problems of practical interest consist in finding the extrema of simple functions, mostly linear or quadratic, subject to a large set of constraints, mostly linear (in)equalities.

In the following I will consider three different problems and touch superficially upon a few others. These problems are: linear programming, the Bayes perceptron, and the mixture inference problem. The presentation is mostly geometrical: once a problem and its ramifications are understood geometrically, it is easy to write it down in algebraic form, and ultimately, to devise appropriate algorithms.

One of the simplest Hamiltonian systems is a free particle moving in a closed container of given shape. Assuming that collisions between the particle and the walls are perfectly elastic, we are led to a *billiard*. The analysis of all possible phase space trajectories is exciting physics, and the quantization of such systems even more so. This paper is, however, not about new results concerning dynamic systems and ergodic theory. Our goal is to use physical knowledge in order to devise working algorithms for effectively solving difficult problems.

Consider the convex polygon shown in Fig. 3.1. In general terms, I will deal here with convex polyhedra living in d-dimensions. Such polyhedra can be defined in different ways.

1. Give the set of its vertices (here A, B, C, D and E)

$$x_A, \; x_B, \ldots, x_E \; \in \; \mathcal{R}^d \; . \tag{3.1}$$

2. Give a set of normal vectors pointing inside the polyhedron (n_A, n_B, etc.) as a set of inequalities:

$$n_A^T x \geq b_A \, ,$$
$$n_B^T x \geq b_B \, ,$$
$$n_C^T x \geq b_C \, ,$$
$$n_D^T x \geq b_D \, ,$$
$$n_E^T x \geq b_E \, ,$$

(3.2)

where $a^T b$ denotes here the scalar product of vectors a and b.

3. Give a set of linear equations as

$$Ax = b \, , \qquad x \geq 0 \, .$$

(3.3)

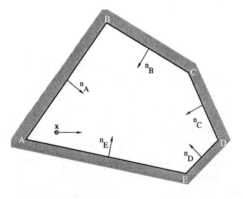

Fig. 3.1. A convex polygon

Form (2) is equivalent to form (3), as explained below. In low dimensions, we can easily transform (1) into (2) and vice versa. Since the number of vertices, lines, facets, etc., grows exponentially with the dimensionality, this transformation becomes unfeasible in high dimensions. The difficulty of the problem we have to solve therefore depends crucially on the available representation of the convex polyhedron.

Any point of a convex polyhedron can be expressed as a convex combination of its N vertices $\{\xi^{(i)}\}_1^N$:

$$x = \sum_{i=1}^{N} a_i \xi^{(i)} \, ,$$

(3.4)

where

$$\sum_{i=1}^{N} a_i = 1 \quad \text{and} \quad a_i \geq 0 \, , \quad i = 1, \dots, N \, .$$

(3.5)

Note that a polyhedron might have many more vertices than the space dimensionality, so that the linear combination (3.4) is in general not unique. A vertex of the polyhedron i has by definition the form $a_i = 1$, $a_{j \neq i} = 0$, a segment-like facet is defined by $a_i + a_j = 1$, $a_{k \neq i, j} = 0$, etc.

As exposed here, it might not be clear why this type of problem is *combinatorial*. In reality, a typical problem looks more like Fig. 3.2, where we do not know a priori which of the constraints are stringent and which ones are shadowed by others.

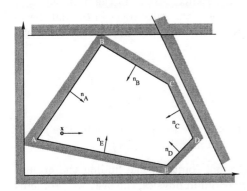

Fig. 3.2. A set of redundant linear inequalities

One aspect of the combinatorial problem is to determine the *stringent* inequalities, those which form in our case the faces of the internal polyhedron (polygon). Once the stringent inequalities and the corresponding polyhedron have been found, we must find at least one *feasible* solution, a vector x satisfying all constraints. The space of all such vectors is called the feasibility space in mathematical programming or *version* space in pattern recognition and Artificial Intelligence. Another aspect of the combinatorial problem is the sheer number of constraints. Many difficult optimization problems, like for instance the traveling salesman problem, lead to an exponential number of constraints. An example will be given below.

In the rest of this paper we will consider three different problems and discuss various algorithms for solving them.

1. **The Linear Programming Problem**
   Given a cost vector c with N components, find

   $$\max_{x} c^{\mathrm{T}} x \tag{3.6}$$

   satisfying the set of M inequalities (3.2) and $x \geq 0$.

2. **The Perceptron Bayes Point Problem**
   Find the theoretically best classifier for a linear classification machine. This is equivalent to the following. Given a set of inequalities (3.2), compute the center of mass of the polyhedron. As discussed later, the polyhedron is in this case either open (a polyhedral cone) or defined on the intersection of a polyhedral cone with the unit hypersphere.

We will also discuss the main problems occurring when kernels are introduced and the space dimensionality becomes infinite.

3. **The Bayesian Inference Problem**
   for the Mixture Assignment Problem

   Consider M substances defined through their (normalized) spectral decomposition. Maximally, N spectral lines are considered. A large database of such substances is built up, with $M > N$. Next, the spectrum of an unknown substance is measured. We are asked to make a scientifically sound estimate of its origin in terms of the substances stored in the database.

   Define a set of M vertices in N dimensions (the substances in the database) as $\{\boldsymbol{\xi}^{(\alpha)}\}_1^M$, an additional set of N equalities, and a point $\boldsymbol{x}_0$ (the unknown substance). Then,

$$\boldsymbol{x}_0 = \sum_{\alpha=1}^{M} a_\alpha \boldsymbol{\xi}^{(\alpha)} , \tag{3.7}$$

   where $a_\alpha \geq 0$ and $\sum_{\alpha=1}^{M} a_\alpha = 1$. Estimate the marginal probabilities $p(a_\alpha)$, $\alpha = 1, \ldots, M$ over all allowed convex decompositions (3.7).

While the first problem asks for an extremal solution, like finding the ground state of a system with many degrees of freedom in physics, the last two problems require finding optimal solutions based on posteriori probabilities. I call such problems *statistical* combinatorial optimization problems, since they require the computation of certain averages over the whole feasibility (version) space.

3.2 Billiards

A *billiard* is defined as a closed space region (compact set) $\mathcal{P} \in \mathbb{R}^N$ of the N-dimensional vector space. The boundaries of a billiard are piecewise smooth functions. Within these boundaries, a point mass (ball) moves freely, except for elastic collisions with the enclosing walls. Hence, the absolute value of the momentum is preserved and the phase space $\mathcal{B} = \mathcal{P} \times \mathcal{S}^{N-1}$, is the direct product of $\mathcal{P}$ and $\mathcal{S}^{N-1}$, the surface of the N-dimensional unit velocity sphere. Such a simple Hamiltonian dynamics defines a flow and its Poincaré map an automorphism. The mathematicians have defined a finely tuned hierarchy of notions related to such dynamic systems. For instance, simple ergodicity as implied by the Birkhoff–Hincsin theorem means that the average of any integrable function defined on the phase space over a single but very long trajectory equals the spatial mean, except for a set of zero measure. Furthermore, integrable functions invariant under the dynamics must be constant. From a practical point of view this means that almost all very long trajectories will cover the phase space uniformly.

Properties like mixing (Kolmogorov mixing) are stronger than ergodicity. They require the flow to mix uniformly different subsets of $\mathcal{B}$. In hyperbolic systems, one can go even further and construct Markov partitions defined on symbolic dynamics and eventually prove related central limit theorems.

Not all convex billiards are ergodic. Notable exceptions are ellipsoidal billiards, which can be solved by a proper separation of variables [1]. Jacobi already knew that a trajectory starting close to and along the boundaries of an ellipse cannot reach a central region bounded by the so-called caustics. This simple fact seems to be ignored by the many architects building round concert halls.

In addition to billiards which can be solved by a separation of variables, there are a few other exactly soluble polyhedral billiards [2]. Such solutions are intimately related to the reflection method – the billiard perfectly tiles the entire space. A notable example is the equilateral triangle billiard, first solved by Lamé in 1852. Other examples of such integrable billiards can be obtained by mapping an exactly soluble 1-dimensional many-particle system into a one-particle high-dimensional billiard [3].

Apart from these exceptions, small perturbations in the form of the billiard usually destroy integrability and lead to chaotic behavior. For example, the stadium billiard (two half-circles joined by two parallel lines) is ergodic in a strong sense: the so-called metric entropy is non-vanishing – for definitions and details see [4]. The dynamics induced by the billiard is hyperbolic if at any point in phase space there are both expanding (unstable) and shrinking (stable) manifolds. A famous example is Sinai's reformulation of the Lorentz gas problem. Deep mathematical methods were needed to prove the Kolmogorov mixing property and in constructing the Markov partitions for the symbolic dynamics of such systems [5].

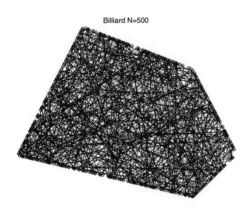

Billiard N=500

Fig. 3.3. A trajectory after 500 collisions

The question as to whether a particular billiard is ergodic or not can be decided in principle by solving the Schrödinger problem for a free particle trapped in the billiard box. If the eigenfunctions corresponding to the high energy modes are roughly constant, then the billiard is ergodic. Only a few general results are known for such quantum problems. In fact, there are not many theoretical results concerning the ergodic properties of convex polyhedral billiards in high dimensions. If all angles of the polyhedra are rational, then the billiard is weakly ergodic in the sense that the velocity direction will reach only rational angles (relative to the initial direction). In general, as long as two neighboring trajectories collide with the same polyhedral faces, their distance will grow only linearly. Once they are far enough apart to collide with different faces of the polyhedron, their distance will abruptly increase. Hence, except for very special cases with high symmetry, it seems unlikely that high-dimensional convex polyhedra such as those generated by the training examples will fail to be ergodic.

Two pictures are given to illustrate this concept: Fig. 3.3 shows the space covered by a trajectory after 500 collisions while Fig. 3.4 shows the collision points only. Both the volume and the surface of the polygon are already well covered by the trajectory.

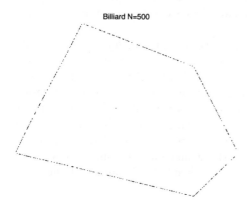

Billiard N=500

Fig. 3.4. The 500 collision points

3.3   Linear Programming

The *general* linear programming problem (LP) is defined as follows. Find

$$\max_{\boldsymbol{x}} \boldsymbol{c}^{\mathrm{T}} \boldsymbol{x} \, , \tag{3.8}$$

subject to m_1 inequalities of the form

$$\boldsymbol{\xi}^{(\alpha_+)} < b_{\alpha_+} \, , \quad \alpha_+ = 1, \, \ldots, \, m_1 \, , \tag{3.9}$$

m_2 equalities

$$\boldsymbol{\xi}^{(\alpha)} = b_\alpha \, , \quad \alpha = m_1 + 1, \, \ldots, \, m_1 + m_2 \, , \tag{3.10}$$

and m_3 inequalities

$$\boldsymbol{\xi}^{(\alpha_-)} > b_{\alpha_-} \, , \quad \alpha_- = m_1 + m_2 + 1, \, \ldots, \, m_1 + m_2 + m_3 \, . \tag{3.11}$$

Consider the following simple elementary transformations.

- If $a < b$ then, from $a + s = b$, it follows that $s \geq 0$. The auxiliary variable s is called a *slack* variable.

- If x is arbitrary, write x as $x = x_+ - x_-$, where now x_+, $x_- \geq 0$.
- If a system of linear equations has N unknowns but only $M < N$ linearly independent variables, parameterize the solution in terms of $N - M$ arbitrary variables.

Using these transformations, one can rewrite the general LP format in either the *canonical* format with constraints of the form

$$\mathbf{A}x \geq r \,, \tag{3.12}$$
$$x \geq 0 \,, \tag{3.13}$$

or the *standard* form

$$\mathbf{A}x = b \,, \tag{3.14}$$
$$x \geq 0 \,. \tag{3.15}$$

For the geometrical approach, the appropriate form is the canonical form (3.12) and (3.13). For example, consider the constraints

$$\begin{aligned}
x_1 + x_2 + x_3 &\leq 4 \,, \\
x_1 &\leq 2 \,, \\
x_3 &\leq 3 \,, \\
3x_2 + x_3 &\leq 6 \,, \\
x_1 &\geq 0 \,, \\
x_2 &\geq 0 \,, \\
x_3 &\geq 0 \,,
\end{aligned} \tag{3.16}$$

which we will use to illustrate the various algorithms. The corresponding polytope is shown in Fig. 3.5.

The cost vector c is shown as a dotted line. It could point in either of the two directions, depending on whether we seek a minimum or a maximum of the linear function (3.8). All algorithms described below must start from a feasible point belonging to the convex hull. There are only three possibilities: either the feasibility (version) space is empty, the polytope is open and c has its extremum at infinity, or (as in our case) the solution is one of the faces of the polytope (most probably a vertex). However, if c points in the $-z$ direction, then the upper triangle corresponds to the minimal cost.

3.3.1 The Simplex Algorithm

The simplex algorithm was developed by Dantzig and implements the following simple but effective idea: once we have a feasible starting solution (usually a vertex), we move from one vertex to the next in such a way that, in each step, the cost does not increase. Eventually, the optimal vertex or facet is reached. The starting vertex can always be defined by extending the dimension of the problem (the number of variables in x) so that all (hyper)planes will pass through the

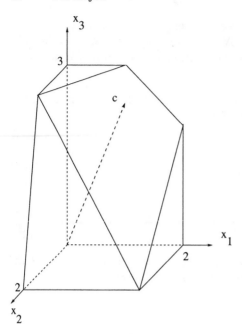

Fig. 3.5. The polytope defined by the constraints (3.16). The vector c indicates a possible cost vector

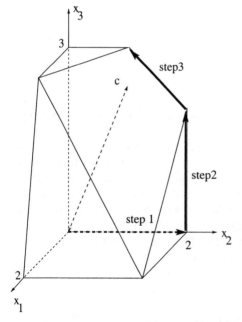

Fig. 3.6. The simplex algorithm moves from one vertex to a neighboring vertex

origin, which is then a good starting point. The first phase of the algorithm then tries to remove these auxiliary variables by maximizing an 'artificial' cost function $c = \sum_{\mathrm{aux}} x_{\mathrm{aux}}$. This type of solution is illustrated in Fig. 3.6.

In practice, the simplex algorithm runs quite fast (mostly linear in N or M), but one can construct examples when it has to visit a finite fraction of all vertices, and this leads to exponential run times [6]. Since mathematicians are mostly interested in worst-case analysis, the question is whether the LP problem is worst-case solvable in polynomial time (class $\mathcal{P}$) or not. The breakthrough came with the work of L.G. Khachian, who published the ellipsoid algorithm in 1979, showing that linear programming is in the worst-case polynomial on N, M, and the number of significant digits of the solution vector. For more details, see [6].

3.3.2   The Internal Point Algorithms

The ellipsoid algorithm was the first among a family of algorithms called interior point algorithms. While the simplex and dual–primal algorithms move on the 'surface' of the version polytope, the ellipsoid and other algorithms move 'inside' the feasibility space. The ellipsoid algorithm is rather complex but the main idea is simple: given a polytope, fit an ellipsoid to it in such a way that, by cutting the space with a plane perpendicular to the cost vector, the feasibility space is roughly halved. In this way, given a desired numerical precision, the optimal solution will be localized in a polynomial number of steps.

The best worst-case algorithms known today use the concept of the 'analytic center of a set of inequalities' introduced by Sonnevend [7]. See for instance the article by Vaidya [17].

3.3.3   The Billiard Algorithm I

Instead of these complex methods let us consider a simpler approach. Each inequality defines a hyperplane (halfspace), whose normal always points towards the 'inside' of the polytope. Let us consider these hyperplanes as made from a material behaving like a half mirror. Each hyperplane would then specularly reflect a light ray incident against the normal, but let it pass through when incident from the other side. If such a ray enters the polytope, it will remain trapped inside. The most difficult part of the algorithm is the initialization (finding an interior point), because we do not know a priori if the problem has a solution at all, or if that solution is bounded. However, this difficulty is common to all algorithms described here and, for 'well behaved' problems, it is not that difficult to 'shoot' a ray into the feasible polytope.

Once this is achieved, we compute the trajectory's collision points and their cost. Once a collision point gives a better cost, we move a hyperplane whose normal is parallel to the cost vector up to the collision point and restart the billiard. We measure the projections of the trajectory on the cost vector. The algorithm converges when its fluctuations become smaller than the required precision.

3.3.4   The Billiard Algorithm II

Note that, when solving the LP or similar optimization problems, we do not really need ergodic dynamics. In fact, we would like to ensure that all trajectories have an attractive sink at the optimum. Therefore, instead of specular reflections, we bend the reflection angle as much as allowed towards the cost vector. This strategy is illustrated in Fig. 3.7.

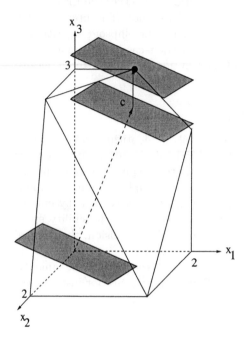

Fig. 3.7. The billiard algorithm moves from facet to facet. It starts from the origin and first runs parallel to the cost vector (here we seek the maximum cost, not the minimum). When a collision point has a better cost, a large part of the feasibility region will be cut by a plane whose normal is the cost vector (shown schematically)

3.3.5 Difficult (Integer) Linear Programming Problems

If in addition to the usual conditions one requires the solution vector x to have integer components, the problem is called *integer linear programming* (ILP) and is $\mathcal{NP}$-hard. The naive expectation that we might solve the problem as if it were a normal LP ('relaxed problem') and then look for the nearest integer x can go very wrong, as illustrated in Fig. 3.8.

One practical solution to this problem is to introduce *cuts*, additional inequalities reducing the feasibility region, but leaving all feasible integer solutions intact. It is not yet clear how the billiard method could accommodate the introduction of such additional Gomory cuts. We need a clever trick forcing the trajectory to run only along paths whose normals are Miller indices.

All $\mathcal{NP}$-hard problems are related through polynomial time algorithms. For example, it is shown below how to formulate the traveling salesman problem as an ILP problem. Assume we have $N + 1$ cities $0, 1, 2, \ldots, N$ and intercity distances $[c_{ij}]$. Let the variable x_{ij} be equal to 1, if that edge is in the tour and

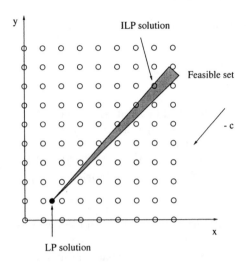

Fig. 3.8. The relaxed LP (condition on integer solution removed) and the ILP solutions can be very far apart

0, if not. We then impose

$$\min z = \sum_{i,j=0}^{N} c_{ij}x_{ij} \ , \tag{3.17}$$

$$x_{ij} = \{0,1\} \ , \quad \forall \, i,j \ , \tag{3.18}$$

$$\sum_{i=0}^{N} x_{ij} = 1 \ , \quad j = 0, \dots, N \ , \tag{3.19}$$

$$\sum_{j=0}^{N} x_{ij} = 1 \ , \quad i = 1, \dots, N \ , \tag{3.20}$$

requiring only one edge to be incident and outward bound from each node. As formulated above, however, this corresponds to the *assignment* problem, which allows for disjoint (cycles) subtours. In order to exclude this possibility, consider all possible decompositions of the tour S into two disjoint subsets, S_1 and S_2, $S = S_1 \cup S_2$. For each pair of such city subsets, one must satisfy

$$\sum_{i \in S_1, \, j \in S_2} x_{ij} = 1 \ . \tag{3.21}$$

From this formulation, it becomes clear where time is needed when trying to solve the TSP exactly. Of course, there are more economical formulations of the TSP, leading to mixed LPs (with both real and integer variables) [19,20]. These are the methods used together with a bound and cut strategy to solve quite large TSP problems exactly (over 20 000 cities).

3.4   The Bayes Perceptron

As shown in Fig. 3.9, a *perceptron* [8] is a network consisting of N (binary or real) inputs x_i and a single binary output neuron. The output neuron first sums all inputs with the corresponding synaptic weights w_i, then performs a threshold operation according to

$$\sigma = \text{sign}\left(\sum_{i=1}^{N} x_i w_i - \theta\right) = \text{sign}(x^{\text{T}}w - \theta) = \pm 1 . \tag{3.22}$$

In general, the synaptic weights and the threshold can be arbitrary real numbers. A network where w has only binary components is called a *binary* perceptron. The output unit has binary values $\sigma = \pm 1$ and labels the class to which the input vector belongs. Both the weight vector $w = (w_1, w_2, \ldots, w_N)$ and the threshold θ should be determined from a predefined set of M labeled examples (the training set).

Our protocol assumes that both the training and test examples are drawn independently and identically from the distribution $P(\xi)$. The simplest way to generate such a rule is to define a *teacher* network with the same structure, as shown in Fig. 3.9, providing the correct class label $\sigma = \pm 1$ for each input vector. Given a set of M examples $\{\xi^{(\alpha)}\}_{\alpha=1}^{M}$, $\xi^{(\alpha)} \in \mathbb{R}^N$ and their corresponding binary class label $\sigma_\alpha = \pm 1$, the task of the learning algorithm is to find a *student* network which mimics the teacher, by choosing the network parameters in such a way as to classify the training examples correctly. Equation (3.22) implies that the learning task consists in finding a hyperplane $w^{\text{T}}x = \theta$ separating the positive from the negative labeled examples:

$$\begin{aligned}
w^{\text{T}}\xi^{(\alpha)} &\geq \theta_1 , \quad \sigma_\alpha = 1 , \\
w^{\text{T}}\xi^{(\beta)} &\leq \theta_2 , \quad \sigma_\beta = -1 , \\
\theta_1 &\geq \theta \geq \theta_2 .
\end{aligned} \tag{3.23}$$

The typical situation is that either zero or infinitely many such solutions exist. The vector space whose points are the vectors (w, θ) is called the *version space* and its subspace satisfying (3.23) is known as the *solution polyhedron*.

In theory, one knows that for a given training set the optimal Bayes decision [9] implies an average over all solutions $\{w, \theta\}$ satisfying (3.23). Since each solution is perfectly consistent with the training set, in the absence of any other a priori knowledge, one must consider them as equally probable. This is not the case, for instance, in the presence of noise. The best strategy is then to associate with each point in the version space a Boltzmann weight whose energy is the squared error function and whose temperature depends on the noise strength [10].

For examples independently and identically drawn from a constant distribution $P(\xi) = \text{Const.}$, Watkin has shown that, in the thermodynamic limit, the single perceptron corresponding to the center of mass of the solution polyhedron has the same generalization probability as the Bayes decision [11]. On the practical side, known learning algorithms like Adaline [12,13] or the maximal stability (or maximal margin) perceptron (MSP) [14–16] are not Bayes-optimal. In fact, if all input vectors have the same length, then the maximal stability perceptron network corresponds to the center of the largest hypersphere inscribed in the solution polyhedron.

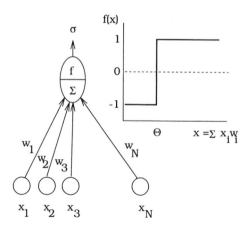

Fig. 3.9. The perceptron as a neural network

3.4.1 The Geometry of Perceptrons

Consider a set of M training examples, consisting of N-dimensional vectors $\boldsymbol{\xi}^{(\nu)}$ and their class σ_ν, $\nu = 1, \ldots, M$. Now let us introduce the $(N+1)$-dimensional vectors $\boldsymbol{\zeta}^{(\nu)} = (\sigma_\nu \boldsymbol{\xi}^{(\nu)}, -\sigma_\nu)$ and $\boldsymbol{W} = (\boldsymbol{w}, \theta)$. In this representation, (3.23) becomes equivalent to a standard set of linear inequalities

$$\boldsymbol{W}^{\mathrm{T}} \boldsymbol{\zeta}^{(\nu)} \geq \Delta > 0 \,. \tag{3.24}$$

The parameter $\Delta = \min_{\{\nu\}} \boldsymbol{W}^{\mathrm{T}} \boldsymbol{\zeta}^{(\nu)}$ is called the *stability* of the linear inequality system. A bigger Δ implies a solution which is more robust against small changes in the example vectors.

The vector space whose points are the examples $\boldsymbol{\zeta}^{(\nu)}$ is the *example space*. In this space, $\boldsymbol{W}$ corresponds to the normal vector of an $(N+1)$-dimensional hyperplane. The *version space* is the space of vectors $\boldsymbol{W}$. Hence, a given example vector $\boldsymbol{\zeta}$ corresponds here to the normal to a hyperplane. The inequalities (3.24) define a convex polyhedral cone whose boundary hyperplanes are determined by the training set $\{\boldsymbol{\zeta}^{(\nu)}\}_{\nu=1}^M$.

How can one use the information contained in the example set to make the best average prediction on the class label of a new vector drawn from the same distribution? Each newly presented example $\boldsymbol{\zeta}^{(\mathrm{new})}$ corresponds to a hyperplane in version space. The direction of its normal is defined up to a factor of $\sigma = \pm 1$, the corresponding class.

The best possible decision for classification of the new example follows the Bayes scheme: for each new (test) example, generate the corresponding hyperplane in version space. If this hyperplane does not intersect the solution polyhedron, consider the normal to be positive when pointing to that part of the version space containing the solution polyhedron. Hence, all perceptrons satisfying (3.24) will classify the new example unanimously. If the hyperplane cuts the solution polyhedron into two parts, point the normal towards the bigger half. Then the decision which minimizes the average generalization error is given by evaluating the average measure of votes for versus the average measure of votes against for all perceptrons making no errors on the training set.

The solution polyhedral cone is either open or else it is defined on the unit $(N+1)$-dimensional hypersphere (see also Fig. 3.10b for an illustration).

The practical problem is now to find simple approximations for the center of mass (Bayes point). One possibility is to consider the perceptron with *maximal stability* (MSP), defined by

$$W_{\mathrm{MSP}} = \arg \max_W \Delta \, , \quad W^{\mathrm{T}} W = 1 \, , \tag{3.25}$$

or, equivalently, by

$$w_{\mathrm{MSP}} = \arg \max_w \{\theta_1 - \theta_2\} \, , \quad w^{\mathrm{T}} w = 1 \, , \tag{3.26}$$

where $\theta = (\theta_1 + \theta_2)/2$ and $\Delta = (\theta_1 - \theta_2)/2$. The quadratic conditions $W^{\mathrm{T}} W = 1$ $[w^{\mathrm{T}} w = 1]$ are necessary because otherwise one could multiply w and $\theta_{1,2}$ by a large number, making $\theta_1 - \theta_2$ and hence Δ arbitrarily large.

Equation (3.26) has a very simple geometric interpretation, depicted in Fig. 3.10. Consider the convex hulls of the vectors $\xi^{(\alpha)}$ belonging to the positive examples $\sigma_\alpha = 1$, and those of the vectors $\xi^{(\beta)}$ in the negative class $\sigma_\beta = -1$. According to (3.26), the perceptron with maximal stability corresponds to the slab of maximal width that can be put between the two convex hulls (the 'maximal dead zone' [18]). Geometrically, this problem is equivalent (dual) to finding the direction of the shortest line segment connecting the two convex hulls (the minimal connector problem). Since the dual problem minimizes a quadratic function subject to linear constraints, it is a *quadratic programming problem*. By choosing $\theta = (\theta_1 + \theta_2)/2$ (dotted line in Fig. 3.10), one obtains the maximal stability perceptron.

The direction of w is determined by at most $N+1$ vertices taken from both convex hulls, called *active constraints*. Figure 3.10a shows a simple 2-dimensional example, with active constraints labeled by A, B, and C, respectively. The version space is 3-dimensional, as shown in Fig. 3.10b. The three planes represent the constraints imposed by the examples A (left plane), B (right plane), and C (lower plane). The bar points from the origin to the point defined by the MSP solution. The sphere corresponds to the normalization constraint. If the example vectors $\xi^{(\nu)}$ all have the same length $\mathcal{L}$, then (3.24) and (3.25) imply that the distance between the W_{MSP} and the hyperplanes corresponding to active constraints are all equal to $\Delta_{\max}/\mathcal{L}$. All other hyperplanes participating in the polyhedral cone are further away. Accordingly, the maximal stability perceptron corresponds to the center of the largest circle inscribed in the spherical triangle defined by the intersection of the unit sphere with the solution polyhedron, the point where the bar intersects the sphere in Fig. 3.10b.

A fast algorithm for computing the minimal connector, requiring on average $O(N^2 M)$ operations and $O(N^2)$ storage place, can be found in [16].

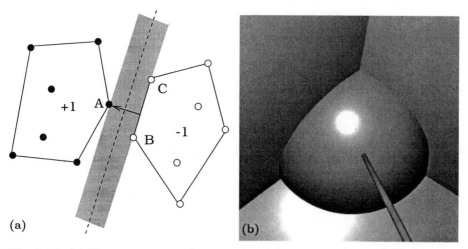

Fig. 3.10. (a) The perceptron with maximal stability in example space. (b) The solution polyhedron (only the bounding examples A, B, and C are shown). See text for details

3.4.2   How to Play Billiards in Version Space

Each billiard game starts by first placing the ball(s) on the pool. As already discussed for the LP, this is not a trivial task. In our case, the maximal stability perceptron algorithm [16] either does this or signals that a solution does not exist. Alternatively, one can use the half-rejecting mirror method (see the LP algorithms) to find an interior point. The trajectory is initiated by generating at random a unit direction vector v in version space.

The basic step consists in finding out where – on which hyperplane – the next collision will take place. The idea is to compute how much time the ball needs until it eventually hits each of the M hyperplanes. Given a point $W = (w, \theta)$ in version space and a unit direction vector v, let us denote the distance along the hyperplane normal ζ by d_n and the component of v perpendicular to the hyperplane by v_n. In this notation, the flight time needed to reach this plane is given by

$$d_n = W^T \zeta \,, \qquad v_n = v^T \zeta \,, \qquad \tau = -\frac{d_n}{v_n} \,. \tag{3.27}$$

After computing all M flight times, one looks for the smallest positive $\tau_{min} = \min_{\{v\}} \tau > 0$. The collision will take place on the corresponding hyperplane. The new point W' and the new direction v' are calculated as

$$W' = W + \tau_{min} v \,, \tag{3.28}$$
$$v' = v - 2v_n \zeta \,. \tag{3.29}$$

This procedure is illustrated in Fig. 3.11a.

In order to estimate the center of mass of the trajectory, one has first to normalize both W and W'. By assuming a constant line density, one assigns

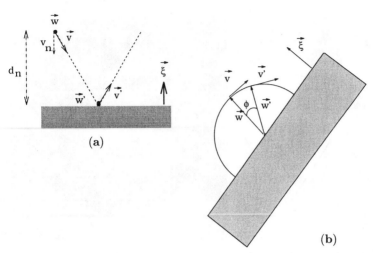

Fig. 3.11. Bouncing in version space. (a) Euclidean, (b) spherical geometry

the length of the vector $W' - W$ to the (normalized!) center of the segment $(W' + W)/2$. This is then added to the actual center of mass – as when adding two parallel forces of different lengths. In high dimensions ($N > 5$), however, the difference between the mass of the full ($N + 1$)-dimensional solution polyhedron and the mass of the bounding N-dimensional boundaries becomes negligible. Hence, we could just as well record only the collision points, assign them the same mass density, and construct their average.

Note that, by continuing the trajectory beyond the first collision plane, one can also sample regions of the solution space where the W makes one, two, etc., mistakes (the number of mistakes equals the number of crossed boundary planes). The additional information can be used for taking an optimal decision when the examples are noisy [10]. This method, which we call the 'onion algorithm', has been shown to be superior to all others. This is not surprising, since we have here some free parameters to adjust on the test set.

Since the polyhedral cone is open, the implementation of this algorithm must take into account the possibility that the trajectory might escape to infinity. The minimal flight time then becomes very large, $\tau > \tau_{\max}$. When this exception is detected, a new trajectory is started from the maximal stability perceptron point in yet another random direction. Hence, from a practical point of view, the polyhedral solution cone is closed by a spherical shell with radius $\tau_{\max}$ acting as a special 'scatterer'. This *flipper* procedure is iterated until enough data is gathered.

If we are class conscious and want to remain in the billiard club, we must do a bit more. As explained above, the solution polyhedral cone can be closed by normalizing the version space vectors. The billiard is now defined on a curved space. However, the same strategy also works here if, between subsequent collisions, one follows geodesics instead of straight lines. Figure 3.11b illustrates the change in direction for a small time step, leading to the well known geodesic differential equation on the unit sphere:

$$\dot{W} = v , \qquad \dot{v} = -W .\tag{3.30}$$

The solution of this equation has an extra cost. Actually, the solution of the differential equation is strictly necessary only when there are no bounding planes on the actual horizon.[1] Once one or more boundaries are 'visible', the choice of the shortest flight time can be evaluated directly, since in the two geometries the flight time is monotonically deformed. Even so, the flipper procedure is obviously faster.

Both variants deliver interesting additional information, such as the mean escape time of a trajectory or the number of times a given border plane (example) has been bounced upon. The collision frequency classifies the training examples according to their 'surface' area in the solution polyhedron – a good measure of their relative 'importance'.

Note that the billiard algorithm has been successfully generalized to kernel classifiers [22] by ourselves and others. Kernel classifiers *implicitly* perform a nonlinear transformation from $x \rightarrow \boldsymbol{\Phi}(x)$, where the $\boldsymbol{\Phi}$ are high- or infinite-dimensional vectors. In order to generalize the learning method described above to kernels, one must rewrite the algorithm in terms of dot products of the form $q_{ij} = \boldsymbol{\xi}^{(i)}\boldsymbol{\xi}^{(j)}$. If this is possible, one makes the substitution

$$q_{ij} \leftarrow \boldsymbol{\Phi}(\boldsymbol{\xi}^{(i)})\boldsymbol{\Phi}(\boldsymbol{\xi}_{(j)}) = k(\boldsymbol{\xi}^{(i)}, \boldsymbol{\xi}^{(j)}) ,$$

where $k(x, y)$ is a (positive definite) Mercer kernel. The output function of the perceptron now becomes an operator

$$w^{\mathrm{T}}x = \sum_{\alpha} a_\alpha k(x, \boldsymbol{\xi}^{(\alpha)}) ,\tag{3.31}$$

where the α index runs over all examples which contributed to the Bayes point w.

The interesting fact about the kernel classifiers is that, although the dimension where the separation might be infinite (if the kernel is a Gauss function, for example), their estimated error is bounded (in the case of errorless learning) by $O(R/\Delta)$, where R is the radius of the smallest circumscribed circle around all examples and Δ is the distance between the two convex hulls of *active* positive and negative examples. More details, results and performance tests can be found in [21,22].

3.5   Bayesian Inference
for the Mixture Assignment Problem

This problem was originally considered by Brumsack at ICBM in Oldenburg, for identifying sources of industrial air pollution. It can be generally formulated

---

[1] Assuming light travels along Euclidean straight lines.

as follows. Assume we create a database containing M substances generated by different processes. For example, these substances are particles polluting the air, produced by different industrial processes. Each substance is analyzed with a mass spectrograph and leads to a (discrete) set of normalized spectrograms containing N lines each. Hence, each substance is a vector

$$\boldsymbol{\xi}^{(\alpha)} = (\xi_1^{(\alpha)}, \ldots, \xi_N^{(\alpha)}), \tag{3.32}$$

normalized as $\sum_{i=1}^{N} \xi_i^{(\alpha)} = 1$, for all $\alpha = 1, \ldots, M$.

Now we are faced with an unknown substance (particle) and its spectrogram, $\boldsymbol{x} = (x_1, \ldots, x_N)$. We are asked to estimate, using the database, how likely it is that this particle was produced by a particular process (or industrial plant).

One way to proceed is to compute the minimal distance between the convex hull of 'example' substances in our catalog and $\boldsymbol{x}$. If this distance is positive (larger than, say, epsilon) we can uniquely identify a point belonging to the convex hull of the catalog examples. As explained in the case of a perceptron, the shortest distance between this point and the database convex hull results in a unique convex combination of the form

$$\boldsymbol{x} = \sum_{\alpha=1}^{M} a_\alpha \boldsymbol{\xi}^{(\alpha)} . \tag{3.33}$$

Here, we can interpret a_α as the probability that substance α contributed to the pollution. If our catalog is complete, however, the unknown substance will lie inside the database convex hull, in which case the minimal distance to the convex hull is zero and the convex combination (3.33) is not unique. Assuming that this is the most frequent case, the interesting question is to rank the possible catalog substances as possible 'sources' of the 'mixture' $\boldsymbol{x}$. It is not hard to see why this problem is important (especially in culinary art!) but also why it is very difficult to make scientific statements about it.

First let us remark that, due to mass conservation,

$$\sum_{\alpha=1}^{M} a_\alpha \xi_i^{(\alpha)} = x_i , \qquad i = 1, \ldots, N , \tag{3.34}$$

and that the coefficients a are convex,

$$\sum_{\alpha=1}^{M} a_\alpha = 1 , \qquad a_\alpha \geq 0 . \tag{3.35}$$

We are interested in the joint distribution $P(a_1, \ldots, a_M)$. It seems very difficult to solve that task. For Bayesian classification it might be enough to estimate the marginal probabilities $p_i = P(a_i)$. This we can do by considering the version space of $\boldsymbol{a}$ points. If $M > N + 1$, we eliminate the equality constraints and play billiard in the remaining $(M - N - 1)$-dimensional space, accumulating statistics from the collision points. That allows us to build a histogram approach to p_i and estimate the average, the most probable value, and the variance of a_i. I mention this problem here because it is a very interesting one from both a theoretical and a practical point of view. Moreover, a lot of questions remain open, for example, how to account correctly for a priori probabilities of finding catalogued substances.

3.6   Conclusion

Although the billiard method is in practice slower than other methods, like the simplex or the primal–dual implementation I developed for finding the maximal margin perceptron [16], it has some evident advantages:

- it is fully parallelizable, since we can gather data from two or more billiard trajectories running in parallel,
- it is conceptually simple to understand and to program, and hence can be easily adapted to special situations,
- it provides a powerful method for sampling (posterior) probability distributions for many optimization problems. This is illustrated by the Bayesian inference problem for mixture assignment, a kind of inverse diet problem.

With computing power becoming a mass-market commodity, it is not hard to see that statistical (or Bayesian) combinatorial optimization problems will be solved routinely on PCs or even handheld devices. Although the above considerations might seem rather academic, such algorithms can solve real life problems surprisingly well. I have no doubt that they will become the core of a new era in information technology.

References

1. J. Moser: 'Geometry of quadrics and spectral theory'. In: *The Chern Symposium*, ed. by Y. Hsiang et al. (Springer, Berlin, Heidelberg 1980) p. 147
2. J.R. Kuttler, V.G. Sigillito: SIAM Review **26**, 163 (1984)
3. H.R. Krishnamurthy, H.S. Mani, H.C. Verma: J. of Phys. A **15**, 2131 (1982)
4. L.A. Bunimovich: Comm. Math. Phys. **65**, 295 (1979)
5. L.A. Bunimovich, Y.G. Sinai: Comm. Math. Phys. **78**, 247 (1980)
6. C. H. Papadimitriou, K. Steiglitz: *Combinatorial Optimization* (Prentice Hall, Englewood Cliffs 1982)
7. Gy. Sonnevend: 'New algorithms based on a notion of 'centre' (for systems of analytic inequalities) and on rational extrapolation.' In: *Trends in Mathematical Optimization – Proceedings of the 4th French–German Conference on Optimization, Irsee 1986*, ed. by K.H. Hoffmann, J.-B. Hiriart-Urruty, C. Lemarechal, J. Zowe. ISNM, Vol. **84** (Birkhauser, Basel 1988)
8. F. Rosenblatt: Psychological Review **65**, 386 (1961)
9. M. Opper, D. Haussler: Phys. Rev. Lett. **66**, 2677 (1991)
10. M. Opper, D. Haussler: 'Calculation of the learning curve of Bayes optimal classification algorithm for learning a perceptron with noise.' In: *IVth Annual Workshop on Computational Learning Theory (COLT91), Santa Cruz 1991* (Morgan Kaufmann, San Mateo 1992), pp. 61–87

11. T. Watkin: Europhys. Lett. **21**, 871 (1993)
12. B. Widrow, M.E. Hoff: *Adaptive switching circuits* (1960 IRE WESCON Convention Record, New York, IRE, 96–104)
13. S. Diederich, M. Opper: Phys. Rev. Lett. **58**, 949 (1987)
14. D. Vapnik: *Estimation of Dependencies from Empirical Data* (Springer, Berlin 1982) – see Addendum I
15. J. Anlauf, M. Biehl: Europhys. Lett. **10**, 687 (1989)
16. P. Ruján: J. de Physique (Paris) I **3**, 277 (1993)
17. P.M. Vaidya: 'A new algorithm for minimizing a convex function over convex sets.' In: *Proceedings of the 30th Annual FOCS Symposium, Research Triangle Park, NC 1989* (IEEE Computer Society Press, Los Alamitos, CA 1990), pp. 338–343
18. P.F. Lampert: 'Designing pattern categories with extremal paradigm information.' In: *Methodologies of Pattern Recognition*, ed. by M.S. Watanabe (Academic Press, New York 1969), p. 359
19. E.L. Lawler, J.K. Lenstra, A.H.G. Rinnoy Kan, D.B. Shmoys (Eds.): *The Traveling Salesman Problem* (John Wiley, New York 1984)
20. M. Padberg, G. Rinaldi: Oper. Res. Lett. **6**, 1 (1987) and IASI preprint R 247 (1988)
21. P. Ruján: Neural Computation **9**, 197 (1997)
22. P. Ruján, M. Marchand: 'Computing the Bayes Kernel Classifier.' In: *Advances in Large Margin Classifiers*, ed. by A. Smola et al. (MIT Press, Cambridge, London 2000), pp. 329–348

4 The Statistical Physics of Energy Landscapes: From Spin Glasses to Optimization

Karl Heinz Hoffmann

Summary. The concept of energy 'landscapes' leads to a unified understanding of phenomena in a number of different complex physical systems. All these systems are characterized by an energy function which possesses many local minima separated by barriers as a function of the state variables. If depicted graphically, such an energy function looks very much like a mountainous landscape. Typical examples of such complex systems are spin glasses which show a wealth of interesting relaxation phenomena, but also a number of industrially important minimization problems, which have a mountainous cost function landscape. These problems are intimately related by the thermally activated relaxation dynamics on complex energy landscapes.

4.1 An Introduction to Energy Landscapes

The concept of energy landscapes is a powerful tool for describing phenomena in a number of different physical systems. All these systems are characterized by an energy function which possesses many local minima, separated by barriers as a function of the state variables. If depicted graphically, the energy function thus looks very much like a mountainous landscape. This picture is a good guide to understanding physical phenomena which are connected to thermal relaxation processes, because these can be visualized as moving about on such an energy landscape. A good example are metastable systems, characterized by states which decay very slowly over long periods of time. These metastable states are the deep wells in the energy function, from which escape can take a very long time.

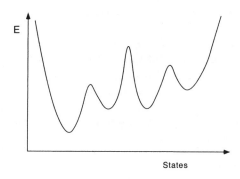

Fig. 4.1. A sketch of a complex state space. The energy is depicted as a function of a sequence of neighboring states. It resembles a cut through a mountainous landscape, and the energy function is thus sometimes referred to as the energy landscape

Systems with this type of energy function are often called complex systems. Figure 4.1 shows a sketch of a such a complex state space. It is important to note that, in order to draw a figure like this, one needs a definition of what constitutes neighboring states. Otherwise one would not know what the abscissa means. From a mathematical point of view one needs this definition to define a local minimum (a state where all neighbors have higher energy).

In this chapter, we show the power of the energy landscape concept by means of two examples: spin glasses and stochastic optimization schemes.

First we will present the progress made in the modeling of spin glasses. In particular, we will look at coarse-graining methods developed to describe the so-called aging phenomena, in which the slow relaxation of such systems becomes very apparent. With the insight gained from this physical example, we then move on to stochastic optimization schemes such as simulated annealing.

Stochastic optimization procedures aim to provide solutions to optimization problems which have many local minima in their objective function. For these optimization problems, the usual steepest descent algorithms fail, as they are easily caught in local minima. For such problems, stochastic optimization procedures, and especially simulated annealing, have been used with growing success – not to determine the global minimum, but to provide 'good' solutions with values of the objective which are not too far removed from the desired global minimum.

We will see that the concept of an energy landscape is particularly helpful in describing the thermal relaxation dynamics important in both areas, i.e., the dynamics which describes the equilibration of the system in contact with a heat bath. Below, we present two major tools for analysing thermal relaxation. One comes from the realm of theoretical physics, namely the theory of Markov processes [1], and the other comes from computational physics, namely the Metropolis algorithm.

4.2   Thermal Relaxation on Energy Landscapes

4.2.1   Markov Processes and the Metropolis Algorithm

Thermal relaxation in contact with a heat bath can be modelled as a discrete time Markov process. The thermally induced hopping process induces a probability distribution in the state space.

Let $\Omega = \{\alpha\}$ represent this state space and let $E : \Omega \to \mathbb{R}$ be the energy defined on the state space. Let T be the temperature of the heat bath in which the physical system is immersed. In addition, a so-called neighborhood relation or move class is needed, which typically takes the form of an undirected graph structure on the state space. We will denote by $N(\alpha)$ the set of neighbors of a state α in this graph. As an example, remember the above-mentioned case of an Ising spin glass, where two states (= spin configurations) are usually neighbors if they differ by one spin flip.

The time development of $P_\alpha(k)$, the probability of being in state α at step k, can then be described by a master equation [1]

$$P_\alpha(k+1) = \sum_\beta \Gamma_{\alpha\beta}(T) P_\beta(k) \ . \tag{4.1}$$

The transition probabilities $\Gamma_{\alpha\beta}(T)$ depend on the temperature T. They have to ensure that the stationary distribution is the Boltzmann distribution

$$P_\alpha^{\mathrm{eq}}(T) = g_\alpha \exp(-E_\alpha/T)/Z \ ,$$

where $Z = \sum_\alpha P_\alpha^{\mathrm{eq}}$ is the partition function and g_α the degeneracy of state α. The latter is needed if the states α already represent quantities which include more than one micro state.

Possible choices for the transition probabilities are the Glauber dynamics [2], and the Metropolis dynamics [3]. The latter describes the process induced by the Metropolis algorithm described below. Its transition probabilities are defined as follows.

The first step is to define the infinite-temperature transition probabilities $\Pi_{\beta\alpha} = \Gamma_{\beta\alpha}(\infty)$ from state α to β by

$$\Pi_{\beta\alpha} = \begin{cases} 0 & \text{if} \quad \beta \notin N(\alpha) \ , \\ 1/|N(\alpha)| & \text{if} \quad \beta \in N(\alpha) \ , \end{cases} \tag{4.2}$$

where $|N(\alpha)|$ is the number of neighbors of α. These are the transition probabilities if the algorithm automatically accepts each attempted move, i.e., if $T = \infty$.

At finite temperature, the acceptance decision is superimposed on Π in (4.2) to give $\Gamma(T)$ defined by

$$\Gamma_{\beta\alpha} = \begin{cases} \Pi_{\beta\alpha} \exp(-\Delta E/T) & \text{if} \quad \Delta E > 0, \ \alpha \neq \beta \ , \\ \Pi_{\beta\alpha} & \text{if} \quad \Delta E \leq 0, \ \alpha \neq \beta \ , \\ 1 - \sum_{\xi \neq \alpha} \Gamma_{\xi\alpha} & \text{if} \quad \alpha = \beta \ , \end{cases} \tag{4.3}$$

where $\Delta E = E(\beta) - E(\alpha)$.

We now turn to the algorithmic version of the Metropolis algorithm [3]. Its aim in simulating a system in contact with a heat bath is to create a Boltzmann distribution in the state space according to the temperature of the heat bath.

To start the algorithm, a state ω_0 is chosen at random. Then, at each step of the algorithm, a neighbor ω' of the current state ω_k is selected at random to become the candidate for the next state. It actually becomes the next state only with probability

$$P_{\mathrm{TA}} = \begin{cases} 1 & \text{if} \quad \Delta E \leq 0 \ , \\ \exp(-\Delta E/T) & \text{if} \quad \Delta E > 0 \ , \end{cases} \tag{4.4}$$

where $\Delta E = E(\omega') - E(\omega_k)$. If this candidate is accepted, then $\omega_{k+1} = \omega'$, otherwise the next state is the same as the old state, $\omega_{k+1} = \omega_k$.

This probabilistic decision rule is implemented by choosing a random number r, uniformly distributed between zero and one, and comparing it with the Boltzmann factor $\exp(-\Delta E/T)$. If $r < \exp(-\Delta E/T)$, then $\omega_{k+1} = \omega'$, otherwise $\omega_{k+1} = \omega_k$.

With the metropolois algorithm and the Markov process description, we are now able to model thermal relaxation on energy landscapes.

4.2.2   Coarse-Graining Mountainous Landscapes

In this section, we want to study the thermal relaxation of a complex system with a mountain-like energy function by means of a Markov process. This leads us to a major problem, common to macroscopic physical systems described on a microscopic level: the enormous number of states.

To exemplify this problem, consider an Ising spin glass model [4]. This will serve as a convenient example throughout the next few sections. Its energy is given by

$$E = \sum_{i,j} J_{ij} s_i s_j \; , \tag{4.5}$$

where the Ising spins s_i can only take the values $+1$ or -1, and the coupling constants J_{ij} are random quantities, which can take positive and negative values. Here the states are defined by the configuration of all spins $\{s_i\}$, and neighboring states are obtained from each other by flipping one of the spins.

The number of states is 2^N if N is the number of spins and grows exponentially with the system size, a feature common to the systems we are interested in. To overcome this problem, the concept of an energy landscape and the rough qualitative picture of the thermal relaxation process, treated as a random hopping process on this mountainous landscape, serves as a good starting point.

It suggests that – as in real mountain areas – movement within a valley is relatively easy compared with movement from one valley to another, which involves crossing a pass. In terms of thermal relaxation, this means that, within such a valley in state space, relaxation proceeds quite fast and local equilibrium is obtained after a short while. Then crossing the barrier between two valleys takes much longer and, as there are valleys inside larger valleys inside larger valleys, and so on, we have to expect a whole spectrum of longer and longer relaxation times.

Once again, as in real mountain areas, the crossing of a pass might not only be influenced by the height (the energy) of the pass alone. There might also be a dynamical restriction (for instance the width of a pass) which requires its own model. It is nevertheless important to realize that the complex energy landscape already leads to a slow-down of the relaxation.

Summarizing the above insight, we get the overall picture of a sequence of quasi-equilibria in larger and larger regions of the state space, or – in terms of an energy landscape – in larger and larger valleys.

Structural Coarse Graining. In the following the above idea is taken further and a simple algorithm is presented which shows how a coarse graining – a simplification of the state space – can be obtained on the basis of the above ideas. The only input is structural information taken from the energy landscape.

The aim of this structural coarse graining is to collect those sets of microscopic states into larger clusters which easily get into internal equilibrium. In order to obtain a good approximation for the dynamical properties of the system on macroscopic time scales, it is important that the inner relaxation in a cluster should be faster than the interaction with the surrounding clusters. In this case, the clusters can be considered to be in internal equilibrium on larger time scales.

The algorithm for constructing clusters from a microscopic state space is not completely defined by the above conditions. In fact, several different possibilities exist. Below, we show results from Klotz et al. [5], who presented an algorithm producing clusters without any internal barrier.

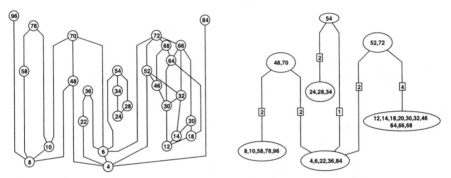

Fig. 4.2. *Left*: Microscopic states with connections. *Right*: Coarse-grained state space obtained from the microscopic states using the algorithm. The *number* in a *square* at a connection is the number of corresponding microscopic connections

In Fig. 4.2 (left), structural coarse graining has been applied to a small subsystem with 28 states for demonstration purposes. The subsystem shown is actually a low energy subset of the states of an Ising spin glass. The resulting coarse-grained system is shown in Fig. 4.2 (right), where the numbers inside a cluster are the numbers of the states which are lumped into the cluster.

In Fig. 4.3, the resulting total number of microscopic states and of clusters is plotted versus $E - E_0$, where E is the energy of the microscopic state and E_0 is the energy of the global minimum. In the considered energy range, the number of clusters increases more slowly than that of the microscopic states. The mean number of microstates lumped in one cluster increases exponentially, as we can see from the dotted curve. Thus the coarse-graining algorithm becomes more and more effective for larger energies and hence for larger systems. Furthermore, it can be shown that the maximum energy difference of the microscopic states inside a cluster is large, i.e., clusters are 'long' on the energy axis.

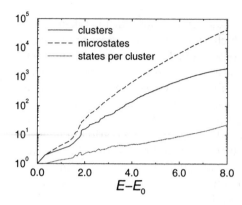

Fig. 4.3. Number of microscopic states with an energy below $E - E_0$ (*dashed line*) and number of clusters resulting from the coarse-graining procedure which start at energies below $E - E_0$ (*solid line*). The *dotted curve* displays the ratio of the number of microstates below $E - E_0$ to the number of clusters starting below this energy

Dynamical Coarse Graining: Transition Rates. Consider now thermal hopping on the complex energy landscape. Rather than calculating the full time dependence of the probability distribution in the state space, we can choose to monitor the presence or absence from a cluster of the coarse-grained system. We thereby define a stochastic process, which in general will not be a Markov process [6], because the induced transition probability from cluster to cluster might depend on the internal (microscopic) distribution within one cluster. However, it turns out [7] that inside a coarse-grained area, a kind of local equilibrium distribution is very quickly established, which then makes the coarse-grained relaxation process (at least approximately) Markovian. The result is that Markov processes on coarse-grained systems are good tools for modelling the thermal relaxation of complex systems [8–10].

Using this insight, the proper transition rates of the coarse-grained system can be determined. To find the structure of these rates, let us begin with the exact calculation of the transition probability between two neighboring clusters, starting from the microscopic picture. The probability flux from the states belonging to cluster C_ν to the states belonging to cluster C_μ is given by

$$J_{\mu\nu} = G_{\mu\nu} P_\nu = \sum_{\alpha \in C_\mu, \beta \in C_\nu} \Gamma_{\alpha\beta} p_\beta , \qquad (4.6)$$

where P_ν is the total probability of being in cluster C_ν, i.e., the sum of the probabilities of all states in cluster C_ν, and $G_{\mu\nu}$ is the transition rate from cluster C_ν to cluster C_μ. Once again, we assume that the internal relaxation inside the clusters is fast compared with the relaxation between different clusters. For the time scale of interest, all clusters are in internal equilibrium, i.e., $p_\beta \propto \exp(-E_\beta/T)$. As the microscopic transition rates have the form $\Gamma_{\alpha\beta} \propto \exp[-\max(E_\alpha - E_\beta, 0)/T]$, the coarse-grained transition rate is

$$G_{\mu\nu} = \frac{\sum_{\alpha \in C_\mu, \beta \in C_\nu} t_{\alpha\beta} \exp[-\max(E_\alpha, E_\beta)/T]}{\sum_{\alpha \in C_\nu} \exp(-E_i/T)} , \qquad (4.7)$$

where $t_{\alpha\beta}$ equals one if the states α and β are neighbors and zero otherwise.

The roughest simplification (here referred to as procedure A) would be to consider all states of a cluster as one state with a certain energy $\hat{E}_\alpha$ which is chosen as the mean energy of the microscopic states. Following this idea, the sums in (4.7) can be simplified to

$$\hat{G}_{\mu\nu} = \frac{\hat{T}_{\mu\nu}\min\left(\exp[-(\hat{E}_\mu - \hat{E}_\nu)/T], 1\right)}{\hat{n}_\nu}, \tag{4.8}$$

where $\hat{T}_{\mu\nu}$ is the number of connections between cluster C_ν and cluster C_μ, and $\hat{n}_\nu$ is the number of states collected in cluster C_ν. A better aproximation can be achieved if each cluster is modelled by a two-level system, which is in internal equilibrium (procedure B). For a detailed description of these procedures, see Klotz et al. [5].

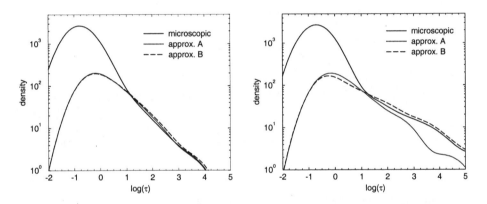

Fig. 4.4. Smoothed relaxation time densities versus the logarithm of the relaxation time for $\beta = 1$ (*left*) and $\beta = 4$ (*right*) for the microscopic system and the two procedures A and B described in the text

In order to check the quality of the approximations made by the coarse graining of the energy landscape, one can for instance look at the density of relaxation times. Figure 4.4 shows this density of relaxation times for $\beta = 1$ and $\beta = 4$. The spectra have been computed with a resolution of 0.2 on the logarithmic τ scale. In the case of high temperatures (Fig. 4.4 left), we find good agreement between the two procedures compared with the original microscopic system in the range of large relaxation times. For short times, the microscopic system has many more eigenvalues, which are neglected in the coarse-grained system. The dynamics in the coarse-grained state space is therefore a good approximation to the dynamics of slow processes in the microscopic system, which are the important ones for the analysis of low temperature relaxation phenomena.

The above algorithm for the coarse graining of complex state spaces shows how the idea of an energy landscape can help in modeling complex systems. This technique is independent of the model and can be used not only for Ising spin glass models, as demonstrated, but also for other complex systems such as Lennard–Jones systems or proteins. Coarse graining not only makes it possible to simulate the dynamics of large complex systems, but it is also of importance for visualizing high-dimensional state spaces associated with complex systems.

4.2.3   Coarse Graining on a Coarser Scale

For many systems the reduction in the size of the state space obtained by the above scheme is not large enough. Then the coarse graining has to take place at a coarser level. Below, we present such an approach which, contrary to the previous scheme, neglects energy barriers up to a certain size. Figure 4.5 depicts the underlying idea [8].

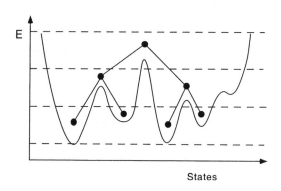

Fig. 4.5. The coarse graining of a complex state space. All connected states within one energy band are lumped into the nodes of a tree

Suppose one introduces hyperplanes with energy

$$E(l) = \Delta E_t \cdot l \tag{4.9}$$

into the state space and calls all states between two cuts an energy band. The connected parts (in the sense of the neighborhood graph introduced above) of the state space within an energy band are then lumped together to represent the nodes of a new coarse-grained structure.

If ΔE_t is large compared with the energy change $\Delta E_{\mathrm{micro}}$ of a transition between microstates, then the connectivity between the microstates induces a unique connectivity between the nodes, as indicated in Fig. 4.5. The resulting structure has a tree topology. Each node has only one connection to a higher energy node, called the mother node. The connected nodes at lower energy are called daughter nodes. The number of daughter nodes will vary from node to node. Moreover, the number of states lumped into a node will vary. In general, we expect that the higher the nodes are in energy, the more states they will include, i.e., the degeneracy of a node will increase with its energy.

There exist other coarse-graining procedures [8] which lead to trees with nodes not seperated by equal energy intervals. The important point is that a hierarchical tree-like structure nevertheless results, and trees can thus be regarded as generic coarse-grained structures for complex state spaces. Consequently, the next section will be devoted to thermal relaxation dynamics on trees.

4.2.4   Tree Dynamics

Thermal relaxation dynamics has been studied for a number of tree structures. We present here one example following [11,12]. The tree is constructed as shown in Fig. 4.6. The states of this tree are organized into a hierarchical structure with the height of the nodes corresponding to their (free) energy. All nodes but those at the lowest level have two 'daughters' connected to their mother by a 'long' and a 'short' edge, such that the energy differences become $\Delta E = L$ and $S < L$, respectively. In this way the degeneracy of nodes belonging to the same level of the tree is broken. In particular, not all metastable states have the same energy. This opens the possibility of competition between thermodynamic and kinetic relaxation mechanisms, as detailed further below. A degeneracy g_μ is assigned to each node μ, describing the state space volume associated with the lumped microstates. The dynamics is given by a nearest neighbor random walk on this structure.

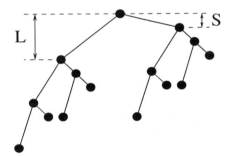

Fig. 4.6. The construction of an LS-tree

The nondiagonal elements of the transition matrix $\Gamma_{\mu\nu}$ are zero, except for states connected by an edge, in which case they can all be expressed in terms of up and down rates along the edge:

$$\Gamma_{\mathrm{up}} = f_j \kappa_j \mathrm{e}^{-\beta \Delta E_j} , \qquad \Gamma_{\mathrm{down}} = f_j . \tag{4.10}$$

The index j distinguishes between L- and S-edges, and κ_j is the ratio between the degeneracy of a node and that of the corresponding daughter node. The diagonal elements of the transition matrix are given by the condition that each column sum vanish to ensure conservation of probability.

The simple but crucial additional feature of our model is the kinetic factors f_j controlling the relaxation speed along each edge. As detailed balance only prescribes the ratios of the hopping rates between any two neighbors, the f_j can be freely chosen without affecting the equilibrium properties of the model. It is less clear, but quite important, that the quasi-equilibrium properties, i.e., the exponents of the slow algebraic relaxation [11], are unaffected by any arbitrary choice of these parameters. This has been numerically demonstrated by Uhlig et al. [12]. However, non-uniform kinetic factors have a decisive effect on the dynamics following temperature steps, which destroy quasi-equilibrium on short time scales. We assume that $f_S > f_L$. This is instrumental in creating competing kinetic and thermodynamic relaxation mechanisms.

To understand qualitatively how the competition comes about, consider the extreme case where the system, initially at high temperature, is quenched to zero temperature. Since upward moves are forbidden, the probability flows downwards through the system, splitting at each node in the ratio $f_L : f_S$, independently of energy differences. Thus, if f_S is larger than f_L, the probability is preferably funneled through the short edges, and ends up mainly in high-lying metastable states. If the system is heated up even very slightly after the quench, thermal relaxation sets in and redistributes the probability. Eventually the distribution of probability becomes independent of the values of f_L and f_S, and the low energy states are favored.

The above description is now easily extended to a many-level tree. An initial quench creates a strongly non-equilibrium situation, mainly determined by the relaxation speeds, whereafter the slow relaxation takes over. At any given time, subtrees of a certain size will have achieved internal equilibration, while larger ones will not.

A temperature cycle – that is, a temperature increase followed by a decrease of the same size or vice versa – always destroys this internal equilibration as it induces a fast redistribution of probability. In positive temperature cycles, probability is pumped up to energetically higher nodes, which leads to a partial reset when the temperature is lowered. By way of contrast, negative temperature cycles push probability down the fast edges, a process which is readily reversed when the temperature is raised again.

4.2.5 A Serious Application: Aging Effects in Spin Glasses

Using the above modeling ideas, based on the concept of an energy landscape, very simple and yet highly successful models can be developed to describe a number of interesting experimental results for spin glasses. These so-called aging effects [13–18] are not confined to spin glasses, but have also been measured in high-T_c superconductors [19] and CDW systems [20] as well. In the so-called Zero Field Cooled (ZFC) experiments, a sample is cooled to a low temperature and 'aged' for a 'waiting time' t_w, without fields. Thereafter, a small field is applied, and the response of the magnetization to the perturbation is measured. Contrary to naive expectations, the response depends both on the time during which the system has been acted upon by the field and on the waiting time. This situation persists through many decades, and indicates that the system never reaches thermodynamical equilibrium during the observation time.

In spin glass ZFC magnetization experiments, the applied field H is kept constant, and the salient feature of the data below the critical spin glass temperature T_g is a kink in the magnetization $M(t, t_w)$ plotted as a function of

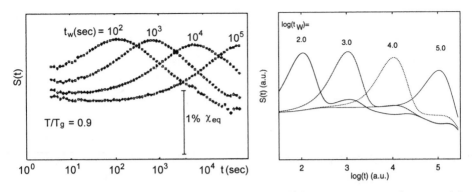

Fig. 4.7. Comparison between experimental data for ZFC experiments and tree model data

logarithmic time at $t = t_w$ or equivalently, a maximum in the derivative $S(t, t_w)$ of the magnetization with respect to the logarithm of the time at $t = t_w$.

Thermal relaxation on a tree model [21] can be shown to have all the important features of an aging system. Figure 4.7 shows a comparison between experimental data and the model result. Note that the latter reproduces very well the maxima in the relaxation rate at times which correspond to the waiting times [22,23].

There are a number of further experiments which measure the response as temperatures are changed during the waiting time. This leads to partial reinitialization effects which can be reproduced very well by the model [24] including the kinetic factors. In the following, we compare our model predictions with the work of Vincent et al. [25], who studied temperature cycling in thermoremanent magnetization experiments. In these experiments, the sample is quenched in a magnetic field, which is turned off after time t_w. Then the decay of the thermoremanent magnetization is measured. The temperature cycling consists of a temperature pulse added during the waiting time t_w.

The data of Vincent et al. [25] are shown in the left-hand part of Fig. 4.8, for $t_w = 30$ and $t_w = 1000$ min. On top of this reference experiment, a very short temperature variation ΔT is imposed on the system 30 min before cutting the field. The important feature is that increasing ΔT shifts the magnetization decay data from the curve corresponding to the 1 000 min curve to the 30 min curve. The corresponding effect in our model is shown in the left-hand part of Fig. 4.9. The parallels between model and experiment are obvious. In both cases, reheating appears to reinitialize the aging process.

The right-hand part of Fig. 4.8 shows the results for a negative temperature cycle. The important feature here is that a temporary decrease in the temperature leads to 'freezing' of the relaxation. In other words, the effect of the time spent at the lower temperature diminishes and eventually disappears as ΔT decreases. This is seen in the experimental data as well as in the model (see Fig. 4.9 right).

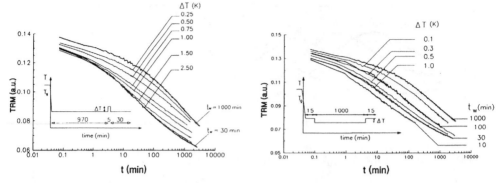

Fig. 4.8. Experimental results for positive (*left*) and negative (*right*) temperature cycles on thermoremanent magnetization (*thin lines*). *Bold lines* are reference curves without temperature cycling. The procedure is shown in the *inset*

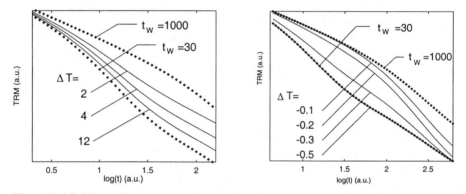

Fig. 4.9. Model results corresponding to the above experimental results

Finally we note that a theoretical understanding of the physical mechanism behind the aging phenomenon of spin glasses has important implications for the more general issue of describing relaxation phenomena in complex systems like glasses and polymers, as well as some nonphysical systems which are currently treated by statistical mechanical methods, e.g., Boltzmann machines and simulated annealing schemata [26]. This is due to the fact that all these systems can be very well understood in terms of a mountainous energy landscape.

4.3   Stochastic Optimization: How to Find the Global Minimum in an Energy Landscape

While in the first part of this chapter the coarse graining of an energy landscape and its use for spin glass modelling was at the center of our interest, we now turn to the problem of finding the global minimum in a multi-mimimum landscape. From a physical point of view, this is equivalent to finding the ground state of the corresponding physical system, which is usually important for low temperature behavior. But the problem of finding the global minimum in a multi-mimimum landscape is of much wider importance. Many industrially important optimization problems are of the kind where the cost-function landscape is as mountainous as the energy landscape for physical systems. By now it is certainly clear that the multi-valley structure with its many local minima makes it quite complicated to find the ground state. Simple descent methods which rely on following the energy function down to states with lower and lower energy will invariably lead into one of the many local minima, but rarely into the global one. Recent investigations [27] show that the number of local minima can grow exponentially with system size, so that just hoping to find the ground state by chance or a few repeated trials does not work. In the following, we will show how this similarity of the energy landscapes can be used to find good solutions to these optimization problems.

4.3.1   Simulated Annealing

One idea for solving this problem was born in the early 1980s: simulated annealing [28,29]. It is based on the observation that a careful annealing of physical systems brings them closer to equilibrium than a quenching process. This is well known and widely used in the preparation of single crystals. Careful annealing of a real physical system brings it into its equilibrium with the ambient temperature T and thus, for $T \to 0$, the system moves into its ground state(s). Simulating the thermal relaxation by the Metropolis algorithm and slowly turning down the temperature parameter which enters in the transition rate leads to the same result in a model system: the ground state or, more generally, the global minimum can be obtained. This method is thus called simulated annealing.

For a general optimization problem, all one needs to do is to introduce a neighborhood relation on the states artificially and then perform a random hopping process on the energy (objective function) landscape according to the Metropolis rules, where the temperature is now just an external parameter which has to be lowered properly.

Simulated annealing has been applied to a wide range of problems with a complex state space structure. Such complex structures are often due to various constraints which have to be satisfied during optimization (see for instance the next chapter). Apart from finding the ground state of a spin glass [30], simulated annealing has also proved a useful tool in the design of integrated circuits [28,31,32] for partitioning, routing, and placement [33]. It has been applied to many other problems including the traveling salesman [34,35], graph partitioning [36], image restoration [37], and parameter estimations [38]. While this list is far from exhaustive, it shows that the problems tackled by simulated annealing are of great scientific and industrial importance.

A thorough analysis has shown [37] that the simulated annealing procedure does indeed find the ground state with probability one. However, the annealing schedule, i.e., the way in which the temperature parameter T is lowered as a function of time, needs to be very slow and turns out to be $T(t) \sim 1/\ln t$. Note that an infinite time is required to reach the ground state with probability one.

Consequently, for any finite time, the ground state is not found with probability one. This raises the problem of finding the schedule that provides the 'best' possible solution under the restriction of finite (computer) time, which in our case translates into a finite number of Metropolis steps. In other words, the question is: what is the 'optimal' schedule?

Before this question can be answered, a yardstick must be defined with respect to which optimality can be determined. Indeed, several criteria are possible. The two most commonly used are:

- the final energy,
- the BSF energy $E_{\text{BSF}}(k) = \min_{0<k'<k} E(k')$, i.e., the lowest energy seen up to a certain step number k.

The final and BSF energies are stochastic quantities [39]. Their probability distribution evolves with time and is induced by the underlying random walk on the energy landscape. The distribution as such cannot be optimized, although certain aspects can, such as its mean, its median, or its mode. The choice between the different criteria has to be made externally. After a choice has been made, the determination of the optimal schedule becomes a new optimization problem, which can be tackled either analytically or numerically.

Optimal schedules have been determined for a variety of systems. The simplest example system consists of only three states [40]. Knowledge of the optimal schedule allows a comparison between different schedules. For instance, it has been shown that the optimal schedule performs much better than any exponential or linear schedule. This simple model also shows that the barrier height enters the optimal schedule as an essential parameter. Later optimal schedules [41,42] for larger systems were determined numerically. It turns out that optimal schedules are dominated by a single barrier during certain time intervals.

Summarizing these studies of optimal schedules, one finds that it is essential to hold the system close enough to equilibrium in order to avoid getting trapped in local minima. On the other hand, one has to maintain a certain lack of equilibrium in order to anneal as quickly as possible.

4.3.2 Threshold Accepting and Tsallis Annealing

When implementing the standard simulated annealing algorithm, computation of the acceptance probability requires evaluation of an exponential function in each step of the random walker. Dueck et al. [43] and Moscato and Fontanari [44] changed the Metropolis acceptance probability when stepping upwards in energy from an exponential to a step function, i.e.,

$$P_{\text{TA}}(\Delta E) = \begin{cases} 1 & \text{if} \quad \Delta E \leq T \,, \\ 0 & \text{if} \quad \Delta E > T \,. \end{cases} \tag{4.11}$$

This algorithm is called threshold accepting. By removing the computation of the exponential function, the algorithm became faster, and it even seems to yield the same if not better solutions than the Metropolis algorithm.

Another technique has come up in the context of the discussion of generalized thermodynamics [45]. Penna [46] and Tsallis and Stariolo [47] introduced an acceptance probability of the form

$$
P_q(\Delta E) = \begin{cases} 1 & \text{if } \Delta E \leq 0 \,, \\[2mm] \left[1 - (1-q)\dfrac{\Delta E}{T} \right]^{1/(1-q)} & \text{if } \Delta E > 0 \quad \text{and} \quad (1-q)\dfrac{\Delta E}{T} \leq 1 \,, \\[2mm] 0 & \text{if } \Delta E > 0 \quad \text{and} \quad (1-q)\dfrac{\Delta E}{T} > 1 \,, \end{cases}
$$

$$(4.12)$$

depending on an additional parameter $q \neq 1$. For $q = 1$, (4.12) is not defined, but one can show that in the limit $q \to 1$, the acceptance probability (4.12) converges to the Metropolis probability. We remark that the Tsallis algorithm can be modified to cover all three schemes (Tsallis, Metropolis, threshold), in such a way that this modification contains the Metropolis case and threshold accepting as limiting cases [48].

Recently Franz, Hoffmann and Salamon [49] have proven that, in a rather wide class of stochastic annealing schemes, including all three of the above-defined optimization algorithms, threshold accepting is the best strategy. It is thus especially important to look into the problem of optimal schedules for threshold accepting.

4.3.3   Adaptive Schedules and the Ensemble Approach

Investigations of truly optimal schedules for simple systems [40,42,50] have shown that the schedule depends critically on the barrier height which has to be over-come to leave a local minimum. In the usual optimization problem, these barrier heights are unknown. Moreover, they differ from problem to problem. The sched-ule must therefore be adapted to the problem.

Adaptive schedules using information gathered during the annealing have already been suggested [51,52,36]. Here we present an adaptive schedule (EBTA – Ensemble-Based Threshold Accepting) which is easy to implement and has only negligible computational overhead. The schedule works as well for simulated annealing as for threshold accepting and we shall present it for the latter case. It is based on the ensemble approach to simulated annealing, where a whole collection of copies of the system (rather than just one) is annealed according to the same schedule [38,39,53,54]. One of the most important advantages of the ensemble approach is that statistical information about the ensemble can be used to adjust the schedule with which the temperature or the threshold is lowered.

The threshold T is lowered in a stepwise fashion by a factor c, $0 < c < 1$, and a certain number of steps is performed at the same threshold. This number of steps can be predetermined or it can be set adaptively during the run, using the statistical data from the ensemble.

The philosophy behind the original version of an adaptive schedule [55] is to hold the ensemble fairly close to the equilibrium corresponding to the threshold. As an indicator for this, we monitor the ensemble average of the energy or cost function $\langle E \rangle$. For an ensemble close to equilibrium $\langle E \rangle$ will fluctuate around the equilibrium corresponding to the threshold, while for an out-of-equilibrium situation $\langle E \rangle$ will move towards that value. As the threshold is lowered during a run, $\langle E \rangle$ should be above its equilibrium value and move down. If it does not do so, this is taken as an indication of a fluctuation around equilibrium and the threshold T is lowered.

Here we have implemented the following somewhat more general schedule, containing an additional freely adjustable parameter γ. The parameter γ is a measure for how much the expected downward movement of $\langle E \rangle$ can be violated without triggering a threshold reduction. For $\gamma \to -\infty$, the threshold is lowered in any case and we thus obtain the exponential schedule, whereas for $\gamma = 0$, we recover the simple adaptive schedule, and for positive values of γ the condition is relaxed further:

> Choose N initial configurations
> Set an initial threshold T
> Compute initial ensemble average $\langle E \rangle$
> **while** end condition is false **do**
> > Perform n TA steps per ensemble member with threshold T
> > Compute new ensemble average $\langle E' \rangle$ and variance $\mathrm{var}(E')$
> > **if** $\langle E' \rangle - \langle E \rangle > \gamma \, \mathrm{var}(E')/\sqrt{N}$
> > > **then**   Set new threshold $T' = c\,T$
> > Set new values as old values
> Stop.

One iteration of the **while**-loop is called a TA sweep. Note that even though the threshold is always lowered by the factor c, the schedule is not in general exponential, as the number of TA sweeps carried out at each threshold varies.

We have implemented the TA ensemble algorithm on high performance workstations and on parallel computers. Depending on the ensemble size used and the number of processors available, each processor of the parallel machine had to handle a certain number of equivalent work processes corresponding to the members of the ensemble. A master process is responsible for evaluating the data received from the ensemble members and for controlling the threshold. The communication required by the algorithm is minimal. At the end of each sweep, each ensemble member has to transmit only its current tour and the corresponding length to the master, and the master transmits back the new threshold. This makes the algorithm well suited for parallel implementations.

Figure 4.10 shows a comparison of different schedules. We present data for the VBSF (very best so far) energy $E_{\mathrm{VBSF}}(k)$, which is the lowest energy seen up to step number k by one of the ensemble members. The performance of EBTA was

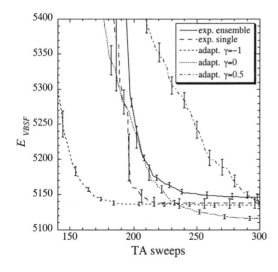

Fig. 4.10. VBSF energies as a function of the TA sweeps for different schedules

tested on a traveling salesman problem (TSP), as a typical example of complex optimization. The TSP for which our results were obtained is the Grötschel drill hole problem [56]. We used an ensemble size of 100, where the initial states were chosen at random. The end condition was given by a maximum number of TA sweeps of 300. For the adaptive schedule, we performed the optimization process with factors $\gamma = -1$, 0, $+0.5$ and $c = 0.9$. To find the best exponential schedule, we tested 100 different factors $0.9 \leq c \leq 0.999$ in preliminary runs and determined the best factor for this schedule to be $c = 0.98$. In order to take advantage of investigating a larger portion of the state space Ω, we also implemented the exponential schedule in an ensemble version of now completely independent random walkers, corresponding to $\gamma \to \infty$ for $c = 0.98$. For a second kind of comparison, we used a single random walker that ran 100 times as long and thus lowered the threshold 100 times, i.e., 100 times slower than each ensemble member above, with a maximum number of TA sweeps of 30 000 and $c = 0.98^{1/100} \approx 0.9998$.

It can be seen that the simple adaptive schedule gets deeper states than the exponential schedule. As expected, the adaptive schedules become slower and better (but not significantly) with increasing γ. Also plotted is the analogous result for the single long-running slow exponential schedule, where $E_{\mathrm{VBSF}} \equiv E_{\mathrm{BSF}}$. Here the sweep numbers were rescaled by a factor of 100, so that they correspond to comparable CPU times for the ensemble-based schedules.

Figure 4.11 shows a typical (cumulative) distribution of E_{BSF} at the end of the runs for the ensemble version of the exponential and for the simple adaptive schedule. As can be seen, the distribution of the adaptive schedule is significantly concentrated towards lower tour lengths compared with the exponential schedule. The error bars indicate the standard error.

In conclusion, we can say that, for the adaptive schedules, the probability of finding states close to the global minimum is significantly higher than it is for the

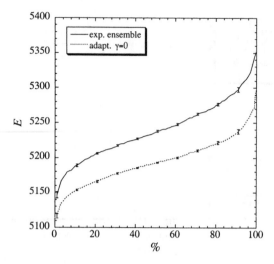

Fig. 4.11. A typical (cumulative) distribution of the final E_{BSF} for the ensemble version of the exponential and for the simple adaptive schedule

exponential schedule. Compared with the ensemble version of the exponential schedule, the simple adaptive schedule is a good choice with respect to both speed and quality.

Finally, we remark that in a number of cases the time dependence of the moments $\langle (E_{\text{BSF}} - E_{\text{gs}})^n \rangle$ of the BSF energy can be reasonably well fitted by power laws for a certain time span [57,58], and this shows once again the connection with thermal relaxation in tree models, where these power laws also occur.

4.4   Conclusion

In this chapter, we have presented the concept of an energy landscape. This allows us to describe a large number of complex systems characterized by an energy function with many local minima. The dynamics in such a state space, which is induced by contact with a heat bath – real or artificial – can be analysed by simulations based on the Metropolis algorithm or by a Markov process description.

Using the concept of an energy landscape, we developed simple relaxation models for spin glasses based on coarse-graining procedures to simplify the state spaces. We then showed how the insights and methods from statistical physics can be put to use in stochastic optimization. We presented effective and easily implemented methods for solving optimization problems with 'mountainous' objective functions, based on adaptive annealing schedules.

References

1. N.G. van Kampen: *Stochastic Processes in Physics and Chemistry* (Elsevier, Amsterdam 1997)
2. K. Binder, D.W. Heermann: *Monte Carlo Simulation in Statistical Physics*, Springer Series in Solid-State Sciences, Vol. 80 (Springer, Berlin 1992)
3. N. Metropolis, A.W. Rosenbluth, M.N. Rosenbluth, A.H. Teller, E. Teller: J. Chem. Phys. **21**, 1087 (1953)
4. K.H. Fischer, J.A. Hertz: *Spin Glasses* (Cambridge University Press, Cambridge 1991)
5. T. Klotz, S. Schubert, K.H. Hoffmann: J. Phys.: Condens. Matter **10**, 6127 (1998)
6. B. Andresen, K.H. Hoffmann, K. Mosegaard, J. Nulton, J.M. Pedersen, P. Salamon: J. Phys. France **49**, 1485 (1988)
7. H.A. Kramers: Physica (The Hague) **7**, 284 (1940)
8. K.H. Hoffmann, P. Sibani: Phys. Rev. A **38**, 4261 (1988)
9. K.H. Hoffmann, S. Grossmann, F. Wegner: Z. Phys. B **60**, 401 (1985)
10. P. Sibani: Phys. Rev. B **35**, 8572 (1987)
11. P. Sibani, K.H. Hoffmann: Europhys. Lett. **16**, 423 (1991)
12. C. Uhlig, K.H. Hoffmann, P. Sibani: Z. Phys. B **96**, 409 (1995)
13. L. Lundgren, P. Svedlindh, P. Nordblad, P. Beckman: Phys. Rev. Lett. **51**, 911 (1983)
14. M. Ocio, H. Bouchiat, P. Monod: J. Phys. Lett. France **46**, 647 (1985)
15. P. Nordblad, P. Svedlindh, J. Ferre, M. Ayadi: J. Magn. Magn. Matter **59**, 250 (1986)
16. P. Svedlindh, P. Granberg, P. Nordblad, L. Lundgren, H.S. Chen: Phys. Rev. B **35**, 268 (1987)
17. N. Bontemps, R. Orbach: Phys. Rev. B **37**, 4708 (1988)
18. J. Hammann, M. Ocio, E. Vincent: 'Attempt at a comprehensive description of the slow spin glass dynamics.' In: *Relaxation in Complex Systems and Related Topics* ed. by I.A. Campbell, C. Giovanella (Plenum Press, New York 1990) pp. 11–21
19. C. Rossel, Y. Maeno, I. Morgenstern: Phys. Rev. Lett. **62**, 681 (1989)
20. K. Biljakovic, J.C. Lasjaunias, P. Monceau: Phys. Rev. Lett. **62**, 1512 (1989)
21. K.H. Hoffmann, T. Meintrup, C. Uhlig, P. Sibani: Europhys. Lett. **22**, 565 (1993)
22. P. Sibani, K.H. Hoffmann: Phys. Rev. Lett. **63**, 2853 (1989)
23. C. Schulze, K.H. Hoffmann, P. Sibani: Europhys. Lett. **15**, 361 (1991)
24. K.H. Hoffmann, S. Schubert, P. Sibani: Europhys. Lett. **38**, 613 (1997)
25. E. Vincent, J. Hammann, M. Ocio: 'Slow dynamics in spin glasses and other complex systems.' In: Saclay Internal Report SPEC/91-080, Centre d'études de Saclay, Orme des Merisiers, 91191 Gif-sur-Yvette Cedex, France, October 1991; also in: *Recent Progress in Random Magnets*, ed. by D.H. Ryan (World Scientific, Singapore 1992)
26. E. Aarts, J. Korst: *Simulated Annealing and Boltzmann Machines*, Wiley Series in Discrete Mathematics and Optimization (Wiley, New York 1989)
27. P. Sibani, C. Schön, P. Salamon, J.-O. Andersson: Europhys. Lett. **22**, 479 (1993)
28. S. Kirkpatrick, C.D. Gelatt, M.P. Vecchi: Science **220** (4598), 671 (1983)
29. V. Černy: J. Optim. Theory Appl. **45**, 41 (1985)
30. R. Ettelaie, M.A. Moore: J. Phys. Lett. France **46**, L893 (1983)
31. P. Slarry, G. Dreyfus: J. Phys. Lett. France **45**, L39 (1983)
32. S.R. White: In *Proceedings of the ICCD 1984* (IEEE 1984) p. 646

33. M.P. Vecchi, S. Kirkpatrick: In *Proceedings of the IEEE Trans. Computer Aided Design CAD-2*, IEEE Trans. CAD (IEEE 1983) p. 215
34. E. Bonomi, S.L. Lutton: SIAM Rev. **26**, 551 (1984)
35. R. Durbin, D. Willshaw: Nature **326**, 689 (1987)
36. P. Salamon, J.D. Nulton, J.R. Harland, J. Pedersen, G. Ruppeiner, L. Liao: Comp. Phys. Comm. **49**, 423 (1988)
37. S. Geman, D. Geman: IEEE Transactions on Pattern Analysis and Machine Intelligence PAMI 6, 721 (1984)
38. M.O. Jakobsen, K. Mosegaard, J.M. Pedersen: In *Model Optimization in Exploration Geophysics 2*, ed. by A. Vogel (Friedrich Vieweg and Son, Braunschweig/Wiesbaden 1988) p. 361
39. K.H. Hoffmann, P. Sibani, J.M. Pedersen, P. Salamon: Appl. Math. Lett. **3**, 53 (1990)
40. K.H. Hoffmann, P. Salamon: J. Phys. A Math. Gen. **23**, 3511 (1990)
41. M. Christoph, K.H. Hoffmann: J. Phys. A Math. Gen. **26** (13), 3267 (1993)
42. K. Ergenzinger, K.H. Hoffmann, P. Salamon: J. Appl. Phys. **77**, 5501 (1995)
43. G. Dueck, T. Scheuer: J. Comput. Phys. **90**, 161 (1990)
44. P. Moscato, J.F. Fontanari: Phys. Lett. A **146**, 204 (1990)
45. C. Tsallis: J. Stat. Phys. **52**, 479 (1988)
46. T.J.P. Penna: Phys. Rev. E **51**, R1 (1995)
47. C. Tsallis, D.A. Stariolo: Physica A **233**, 395 (1996)
48. A. Franz, K.H. Hoffmann: Threshold accepting as limit case for a modified Tsallis statistics. Submitted to Appl. Math. Lett. 2000
49. A. Franz, K.H. Hoffmann, P. Salamon: Phys. Rev. Lett. **86**, 5219 (2001)
50. K.H. Hoffmann, M. Christoph, M. Hanf: 'Optimizing simulated annealing.' In: *Parallel Problem Solving from Nature*, ed. by H.-P. Schwefel, R. Maenner (Springer, Berlin 1991)
51. I. Morgenstern, D. Würtz: Z. Phys. B **67**, 397 (1987)
52. S. Rees, R.C. Ball: J. Phys. A Math. Gen. **20**, 1239 (1987)
53. G. Ruppeiner, J.M. Pedersen, P. Salamon: J. Phys. I **1**, 455 (1991)
54. R. Frost: Ebsa c library documentation, version 2.1 (San Diego Supercomputing Center 1994)
55. K.H. Hoffmann, D. Würtz, C. de Groot, M. Hanf: 'Concepts in optimizing simulated annealing schedules: an adaptive approach for parallel and vector machines.' In: *Parallel and Distributed Optimization*, ed. by M. Grauer, D.B. Pressmar (1991) pp. 154–175
56. M. Grötschel: Preprint 38, Universität Augsburg, Augsburg (1984)
57. R. Tafelmayer, K.H. Hoffmann: Comp. Phys. Comm. **86**, 81 (1995)
58. P. Sibani, J.M. Pedersen, K.H. Hoffmann, P. Salamon: Phys. Rev. A **42**, 7080 (1990)

5 Optimization of Production Lines by Methods from Statistical Physics

Johannes Schneider and Ingo Morgenstern

Summary. During the last few years, simulated annealing [7] and related Monte Carlo optimization algorithms such as threshold accepting [4,5] have become a useful means for finding the ground states of complex physical problems and for optimizing various kinds of economic problems, such as the Traveling Salesman Problem (TSP). In this article, we show how these methods from statistical physics can be adapted to the optimization of a certain type of assembly line, which can be related to a TSP with additional constraints. We concentrate on this problem because most costs in production processes are incurred in this area, and small relative improvements here can lead to large savings.

5.1 A Short Overview over Optimization Algorithms

Both in operations research and in computational physics, one often has to find a good or even optimal solution for a very complex problem. Economists speak of looking for a minimum cost solution, physicists of finding a low energy configuration. These minimization problems are equivalent: the cost function can simply be identified with the energy E.

For some problems, it is possible to find an exact algorithm and to solve the problem analytically. Examples are linear problems, which can be solved by the simplex algorithm [9]. For other problems, one can use a systematic search for finding an optimum, although this approach is limited to small system sizes by the available calculation time. Much effort has been made in the last few years to improve the velocity of these branch & bound algorithms by cutting off branches of the search tree very early on, in accordance with special bound conditions [6]. However, one still has to use heuristics in order to be able to handle complex problems of larger system sizes. This class of heuristics can be divided into two different subclasses: construction heuristics and improvement heuristics.

Starting from a tabula rasa, construction heuristics build a solution for the whole problem out of some pieces or subsolutions. They often use an 'egoistic' approach, in which a solution is built starting from a tabula rasa and adding one piece after the other to the system according to certain rules. These aim to insert the current piece in a quite optimal way, until a complete solution exists at the end. Another way of constructing a solution (an integrated ansatz) consists in putting all parts of the problem to be optimized into a system which is extended by some auxiliary variables in order to guarantee a feasible solution at the beginning. After that, these additional variables are removed in a certain manner until a solution for the original problem is created.

Improvement heuristics can also be divided into two subclasses: old–new and set-based improvement heuristics. Old–new improvement heuristics work according to the following scheme: starting from a random or a constructed solution, they try to improve the given configuration step by step, by changing some parts of the solution. These moves usually use the local search principle, whereby the tentative new solution is not much different from the previous configuration. Consequently, the energies of succeeding configurations are also expected to differ only slightly. In this way, an energy landscape is built up which is very pleasing and shows a nice valley–hill structure without being too rough. Gradient methods are often used to search for the best improvement close to the actual configuration. This approach is also known as the steepest descent method. Another way consists in choosing one of the neighboring configurations randomly.

The various kinds of algorithms only differ in the explicit choice of the probability, i.e., the prescription for deciding whether the tentative new solution is to be accepted or rejected. The simplest variation, the greedy algorithm, accepts every new configuration if its energy is lower or equal to the energy of the previous configuration. More elaborate acceptance probabilities also allow for deteriorations and depend not only on the energies of the participating configurations but also on a control parameter which is changed during the optimization run, such that the size of the deteriorations, or the probability that a deterioration is accepted, is reduced. In case of acceptance, the tentative new solution is set as the current solution; in case of rejection, the previous solution is retained as the current solution. Performing one trial move after another, a Markov chain is built up which ends at a local (or hopefully the global) optimum in the energy landscape.

Whereas old–new improvement heuristics share the feature that a new configuration is created from the actual one without using any other information, genetic algorithms and evolution strategies use a large set of configurations which are considered to be individuals of one or more populations. Besides mutations, which alter one individual and correspond to the moves described above, they use crossover operators, which generate children from two different parent individuals. According to Darwin's principle of the survival of the fittest, it is those individuals with the greatest fitness, i.e., the lowest energy, which survive. Additionally, individuals with greater fitness get more opportunities to reproduce. The various implementations of these algorithms differ in the kinds of mutations and crossover operators and in the choice of configurations which are allowed to mate or which have to commit suicide.

Another type of set-based algorithms is called Tabu search, which stores information about formerly visited configurations in a Tabu list. It therefore belongs to the class of set-based algorithms, although it works in a similar way to old–new heuristics. Tabu search [12] is a mnemonic search strategy which tries to avoid parts of the solution space already visited. One way to achieve this is to forbid configurations which have formerly been accepted. Another way is to place a ban on structures which were common to previous solutions. These configurations or structures are deposited in a Tabu list which is renewed after each move. The rules of this algorithm must guarantee that the optimization run never reaches a solution that has already been visited, that the Tabu list size does not diverge, and that a good solution is reached at the end.

The searching-for-backbones algorithm (SfB) [17] works just the other way round: SfB compares different results of independent optimization runs to find equal parts. These parts are assumed to be optimal, i.e., to be parts of the optimal solution. This information is considered in the next series of optimization runs, in which these parts remain unchanged. The new solutions are assumed to be better than the previous ones because the optimization could concentrate on parts which are more difficult to solve optimally. This algorithm is repeated iteratively until all optimization runs produce the same solution at the end.

In this article, we shall concentrate on physically motivated old–new improvement heuristics.

5.2   Methods from Statistical Physics

Typical methods taken from statistical physics are simulated annealing and its variations, such as threshold accepting. Another approach is the great deluge algorithm.

Using simulated annealing (SA) [1,2,7,10], threshold accepting (TA) [4,5], or the great deluge algorithm (GDA) [3,5] as optimization algorithm, the procedure is as follows. Firstly, an initial solution is chosen randomly. One then has to solve the problem of finding an appropriate value for the start temperature. If it is too high, time may be wasted, whereas if it is too low, the system may be unable to leave the local valley. One approach which can always be used is to perform a random walk while measuring the appearing energy differences ΔE (in the case of SA and TA) or the occurring energies $E(\sigma)$ (in the case of GDA) in order to determine the start temperature. Working with SA, one usually hopes to get an acceptance rate of at least 0.9 at the beginning. This leads to the condition that

$$T_{\text{start}} = -\frac{\max\{|\Delta E|\}}{\ln 0.9} \approx 10 \cdot \max\{|\Delta E|\} \,. \tag{5.1}$$

Similarly, one uses

$$Th_{\text{start}} = \max\{|\Delta E|\} \tag{5.2}$$

for TA, where Th denotes the threshold, and

$$\mathcal{T}_{\text{start}} = \max\{|E(\sigma)|\} \tag{5.3}$$

for GDA, where $\mathcal{T}$ denotes the water level.

The system is then cooled down stepwise in a logarithmic way, i.e., the control parameter is multiplied by a factor f, where usually $0.8 < f < 0.999$. A series of trial moves is performed at each temperature step in order to equilibrate the system at the new value of the control parameter. If required, several measurements can be performed after this equilibration. Several moves should be tried between two measurements in order to get uncorrelated results. The optimization run ends at a low value of the control parameter at which the acceptance rates of the single moves remain zero for a given time. Of course, this condition only works if the local optimum or the ground state, in which the system is caught at the end, is not degenerate.

For each problem, there are some specific observables of interest, which are measured during the optimization run. The mean energy $\langle E \rangle$ and the specific heat

$$C = \frac{\partial \langle E \rangle}{\partial T} \equiv \frac{\mathrm{Var}(E)}{k_\mathrm{B} T^2} \tag{5.4}$$

are defined for every problem and are the most important observables for optimization, where $\mathrm{Var}(E)$ is the variance of the energy. Of course, the equivalence in equation (5.4) is only exact in the case of SA if the system is equilibrated at a certain temperature. However, we also adopt this equivalence for TA because of its close relationship with SA. Looking at $\langle E \rangle$, one can see how cheap the solutions already are. The peak of the specific heat indicates the temperature range in which the largest reordering processes occur inside the system. Most of the available calculation time should be spent inside this range of the control parameter.

5.3   A Simple Example: The Traveling Salesman Problem

The traveling salesman problem (TSP) [12], sometimes called the traveling salesperson problem, consists of the following task: a traveling salesman has to find a closed tour of minimum length through a given set of N cities, going through each city exactly once. All distances $D(i,j)$ between the cities are known.

The simplest way to describe a configuration is to use a permutation σ of the numbers $(1 \ldots N)$. The energy E is given by the length of the configuration,

$$E(\sigma) = D\big(\sigma(N), \sigma(1)\big) + \sum_{i=1}^{N-1} D\big(\sigma(i), \sigma(i+1)\big) . \tag{5.5}$$

For a symmetric distance matrix $D(i,j) = D(j,i)$, each energy level is at least two-fold degenerate because the length does not change if the traveling salesman travels in a clockwise or anticlockwise direction.

There are several ways to compare different algorithms for the traveling salesman problem. Some authors use nodes on a two-dimensional quadratic lattice to represent the different cities. Let n be the number of rows and columns of the lattice. Then the number of cities is $N = n^2$. The optimal length in such a setup is simply given by

$$E_{\mathrm{opt}}(N) = \begin{cases} d \cdot N & \text{if } n \text{ is even ,} \\ d \cdot (N - 1 + \sqrt{2}) & \text{if } n \text{ is odd ,} \end{cases} \tag{5.6}$$

where d is the distance between neighboring rows. Another synthetic problem uses instances with randomly distributed cities. For some fixed numbers of cities, several instances are created by a uniform random generator. The average is then taken over the results of these instances, and the behavior of the algorithm is discussed for different system sizes.

However, neither approach considers real-life problems, because cities are neither on a quadratic lattice nor purely randomly distributed. There is a third way of investigating well defined real-life problems, for which the optimal energy value is known. A lot of examples introduced by several authors have been collected by Reinelt in his library TSPLIB95 [13]. One of these is the problem of the 127 beer gardens in the area of Augsburg, which was introduced by Jünger and Reinelt. This BIER127 problem is shown in Fig. 5.1.

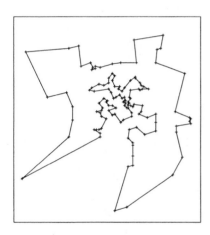

Fig. 5.1. Optimal solution for the problem of the 127 beer gardens in the area of Augsburg

The example of the 127 beer gardens is a relatively simple one, although it consists of a central part with many beer gardens in the city of Augsburg and only a few in the villages further out. Using SA or TA, one generally finds the optimum configuration with a length of 118293.52... according to the Euclidean metric in REAL*8-format. Of course, there are many other benchmark instances like the 16 stations of the Odyssey or the 4461 townships in the five newly-formed German states. However, we restrict ourselves to the BIER127 problem in this article, giving results that are relevant for the sections about production lines.

The first question arising is: which moves should be used for the TSP? The classic move is exchange (EXC), also called transposition or swap. It simply exchanges two cities in the tour. Lin invented the Lin-2-Opt (L2O) (shown in the left half of Fig. 5.2), which turns round a part of the tour. Stadler and

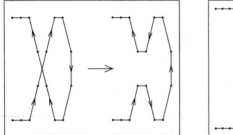 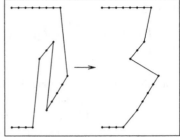

Fig. 5.2. Moves for the TSP. *Left*: Lin-2-Opt, *right*: Lin-3-Opt

Schnabl showed in their paper [18] that the Lin-2-Opt leads to better results than exchange because it cuts only two edges whereas EXC removes four edges from the system.

Cutting three edges, there are four possibilities for creating a new tour. Following Stadler and Schnabl, these Lin-3-Opts (L3O) would lead to worse results than L2O. Using only very short calculation times, this is surely correct. However, spending more calculation time, one gets better results with L3O, a result which is in accordance with Kirkpatrick and Toulouse [8]. One of these Lin-3-Opts (L3O), shown in the right half of Fig. 5.2, exchanges two successive parts of the tour without changing their directions. We prefer to use both L2O and this kind of L3O for our simulations.

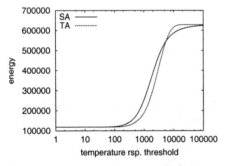 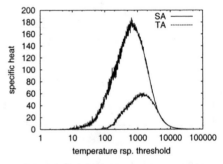

Fig. 5.3. Mean energy and specific heat of the BIER127 problem using SA and TA

Figure 5.3 shows the mean energy and the specific heat of this problem using SA or TA. $\langle E \rangle$ shows a sigmoidal decrease for SA. Using TA, the system accepts every move at the beginning, so that the mean energy stays constant at high thresholds. Lowering the threshold, the system loses energy to a large extent. The decrease in $\langle E \rangle$ is steeper than for SA, so that the curves of SA and TA cross each other. Looking at the specific heat, one finds that the curves of SA and TA are nearly identical for high temperatures or thresholds. The peak in the specific heat is much smaller for TA and lies at higher values of the control parameter.

The results of this comparison between SA and TA are also valid for all other optimization problems studied by the authors [14].

5.4   Optimization with Constraints

Optimization problems generally suffer from more or less severe constraints, which have to be considered during the optimization run in order to result in a feasible solution. There are 'hard' constraints which must definitely be fulfilled, while 'soft' constraints may be violated, although the magnitude of these violations should be as small as possible.

Modelling the optimization problem, the best way to deal with such a constraint is to integrate it into the system, in such a way that each state in the configuration space automatically fulfills the constraint and every move between these states is constructed so that each possible state can be reached from any other state. Furthermore, moves are not allowed to create configurations which are not members of this configuration space. This approach is especially recommended for hard constraints. An example is the modeling of the traveling salesman problem described above which automatically fulfills the constraints that each city has to be visited exactly once and that there is one closed tour without any closed subtours. However, this approach cannot be used for all constraints.

If it is possible to find a feasible solution very easily, another ansatz is often used which only allows for feasible solutions during the whole optimization run, i.e., if a move led to a violation of a constraint it would not be accepted. However, such a procedure seldom leads to good results for most problems because the set of feasible configurations decomposes into many small islands. Consequently, the optimization run searches for the global optimum inside one of these islands, but cannot reach other islands and may fail to end up at the global optimum.

Because of this insight, it is better to work with penalty functions [11]: a configuration which violates some constraints is not forbidden by barriers. Instead, the configuration is penalized. The penalty is determined by the size of the violations: the worse the violations are, the larger is the penalty. An unfeasible solution is penalized by raising its energy. The optimization run, which simply searches for the minimum in the energy landscape, should end at a low-energy feasible solution, if the penalties are chosen correctly.

The energy E does not only consist of the real costs E_0 but also of additional pseudo-costs E_i. At the end of the optimization run, all E_i ($i \geq 1$) should vanish, in such a way that all constraints are fulfilled.

There are many different kinds of penalty functions. Generally, $E_i = 0$ if the constraint i is fulfilled. If there is a violation of the constraint, the penalty function often increases linearly or quadratically with or without an offset for the size of the violation.

One of the best examples in the area of operations research for a problem extended by some constraints is the vehicle routing problem with time windows. This is based on a multi-traveling salesman problem: several trucks have to deliver goods to customers on closed tours, which start and end at a depot. Additionally, there is the constraint that the vehicles have a finite capacity, i.e., the trucks must not be overloaded by the amount of goods they deliver to the customers. Furthermore, customers tell the forwarding agent a time interval in which the truck should arrive at their factory or home. Of course, the real costs are still given by the sum of the tour lengths. Violations of the two constraints are assumed to produce additional pseudo-costs (e.g., disgruntled customers when time constraints have not been fulfilled or fines due to overloaded trucks), which are simply added to the real costs. In addition to the moves used for the TSP, which serve as intra-tour moves, one has to introduce inter-tour moves, which exchange one or more customers between the trucks. Besides the move EAT, which transfers one customer to another truck, and inter-tour exchange, which exchanges two customers between their trucks, it has been shown to be useful also to use cluster moves, which exchange sequences of customers between the tours [15].

An example from physics is the Ising model with an applied magnetic field H. The energy function of the model is given by

$$E = -\sum_{\langle i,j \rangle} J_{ij} S_i S_j - H \sum_{i=1}^{N} S_i \,, \tag{5.7}$$

with N Ising spins $S_i = \pm 1$ and the interaction matrix J_{ij} [2], where $\langle i, j \rangle$ indicates that only neighboring spins interact with each other. The energy therefore consists of two parts. The first part specifies the interaction between single spins. The additional Zeeman term describes the coupling of the spins to an external magnetic field and can be considered as a penalty function; according to the size of H (which may be considered as a Lagrange multiplier), the single spins are more or less forced to lie parallel to H. The magnetization M is given by

$$M = \frac{1}{N} \sum_{i=1}^{N} S_i \,. \tag{5.8}$$

The susceptibility is defined as

$$\chi = \frac{\partial \langle M \rangle}{\partial H} \equiv \frac{N \cdot \mathrm{Var}(M)}{k_B T} \,. \tag{5.9}$$

This equivalence is again only true for SA, but we shall continue to use it for TA. The peak of the susceptibility indicates the temperature range in which the system orders itself according to the corresponding constraint.

If an optimization problem underlies several constraints which are modeled by partial energies E_i, then a susceptibility χ_i may be defined for each i by

$$\chi_i = \frac{\mathrm{Var}(E_i)}{k_B T} \,. \tag{5.10}$$

In this case, the peaks of the susceptibilities are of great importance since they should be in the same temperature range, in such a way that the system is optimized according to all constraints at the same time. This generally leads to better results than first optimizing according to one constraint and then according to another constraint, and so on.

We now have all requisites to start optimizing assembly lines.

5.5   The Linear Assembly Line Problem

5.5.1   The Problem

Nowadays, the production of complex goods, like automobiles, is usually performed on an assembly line, using the fact that workers work faster when they maximally perform a dozen or so tasks. The optimization of assembly lines can be pushed a long way. Usually, the production process is divided up into small steps of equal duration. While one group of workers or robots perform a series of tasks on one product at a certain station on the assembly line, another group located elsewhere performs another series of tasks in exactly the same amount of time. After each time step, the products are shifted to the next station. In this way, the production can be planned exactly on a drawing board for many products.

However, some industries, like the automobile industry, face the problem that the products may differ due to special equipment, which is not introduced in every final product. For example, car purchasers today have many possibilities for 'creating' their own car. They can choose among various devices like air conditioning, a sun roof, or a towbar. Because of seasonal fluctuations and trends, customer orders cannot be predicted at all, and car producers cannot plan their production over a time period of months. They have to construct a continuous sequence of cars, produced according to these options. Building a special station for each different piece of equipment would either waste resources, if the station were fully equipped, or lead to a loss of quality if the workers had to hurry. Introduction of options must therefore be performed 'on the fly' while other workers at conventional stations build the usual equipment into the cars.

In order to burden the workers evenly and to use all resources continuously, the products including a certain option have to be uniformly distributed over the whole production process. Usually, more than one special device is offered by a car producer, and uniform distribution according to all options is not trivial. Because of the finite length of the assembly line, it is essential to introduce the conditions of a maximum pulk (maximum number of cars with the same option, important for options like air conditioning, which are introduced into almost every car) and a minimum distance between two cars that both include the same special equipment (for options like a towbar, which are relatively seldomly ordered), whereas the other cars between them must not have this special equipment. We must therefore represent the basic problem of the uniform distribution by an energy E_0, and the additional hard constraints of the maximum pulk and the minimum distance by additional penalties E_1 and E_2.

5.5.2 The Model

More precisely, we wish to examine the following task. Car bodies are taken one after the other from a shelf in front of the assembly line, in which there is no car at the beginning. We have to determine the sequence in which the car bodies are put onto the assembly line according to the constraints mentioned above. Let us now describe the model used to solve this problem.

Let M be the number of options and N the number of final products. Define a matrix $A(k,i)_{k=1,...,M,\,i=1,...,N}$ with

$$A(k,i) = \begin{cases} 1 & \text{if option } k \text{ is introduced in final product } i , \\ 0 & \text{otherwise} , \end{cases} \tag{5.11}$$

i.e., each final product is represented as a vector of bits. The components of the vectors are the various options. Then the objective functions can be defined as follows.

Uniform Distribution of Final Products According to Options. Define a 'distance' $D(i,j)$ between two final products i and j by

$$D(i,j) = \sum_{k=1}^{M} \lambda_k A(k,i) \cdot A(k,j) , \tag{5.12}$$

with Lagrange multipliers λ_k. For a configuration σ, we get

$$E_0(\sigma) = \sum_{i=1}^{N-1} D\big(\sigma(i), \sigma(i+1)\big) . \tag{5.13}$$

This condition of a uniform distribution in the assembly line, which is modelled by a repulsive short-range Coulomb interaction, leads in effect to a mapping of the given optimization problem onto the well-known traveling salesman problem with open end points.

Maximum Pulk. Secondly, only α_i final products with option i may be produced on the assembly line one after the other. Then a final product without i has to come onto the line. This condition is easily mapped on the penalty function

$$E_1(\sigma) = \sum_{i=1}^{M} \left[\mu_i \Theta(\alpha_i) \sum_{j=1}^{N-\alpha_i} \prod_{k=0}^{\alpha_i} A\big(i, \sigma(j+k)\big) \right] \tag{5.14}$$

with Lagrange multipliers $\mu_i > 0$ and the slightly changed Heaviside function

$$\Theta(x) = \begin{cases} 1 & \text{if } x > 0 , \\ 0 & \text{if } x \leq 0 . \end{cases} \tag{5.15}$$

Minimum Distance. This condition requires that, after one final product with the option i, there must be $\beta_i - 1$ final products without i before there may be another one with option i. We get

$$E_2(\sigma) = \sum_{i=1}^{M} \left[\nu_i \sum_{j=1}^{N-1} A\big(i, \sigma(j)\big) \times \sum_{k=1}^{\min\{\beta_i-1, N-j\}} A\big(i, \sigma(j+k)\big) \right] , \qquad (5.16)$$

with Lagrange multipliers $\nu_i > 0$.

The energy $E = E_0 + E_1 + E_2$ is to be minimized. The Lagrange multipliers have to be chosen in such a way that $E_1 = E_2 = 0$ for the resulting configuration and that the 'real costs' are as low as possible.

5.5.3 Computational Results

We generated a test instance for the problem described above with $N = 100$ final products and $M = 10$ options, with a random generator using probabilities as they occur in practice. (This test instance can be received via email from J. Schneider, or see [19].) We chose the Lagrange multipliers as $\lambda_i = 1$, $\mu_i = 1$, and $\nu_i = 10$ in order to get a feasible solution at the end. We wish to study some observables in order to gain insight into the way a physical optimization algorithm like SA or TA processes such an economic problem. Hence we invested a lot of calculation time in order to get graphs with low statistical noise: T was decreased by a factor of 0.95 from 1000 to 0.01, 102 000 measurements were performed in each temperature step (the first 2000 were not used for averaging), and 200 Lin-2-Opt trials and 1000 Lin-3-Opt trials were performed between two measurements. More exactly, each run took roughly 3.5 days, even though a configuration with the optimal energy $E_0 = 70$, $E_1 = E_2 = 0$ can already be reached after a few seconds. Investing one minute, we nearly always get an optimal configuration. Times are given here for a LINUX Pentium 300 using a g77 compiler.

First of all, let us have a look at Fig. 5.4, which shows the development of the energy $\langle E \rangle$ and the specific heat C against T and Th, respectively, for both SA

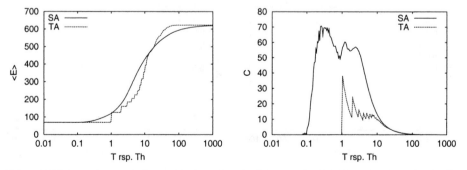

Fig. 5.4. Mean energy and specific heat for the linear assembly line problem, for both SA and TA

and TA. For large T, the system performs a random walk, i.e., every/almost every new configuration is accepted by TA/SA. The energy $\langle E \rangle$ is therefore virtually constant at a high level for large T. Reducing T, the energy $\langle E \rangle$ decreases on a logarithmic T scale until it becomes constant for $Th < 1$ (TA) or $T < 0.1$ (SA), when the system is caught in the ground state. The decrease in energy is much steeper for TA than for SA, so that the curves cross. Interestingly, the decrease is sigmoidal for SA, whereas there are some significant steps in the threshold range $1 \leq Th \leq 10$: if the threshold crosses integer values, the system loses energy to a large extent. This is due to the choice of the Lagrange multipliers because the smallest possible energy difference is 1, so we get the largest decrease at the threshold 1. All other possible energy differences are also integer-valued.

The steps in the energy decrease using TA correspond to significant peaks of the specific heat. The largest peak is at $Th = 1$ as expected. The increase in the height of the peaks is $\propto Th^{-2}$, so that the variance of the length distribution is also constant between two integer thresholds. If we compare the specific heats of SA and TA, we see that they are nearly equal for large T, although the peak is much larger for SA. Additionally, there is a double peak for SA, and the right peak shows a cusp in the middle, so that we might have three peaks all together. This could be explained by the three parts of the energy. However, we cannot significantly exclude this cusp being due to nonequilibrium effects.

A freezing behavior of the system can best be viewed with the acceptance rates, which are shown in Fig. 5.5. For SA, the decrease in the acceptance rates is sigmoidal, but both acceptance rates converge towards a constant greater than zero for $T < 0.1$. This problem thus has a degenerate ground state. Using TA, we again find a decrease in steps at small integer thresholds. The acceptance rates converge towards the same values as for SA.

In addition, we look at the behavior of the individual parts of the system during the optimization run, which is shown in Fig. 5.6: looking at the sigmoidal decrease in the partial energies for SA with decreasing temperature, we find that the system is dominated by $\langle E_2 \rangle$, the penalty for the minimum distance. The system finally freezes for small T, the energy values of $\langle E_1 \rangle$ and $\langle E_2 \rangle$ decrease to 0, as required, and $\langle E_0 \rangle \to 70 = \langle E \rangle$. Moreover, the peak in the susceptibility

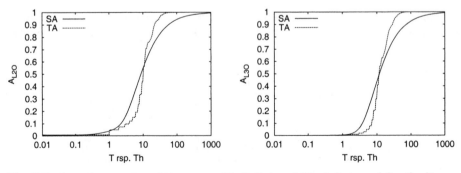

Fig. 5.5. Acceptance rates of the moves Lin-2-Opt and Lin-3-Opt used for the linear assembly line problem, for both SA and TA

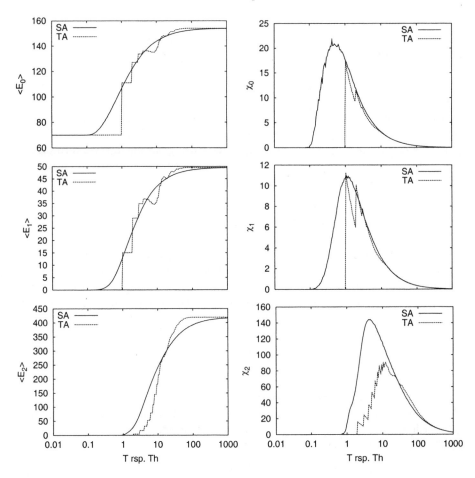

Fig. 5.6. *Left*: decrease in the partial energies of the linear assembly line problem, $\langle E_0 \rangle$ for the uniform distribution, $\langle E_1 \rangle$ for the maximum pulk lengths, and $\langle E_2 \rangle$ for the minimum distance, vs. decreasing T or Th. *Right*: the corresponding susceptibilities, for both SA and TA

χ_2, which corresponds to the minimum distance, is an order of magnitude higher than the peaks for the two other energy terms. Summarizing the information provided by the peaks, we find that the system first orders according to the minimum distance, then according to the maximum pulk, and finally according to the uniform distribution. However, the peaks strongly overlap with each other, so that the optimization is not performed separately for each constraint.

Working with TA, we again find a successive decrease, together with a significant inside knee in $\langle E_0 \rangle$ and $\langle E_1 \rangle$ at $Th \approx 10$, which corresponds to an outside knee in $\langle E_2 \rangle$. The corresponding susceptibilities show a new behavior: χ_2 seems to be typical at first sight, with a peak at a larger value of Th than the corresponding value of SA. However, we find many small peaks on its left leg, which correspond to integer threshold values. The asymmetric appearance of the susceptibilities χ_0 and χ_1 is completely new: they break down at $Th = 1$. Besides these features, we find a double peak structure which corresponds to the steepest decreases in the partial energies. The susceptibilities break down working with TA on systems with discrete energy differences $n \cdot \Delta E_{\min}$, at a threshold value of $T = \Delta E_{\min}$, if the minimum energy difference corresponds to the minimum partial energy difference. This behavior – observed here for the first time as far as we know – should have universal character. It is also observed for other optimization problems with this property [14]. The extent to which the partial energies are correlated with each other during the optimization run is also very important when examining the complexity of the system. The correlation between two partial energies E_i and E_j is given by

$$\varrho(m, n) = \frac{\langle E_m \cdot E_n \rangle - \langle E_m \rangle \cdot \langle E_n \rangle}{\sqrt{\text{Var}(E_m) \cdot \text{Var}(E_n)}} . \tag{5.17}$$

Looking at Fig. 5.7, we find that the maximum pulk and minimum distance constraints are positively correlated with the uniform distribution. The correlation decreases to zero during the optimization run; using TA, $\varrho(0,1)$ mainly decreases in two large jumps. The correlation between maximum pulk and minimum distance is much more interesting: there is a positive total correlation for large T, which then decreases to 0. For TA, we also get a significant anticorrelation in the range between 1 and 10. In this range, the conditions work against each other. However, we expect this behaviour to be general. This is due to the fact that there can only exist either a maximum pulk condition or a minimum distance constraint for an option. However, a correlation consists of the intrinsic partial correlation and additionally of contributions from indirect correlations. For a system of three interacting energies, the partial correlation can be calculated from

$$\varrho(m, n|k) = \frac{\varrho(m, n) - \varrho(m, k)\varrho(n, k)}{\sqrt{[1 - \varrho^2(m, k)][1 - \varrho^2(n, k)]}} . \tag{5.18}$$

$\varrho(1, 2|0)$ shows the anticorrelation, as expected for large T.

In conclusion, we can state that this problem seems to be relatively easily solved by physical optimization algorithms. Because of the overall positive correlation between uniform distribution and maximum pulk, and between uniform distribution and minimum distance, we can neglect the uniform distribution in the energy: if we optimize the system only according to the hard constraints of maximum pulk and minimum distance, the system also optimizes itself according to the uniform distribution condition.

5.6   The Network Assembly Line Problem

5.6.1   The Problem

A production process is not usually as straightforward as in the linear assembly line model. For example, machines may be used several times in the production

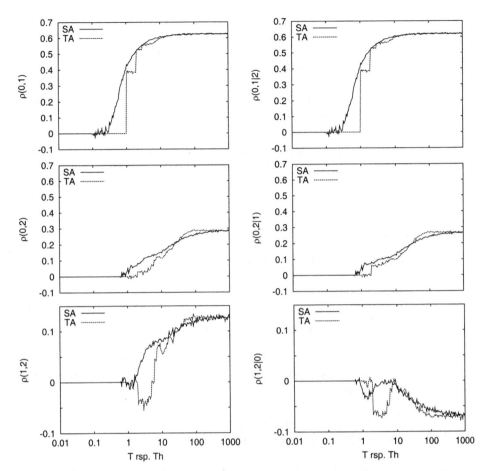

Fig. 5.7. Correlations between partial energies of the linear assembly line problem: E_0 (uniform distribution), E_1 (maximum pulk), and E_2 (minimum distance), shown for both SA and TA. *Left*: total correlations, *right* partial correlations

process for different purposes, intermediate products may be stored for later use in stocks of finite capacity or they may be involved in different ways, so that we get a network of correlated processes. We thus also have to optimize the relative timing of the individual processes when studying the universal case.

However, let us once again concentrate on a specific real-life problem as it occurs in the plant of a Bavarian car manufacturer with five assembly lines. Car bodies are taken one after the other from the shelf onto assembly line No. 1. This assembly line splits at the end into assembly lines Nos. 2 and 3. Nos. 2 and 3 unite at their ends to form assembly line No. 4. No. 4 is continued by No. 5. Nos. 1, 4, and 5 have the same velocity v, whilst Nos. 2 and 3 are run with the velocity $v/2$. Hence car bodies leaving assembly line No. 1 are routed alternately onto No. 2 and No. 3, i.e., the 1st, 3rd, 5th, ... , car bodies in the sequence are supposed to be intermediately handled in assembly line No. 2, whilst the 2nd, 4th, 6th, ... , are handled in assembly line No. 3.

We still have a pseudo-linear problem, since the 'network' only consists of two parallel assembly lines, and the problem does not contain any refluxes. We have to find an optimal sequence of cars taken from the shelf onto assembly line No. 1, regarding some constraints: again we have the constraints of maximum pulk and minimum distance. However, we now have to take into account the fact that some of the options are implemented in the final products in assembly line No. 1, and other special devices on assembly line No. 2 and/or No. 3. Some options can only be integrated into the final products either on line No. 2 or on line No. 3. Only standard equipment is introduced into the final products on lines Nos. 4 and 5.

5.6.2   The Model

Due to these options, we have to introduce a flag $f(i)$ for each final product i:

$$f(i) = \begin{cases} 1 & \text{if } i \text{ can be channelled through No. 2 or No. 3 ,} \\ 2 & \text{if } i \text{ must be channelled through No. 2 ,} \\ 3 & \text{if } i \text{ must be channelled through No. 3 .} \end{cases} \tag{5.19}$$

We may also define an information flag $l_\sigma(i)$ for each final product i, which indicates through which assembly line l the final product i is channelled in the actual configuration σ:

$$l_\sigma(i) = \begin{cases} 2 & \text{if } i \text{ is channelled through No. 2 ,} \\ 3 & \text{if } i \text{ is channelled through No. 3 .} \end{cases} \tag{5.20}$$

We can easily verify whether the final product is channelled through the correct assembly line because, in case of a violation, we get $f(i) \cdot l_\sigma(i) = 6$.

For this new task of simulating a specific real-life problem, we neglect the condition of uniform distribution. Furthermore, according to the original problem, we have a minimum distance condition only for assembly line No. 1 and maximum pulk lengths α_{il} only for the assembly lines $l = 1\text{--}3$ (either for No. 1 or for No. 2 and/or No. 3), so that we get the energy

$$E(\sigma) = \lambda \sum_{i=1}^{N} \delta_{6, f(i) \cdot l_\sigma(i)}$$

$$+ \sum_{i=1}^{M} \left[\sum_{l=1}^{3} \mu_{il} \Theta(\alpha_{il}) \sum_{j=1+\left[\frac{l-1}{2}\right]}^{N+|l-2|-1-\alpha_{il} z_{il}} \prod_{k=0}^{\alpha_{il}} A\big(i, \sigma(j + k z_{il})\big) \right]$$

$$+ \sum_{i=1}^{M} \left[\nu_i \sum_{j=1}^{N-1} A\big(i, \sigma(j)\big) \cdot \sum_{k=1}^{\min\{\beta_i-1, N-j\}} A\big(i, \sigma(j + k)\big) \right] \tag{5.21}$$

where $[x]$ denotes the integer part of x, $\delta_{a,b}$ is the Kronecker symbol defined by

$$\delta_{a,b} = \begin{cases} 1 & \text{if } a = b, \\ 0 & \text{otherwise}, \end{cases} \qquad (5.22)$$

and

$$z_{il} = \begin{cases} 1 & \text{if option } i \text{ is introduced in assembly line } l = 1, \\ 2 & \text{otherwise}. \end{cases} \qquad (5.23)$$

Of course, this energy can easily be extended to cases in which the velocities differ between the individual assembly lines and in which the splitting of the assembly lines is more complicated.

5.6.3   Computational Results

Once again we created a test instance (which can also be received from J. Schneider, or see [19]) with $M = 14$ options and $N = 150$ final products. Fifty of them are already put onto the assembly line according to all constraints, so that we have to find an optimal sequence only for the remaining 100 car bodies. Note that, in contrast to the toy problem in the previous section, we are now working on a real-life problem with which our Bavarian car manufacturer is confronted every day. The parameters for the optimization run were chosen as for the linear assembly line problem.

Because the first car bodies are already put onto the assembly line, we now get symmetry breaking (well-known in many physical systems): turning around the sequence of 100 car bodies will change the energy, because of the interactions with the first 50 car bodies, so one of the two directions is preferred. Of course, there are still many possibilities for degeneracies.

Once again, we wish first to consider the decrease in the mean energy $\langle E \rangle$ with decreasing T, shown in Fig. 5.8. $\langle E \rangle$ decreases to zero, so that all constraints are fulfilled, and we achieve a feasible solution at the end. The decrease in $\langle E \rangle$ is

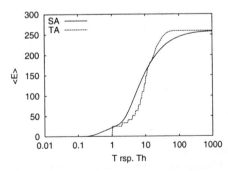
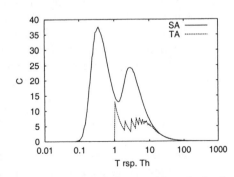

Fig. 5.8. Mean energy and specific heat of the network assembly line problem, for both SA and TA

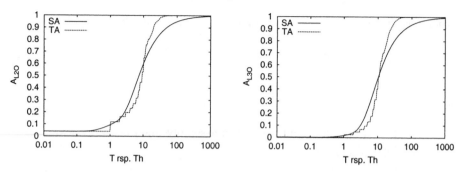

Fig. 5.9. Acceptance rates of the moves Lin-2-Opt and Lin-3-Opt for the network assembly line problem, for both SA and TA

no longer cleanly sigmoidal for SA: we observe a significant knee at $T \approx 2$. For TA, we again find a decrease in steps with an extremely large step at $Th = 1$. This corresponds to the peaks in the specific heat for TA, which is also shown in Fig. 5.8. Here we are able to identify a bell structure also for TA in contrast to Fig. 5.4. The peak of the bell is at $Th \approx 7$. The specific heat of SA possesses a significant double peak structure with a larger left peak.

In contrast to that, in Fig. 5.9, we find the picture we have already seen in Fig. 5.5, namely a sigmoidal decrease in the acceptance rates when using SA and a stepwise decrease at $1 \le Th \le 10$ for TA. Interestingly enough, the value of roughly 4% to which the acceptance rate of the Lin-2-Opt converges, is rather high, i.e., the problem has a highly degenerate ground state. Trivial loops, i.e., moves with $\Delta E = 0$, are also possible using the Lin-3-Opt, whose acceptance rate decreases to 0.34%.

Furthermore, we look at the partial energies and their corresponding susceptibilities, which are shown in Fig. 5.10. We find a sigmoidal decrease in the partial energies for SA, together with several steps and a knee in $\langle E_1 \rangle$ for TA, analogously to Fig. 5.6. The system is once again dominated by the penalty function for violation of the minimum distance. Looking at the susceptibility peaks, we see that for SA the system is ordered firstly according to the minimum distances, secondly by the maximum pulks, and finally by the correct assembly line condition. Using TA, the susceptibilities show the same behavior as for the linear assembly line problem. The peak for χ_2 is at $Th \approx 10$ and has many small peaks on its left flank at integer Th values. After a conventional increase, χ_1 shows a many-peak structure and breaks down at $Th = 1$, whilst χ_0 increases as Th^{-1} (except for some small breakdowns) until $Th = 1$ and vanishes for $Th < 1$.

5.7 Conclusion

In this chapter, we describe the main problem of production planning in the automobile industry: sequencing vehicles in the assembly line. We showed how this problem can be modeled so that physical optimization algorithms can be applied

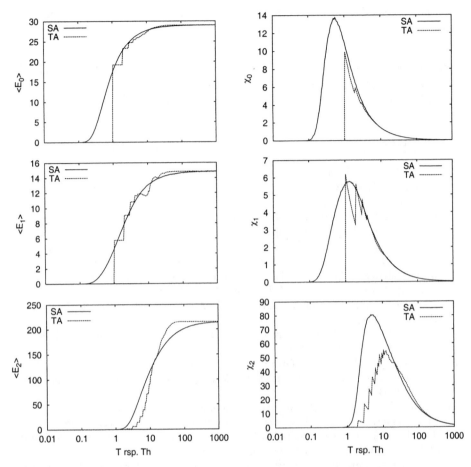

Fig. 5.10. Partial energies and susceptibilities for the network assembly line problem using SA and TA. *Top*: for the correct assembly line, *middle*: for the maximum pulk, and *bottom*: for the minimum distance

to it. Individual constraints are mapped onto different competing penalty functions, which are added to the energy of the system. The latter is then minimized by the optimization algorithms. Through our model, we were able to re-use the moves introduced for the traveling salesman problem. We showed that simulated annealing and threshold accepting are able to solve such problems optimally and reach feasible solutions at the end, even though the individual constraints work against each other and one would expect this problem to be rather hard.

The idea of optimizing assembly lines using methods from statistical physics is well-suited to parallel implementation. Results for difficult instances can be optimally solved on a parallel computer in combination with the searching-for-backbones algorithm [16]. We are continuing our research on this topic and on other areas of supply chain management, enterprise resource planning, and advanced planning and scheduling.

Acknowledgments. We wish to thank directors Hartmann and Pommer (BMW Regensburg, Germany) for providing real-life probabilities for the options. J.S. would like to thank the Swiss National Science Foundation for financial support.

References

1. K. Binder: Rep. Prog. Phys. **60**, 487 (1997)
2. K. Binder, D.W. Heermann: *Monte Carlo Simulation in Statistical Physics*, 2nd edn. (Springer, Berlin 1992)
3. G. Dueck: J. Comp. Phys. **104**, 86 (1993)
4. G. Dueck, T. Scheuer: J. Comp. Phys. **90**, 161 (1990)
5. G. Dueck, T. Scheuer, H.-M. Wallmeier: Spektrum der Wissenschaft 42 (3/1993)
6. M. Grötschel, M. Padberg: Spektrum der Wissenschaft 76 (4/1999)
7. S. Kirkpatrick, C.D. Gelatt Jr., M.P. Vecchi: Science **220**, 671 (1983)
8. S. Kirkpatrick, G. Toulouse: J. Physique **46**, 1277 (1985)
9. R. Mennicken, E. Wagenführer: *Numerische Mathematik 1* (Rowohlt, Reinbek bei Hamburg 1977)
10. N. Metropolis, A.W. Rosenbluth, M.N. Rosenbluth, A.H. Teller, E. Teller: J. Chem. Phys. **21**, 1087 (1953)
11. K. Neumann, M. Morlock: *Operations Research* (Hanser, Munich, Vienna 1993)
12. G. Reinelt, *The Traveling Salesman* (Lecture Notes in Computer Science 840, Springer, Berlin 1994)
13. G. Reinelt, *TSPLIB95*, University of Heidelberg, Germany (1995); http://www.iwr.uni-heidelberg.de/iwr/comopt/software/TSPLIB95
14. J. Schneider: Effiziente parallelisierbare physikalische Optimierungsverfahren. PhD thesis, University of Regensburg, Germany (1999)
15. J. Schneider: Parallelisierung physikalischer Optimierungsverfahren. Diploma thesis, University of Regensburg, Germany (1995)
16. J. Schneider, J. Britze, A. Ebersbach, I. Morgenstern, M. Puchta: IJMPC **11/5** (2000)
17. J. Schneider, Ch. Froschhammer, I. Morgenstern, Th. Husslein, J.M. Singer: Comp. Phys. Comm. **96**, 173 (1996)
18. P.F. Stadler, W. Schnabl: Phys. Letters A **161**, 337 (1992)
19. http://www.tu-chemnitz.de/physik/HERAEUS/2000/Springer.html

6 Predicting and Generating Time Series by Neural Networks: An Investigation Using Statistical Physics

Wolfgang Kinzel

Summary. We give an overview of the statistical physics of neural networks generating and analysing time series. Storage capacity, bit and sequence generation, prediction error, antipredictable sequences, interacting perceptrons and application to the minority game are discussed. Finally, as a demonstration, a perceptron predicts bit sequences produced by human beings.

6.1 Introduction

In the last two decades there has been intensive research on the statistical physics of neural networks [1–3]. The cooperative behaviour of neurons interacting by synaptic couplings has been investigated using mathematical models which describe the activity of each neuron and the strength of the synapses by real numbers. Simple mechanisms change the activity of each neuron receiving signals via the synapses from many others, and change the strength of each synapse according to presented examples on which the network is trained.

In the limit of infinitely large networks and for a set of random examples there exist mathematical tools to calculate properties of the system of interacting neurons and synapses exactly. For many models the dynamics of the network receiving continuously new examples has been described by nonlinear ordinary differential equations for a few order parameters describing the state of the system [4]. If a network is trained on the total set of examples, the stationary state has been described by a minimum of a cost function. Using methods from the statistical mechanics of disordered systems (spin glasses), the properties of the network can be described by nonlinear equations relating a few order parameters.

It turns out that very simple models of neural networks already have interesting properties with respect to information processing. A network with N neurons and N^2 synapses can store a set of order N patterns simultaneously. Such a network functions as a content–addressable, distributed and associative memory.

Even a simple feedforward network with only one layer of synaptic weights can learn to classify high-dimensional data. When such a network (the 'student') is trained on examples which are generated by a different network (the 'teacher'), the student achieves overlap with the teacher network. This means that the student has not only learned the training data but it can also classify unknown input data to some extent, i.e., it generalizes. Using statistical mechanics, the generalization error has been calculated exactly as a function of the number of examples for many different scenarios and network architectures [5].

An important application of neural networks is the prediction of time series. There are many situations where a sequence of numbers is measured and one would like to know the following numbers, without knowing the rule which produces them [6]. There are powerful linear prediction algorithms including assumptions about external noise in the data, but neural networks have proven to provide competitive algorithms compared with other known methods.

The statistical physics of time series prediction has been studied since 1995 [7]. Similarly to the static case, the series is generated by a well known rule, usually a different 'teacher' network, and the student network is trained on these data while moving it over the series. We are interested in the following questions:

- How well can the student network predict the numbers of the series after it has been trained on part of it?
- Has the student network achieved some knowledge about the rule (network) which produced the time series?

It seems straightforward to extend the analytic methods and results of the static classification problem to the case of time series prediction. The only difference appears to be the correlation between input vector and output bit. However, although many experts in this field have looked into the problem, neither the capacity problem nor the prediction problem has been solved analytically up to now, even for the simple perceptron. Furthermore, it turns out that even the problem of the generation of a time series by a neural network is not trivial. A network can produce quasiperiodic or chaotic sequences, depending on the weights and transfer functions. For some models, an analytic solution has been derived, even for multilayer networks [8].

In this chapter I intend to give an overview covering the statistical physics of neural networks which generate and predict time series. Firstly, I discuss the capacity problem: given a random sequence, what is the maximal length a perceptron can learn perfectly? Secondly, in Sect. 6.3, a network generating binary or continuous sequences is introduced and analysed. Thirdly, the prediction of quasiperiodic and chaotic sequences is investigated in Sect. 6.4. In Sect. 6.5, it is shown that, for any prediction algorithm, a sequence can be constructed for which this algorithm completely fails. Section 6.6 considers the problem of a set of neural networks which learn from each other. In Sect. 6.7, this scenario is applied to a simple economic model, the minority game. Finally, in Sect. 6.8, it is shown that a simple perceptron can be trained to predict a sequence of bits entered by the reader, even if he/she tries to generate random bits.

6.2   Learning from Random Sequences

A neural network learns from examples. In the case of time series prediction, the examples are defined by moving the network over the sequence, as shown in Fig. 6.1.

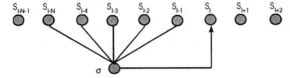

Fig. 6.1. A perceptron moves over a time series

Let us consider the simplest possible neural network, the perceptron. It consists of an N-dimensional weight vector $\boldsymbol{w} = (w_1, \ldots, w_N)$ and a transfer function $\sigma = f(\boldsymbol{w} \cdot \boldsymbol{S}/N)$, where $\boldsymbol{S}$ is the input vector, given by the sequence. We mainly consider two transfer functions, the Boolean and the continuous perceptron:

$$\sigma = \text{sign}(\boldsymbol{w} \cdot \boldsymbol{S}) \,, \tag{6.1}$$

$$\sigma = \tanh\left(\frac{\beta}{N}\,\boldsymbol{w} \cdot \boldsymbol{S}\right) \,, \tag{6.2}$$

where β is a parameter giving the slope of the linear part of the transfer function in the continuous case, $f(x) \simeq \beta x + O(x^3)$.

The aim of our network is to learn a given sequence $S_0, S_1, S_2, \ldots$. This means that the network should find – by some simple algorithms – a weight vector $\boldsymbol{w}$ with the property

$$S_t = f\left(\frac{1}{N}\sum_{j=1}^{N} w_j S_{t-j}\right) \,, \tag{6.3}$$

for all time steps t. For the Boolean function (6.1), this set of equations becomes a set of inequalities

$$S_t \sum_{j=1}^{N} w_j S_{t-j} > 0 \,, \tag{6.4}$$

for all t. If the bits S_t in (6.4) are random, $S_t \in \{+1, -1\}$, instead of being taken from the time series, then the inequalities (6.4) have a solution if the number of inequalities is smaller than $2N$ (with probability one in the limit $N \to \infty$). This is the famous result which was found by Schläfli in about 1850 and was calculated using replica theory by Gardner 140 years later [9,10].

What happens if the bits S_t are not independent but taken from a periodic time series with correlations? Let us assume that we arrange $P = \alpha N$ random bits S_t on a ring or, equivalently, look at P random bits periodically repeated. For this case we ask the question: how long is the typical sequence which a Boolean perceptron can learn perfectly?

Up to now there has been no analytical solution of (6.4) for this scenario, although several experts in this field have tried to solve the problem. However, detailed numerical simulations show that it is harder to learn a random sequence than random patterns: the maximal length of the sequence is $P/N = \alpha_c \simeq 1.7$ [11], which should be compared with $\alpha_c = 2$ for random patterns. Obviously, tiny correlations between input vectors and output bits make the problem harder for a perceptron to learn.

6.3   Generating Sequences

In the previous section the perceptron learned a short random sequence exactly. Consequently, it can also predict it, without errors. If a neural network is able to *predict* a given time series, it can also *generate* the same series. According to Fig. 6.1, generating means that the network takes the last N numbers of the sequence, calculates a new number and moves one step to the right. Repeating this procedure, a sequence $S_0, S_1, S_2 \ldots$ given by (6.3) is generated.

It is therefore interesting to study the structure of sequences generated by a neural network. Here we restrict to the case of *fixed* weights $\boldsymbol{w}$. Adaptive weights are considered in Sects. 6.5–6.8.

Numerical simulations show that, for random weights $\boldsymbol{w}$ and random initial states $\boldsymbol{S}$, the sequence has a transient initial part and finally runs into one of several possible cycles. The structure of these cycles is related to the maxima of the Fourier spectrum of the weights $w_1, \ldots, w_N$. It is thus important to understand the sequence generated by a single Fourier component,

$$w_j = \cos\left(2\pi K \frac{j}{N} + \pi\phi\right) , \tag{6.5}$$

where K is an integer frequency and $\phi \in [-1, 1]$ a phase of the weight vector. For a continuous perceptron, we are looking for a solution $S_0, S_1, \ldots$ of an infinite number of equations

$$S_t = \tanh\left[\frac{\beta}{N} \sum_{j=1}^{N} \cos\left(2\pi K \frac{j}{N} - \pi\phi\right) S_{t-j}\right] . \tag{6.6}$$

In this case, an analytic solution has been derived [8]. For small values of β the attractor is zero, and the sequence relaxes to $S_t = 0$. However, above a critical value of β, which is independent of the frequency K, a nonzero attractor exists. Close to β_c, it is given by

$$S_t = \tanh\left[A(\beta) \cos\left(2\pi(K + \phi)\frac{t}{N}\right)\right] . \tag{6.7}$$

The amplitude $A(\beta)$ increases continuously from zero above a critical value

$$\beta > \beta_c = 2\frac{\pi\phi}{\sin(\pi\phi)} . \tag{6.8}$$

Consequently, the attractor of the sequence is a *quasiperiodic* cycle with frequency $K + \phi$. The phase ϕ of the weights shifts the frequency of the sequence – a result which is not easy to understand without calculating it.

For a multilayer network the situation is similar. Each hidden unit can contribute a quasiperiodic component to the sequence, which has its own critical point. Increasing β, more and more components are activated. This is shown in Fig. 6.2 for a network with two hidden units. For small values of the parameter β, the quasiperiodic attractor is one-dimensional, whereas for large β both components are activated, yielding a two-dimensional attractor, as shown by the return map $S_{t+1}(S_t)$. The attractor dimension is limited by the number of hidden units [12].

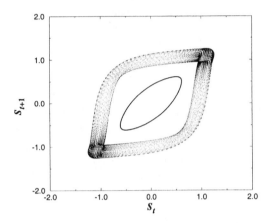

Fig. 6.2. Attractors of a network with two hidden units

If the transfer function is discrete, (6.1), the situation is more complex [7,13]. In this case we obtain a bit generator whose cycle length is limited by 2^N. However, numerical simulations show that the spectrum of cycle lengths has a much lower bound, namely the value $2N$, at least for single component weights with $|\phi| < 1/2$. After a transient part, the bit sequence S_t follows the equation

$$S_t = \text{sign}\left[\cos\left(2\pi(K + \phi)\right)\frac{t}{N}\right] . \tag{6.9}$$

However, the sequence cannot follow this equation forever. In particular, if a window $(S_{t-1}, \ldots, S_{t-N})$ appears a second time, the perceptron has to repeat the sequence. Numerical calculations show that (6.9), in addition to this condition, produces cycles shorter than $2N$. It remains a challenge to show this result analytically.

Figure 6.3 shows the cycle length $L(\phi)$ of the bit generator with weights (6.5). This rather complex figure has a simple origin, it just shows the properties of rational numbers. An integer multiple of the wavelength λ given by (6.9)

$$\lambda = \frac{N}{K + \phi} \tag{6.10}$$

has to fit into the cycle

$$L = n \cdot \lambda . \tag{6.11}$$

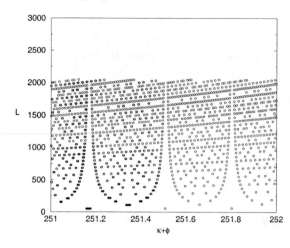

Fig. 6.3. Cycle lengths of the bit generator with cosine weights (6.5). The perceptron has $N = 1024$ input components

Therefore, λ has to be a rational. The pattern $L(\phi)$ shown in Fig. 6.3 turns out to be the numerator as a function of its rational basis. However, this does not explain why this picture is cut for $L > 2N$.

Up to now we have discussed quasiperiodic sequences only. However, time series occurring in applications are generally more complex. We are therefore interested in the question as to whether a neural network can generate a time series with a more complex power spectrum than a single peak and its higher harmonics.

It turns out that a multilayer network cannot generate a sequence with an arbitrary power spectrum. For instance, to generate a sequence with autocorrelations which decay as a power law, one needs a fully connected asymmetric network, a more complex architecture than a feedforward network [14].

However, a simple perceptron can generate a chaotic sequence. When the weights have a bias,

$$b = \frac{1}{N} \sum_i w_i > 0 \,, \qquad (6.12)$$

there are tiny regions in the (β, b)-plane where a chaotic sequence has been observed numerically [15]. Such a scenario has been called *fragile chaos*. The fractional dimension of such a chaotic sequence is between one and two, and in the vicinity of chaotic parameters (β, b), there is always a parameter set with a quasiperiodic sequence.

This situation is different for a nonmonotonic transfer function. If the function $\tanh(x)$ in (6.2) is replaced by $\sin(x)$, there are large compact regions in the parameter space where the sequence is chaotic with a large fractal dimension of the order of N. Neural networks with nonmonotonic transfer functions yield high-dimensional *stable chaos* [15,16]. In this case the attractor dimension can be tuned by the parameter β between the values one and N.

6.4   Predicting Time Series

If a neural network cannot generate a given sequence of numbers, it cannot predict it with zero error. But this is not the whole story. Even if the sequence has been generated by an (unknown) neural network (the teacher), a different network (the student) can try to learn and predict this sequence. In this context we are interested in two questions:

- When a student network with identical architecture to the teacher network is trained on the sequence, how does the overlap between student and teacher develop with the number of training examples (i.e., windows of the sequence)?
- After the student network has been trained on a part of the sequence, how well can it predict the sequence several steps ahead?

Recently these questions have been investigated numerically for the simple perceptron [17]. We have to distinguish several scenarios:

- Boolean versus continuous perceptrons,
- on-line versus batch learning,
- quasiperiodic versus chaotic sequences.

In all cases, we consider only the stationary part of a sequence which was generated by a perceptron. The student network is trained on the stationary part only, not on the transient.

First we discuss the Boolean perceptron of size N which has generated a bit cycle with a typical length $L < 2N$. The teacher perceptron has random weights with zero bias, and the cycle is related to one component of the power spectrum of the weights. The student network is trained using the perceptron learning rule:

$$\Delta w_i = \frac{1}{N} S_t S_{t-i} \quad \text{if} \quad S_t \sum_{j=1}^{N} w_j S_{t-j} < 0 \, ,$$

$$\Delta w_i = 0 \quad \text{otherwise} \, . \tag{6.13}$$

For this algorithm there exists a mathematical theorem [1]: if the set of examples can be generated by some perceptron, then this algorithm stops, i.e., it finds one out of possibly many solutions. Since we consider examples from a bit sequence generated by a perceptron, this algorithm is guaranteed to learn the sequence perfectly. On-line and batch training are identical in this case.

The network is trained on the cycle until the training error is zero. Hence the student network can predict the stationary sequence perfectly. Surprisingly, it turns out that the overlap between student and teacher is small, in fact it is zero for infinitely large networks, $N \to \infty$. The network learns the projection of the teacher's weight vector onto the sequence, but not the complete vector. It behaves like a filter, selecting one of the components of the power spectrum of the weights. Although it predicts the sequence perfectly, it does not gain much information about the rule which generates this sequence.

This situation seems to be different in the case of a continuous perceptron. Inverting (6.3) for a monotonic transfer function $f(x)$ gives N linear equations for N unknowns w_i. If the stationary part of the sequence is either quasiperiodic or chaotic, all patterns are different and the batch training, using N windows, leads to perfect learning.

This holds true for a chaotic time series. However, for a quasiperiodic time series (6.7), the patterns are almost linearly dependent, yielding an ill-conditioned set of linear equations. Without the $\tanh(x)$ in (6.7), one would obtain a two-dimensional space of patterns; with nonlinearity, one obtains small contributions in the other $N - 2$ dimensions of the weight space. Nevertheless, depending on the parameter β, even professional computer routines sometimes fail to solve (6.3) for quasiperiodic patterns generated by a teacher perceptron.

How does this scenario show up in an on-line training algorithm for a continuous perceptron? If a quasiperiodic sequence is learned step by step without iterating previous steps, using gradient descent to update the weights,

$$\Delta w_i = \frac{\eta}{N}[S_t - f(h)] \cdot f'(h) \cdot S_{t-i} \quad \text{with} \quad h = \beta \sum_{j=1}^{N} w_j S_{t-j} , \tag{6.14}$$

then one can distinguish two time scales (time = number of training steps):

- a fast one increasing the overlap between teacher and student to a value which is still far away from the value of one which corresponds to perfect agreement;
- a slow one increasing the overlap very slowly. Numerical simulations for millions of times N training steps yielded an overlap which was still far away from the value of one.

Although there is a mathematical theorem on stochastic optimization which seems to guarantee convergence to perfect success [18], our on-line algorithm cannot gain much information about the teacher network. It would be interesting to know how these two time scales depend on the size of the system. In addition, we cannot exclude the existence of on-line algorithms which can learn our ill-conditioned problem in short times.

This is completely different for a chaotic time series generated by a corresponding teacher network with $f(x) = \sin(x)$. It turns out that the chaotic series appears like a random one: after a number of training steps of the order of N, the overlap relaxes exponentially fast to perfect agreement between teacher and student.

Hence, after training the perceptron with a number of examples of the order of N, we obtain the two cases: for a quasiperiodic sequence the student has not obtained much information about the teacher, while for a chaotic sequence the student's weight vector comes close to that of the teacher. One important question remains: how well can the student predict the time series?

Figure 6.4 shows the prediction error as a function of the time interval T over which the student makes the predictions ($T = \alpha N$). The student network

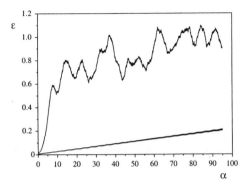

Fig. 6.4. Prediction error ϵ as a function of time steps ahead, for a quasiperiodic (*lower*) and a chaotic (*upper*) series. ϵ is the square of the difference between the series and the output of the network, and α is the number of prediction steps divided by the system size N

which has been trained on the quasiperiodic sequence can predict it very well. The error increases linearly with the size of the interval, even predicting $10N$ steps ahead yields an error of about 10% of the total possible range. On the other hand, the student trained on the chaotic sequence cannot make predictions. The prediction error increases exponentially with time; even after a few steps, the error corresponds to random guessing, $\epsilon \simeq 1$.

In summary, we obtain the following surprising result:

- A network trained on a quasiperodic sequence does not obtain much information about the teacher network which generated the sequence, but the network can predict this sequence over many steps ahead (of the order of N).
- A network trained on a chaotic sequence obtains almost complete knowledge about the teacher network, but this network cannot make reasonable predictions about the sequence.

It would be interesting to find out whether this result also holds for other prediction algorithms, such as multilayer networks.

6.5 Predicting with 100% Error

Consider some arbitrary prediction algorithm. It may contain all the knowledge of mankind. Many experts may have developed it. Now there is a bit sequence $S_1, S_2, \ldots$ and the algorithm has been trained on the first t bits $S_1, \ldots, S_t$. Can it predict the next bit S_{t+1}? Is the prediction error, averaged over a large t interval, less than 50%?

If the bit sequence is random then every algorithm will give a prediction error of 50%. But if there are some correlations in the sequence then a clever algorithm should be able to reduce this error. In fact, for the most powerful algorithm one is tempted to say that for *any* sequence it should perform better than 50% error. However, this is not true [19]. To see this, just generate a sequence $S_1, S_2, S_3, \ldots$ using the following algorithm:

> Define S_{t+i} to be the opposite of the prediction of the algorithm which has been trained on $S_1, \ldots, S_t$.

Now, if the same algorithm is trained on this sequence, it will always predict the following bit with 100% error. Hence there is no general prediction machine; to be successful for a class of problems, the algorithm needs some preknowledge about it.

The Boolean perceptron is a very simple prediction algorithm for a bit sequence, in particular with the online training algorithm (6.13). What does the bit sequence look like for which the perceptron completely fails?

Following (6.13), we just have to take the negative value

$$S_t = -\text{sign}\left(\sum_{j=1}^{N} w_j S_{t-j}\right) , \tag{6.15}$$

and then train the network on this new bit

$$\Delta w_j = +\frac{1}{N} S_t S_{t-j} . \tag{6.16}$$

The perceptron is trained on the opposite (i.e., negative) of its own prediction. Starting from (say) random initial states $S_1, \ldots, S_N$ and weights $\boldsymbol{w}$, this procedure generates a sequence of bits $S_1, S_2, \ldots S_t, \ldots$ and of vectors $\boldsymbol{w}, \boldsymbol{w}(1), \boldsymbol{w}(2),$ $\ldots \boldsymbol{w}(t), \ldots$ as well. Given this sequence and the same initial state, the perceptron which is trained on it yields a prediction error of 100%.

It turns out that this simple algorithm produces a rather complex bit sequence which comes close to a random one. After a transient time the weight vector $\boldsymbol{w}(t)$ performs a kind of random walk on an N-dimensional hypersphere. The bit sequence runs to a cycle whose average length L scales exponentially with N,

$$L \simeq 2.2^N . \tag{6.17}$$

The autocorrelation function of the sequence shows complex properties. It is close to zero up to N, oscillates between N and $3N$, and is similar to random noise for larger distances. Its entropy is smaller than that of a random sequence, since the frequency of some patterns is suppressed. Of course, it is not random, since the prediction error is 100%, whilst it is 50% for a random bit sequence.

When a second (student) perceptron with different initial state $\boldsymbol{w}$ is trained on such an antipredictable sequence, generated by (6.13), it can perform somewhat better than the teacher. The prediction error goes down to about 78% but it is still larger than 50% for random guessing. However, the student obtains knowledge about the teacher. The angle between the two weight vectors relaxes to about 45 degrees.

6.6   Learning from Each Other

In the previous section we discussed a neural network which learns from itself. But more interesting may be the scenario where several networks are interacting, learning from each other. After all, our living world consists of interacting adaptive systems and recent methods of computer science use interacting agents to solve complex problems. Here we consider a simple system of interacting perceptrons as a first step towards developing a theory of the cooperative behaviour of adaptive agents.

Consider K Boolean perceptrons, each of which has an N-dimensional weight vector $\boldsymbol{w}^\nu$, for $\nu = 1, \dots, K$. Each perceptron receives the same input vector $S_1, \dots, S_N$ and produces its own output bit

$$\sigma^\nu = \text{sign}(\boldsymbol{w}^\nu \cdot \boldsymbol{S}) . \tag{6.18}$$

Now these networks receive information from their neighbours in a ring topology: perceptron $\boldsymbol{w}^\nu$ is trained on the output $\sigma^{\nu-1}$ of perceptron $\boldsymbol{w}^{\nu-1}$, and $\boldsymbol{w}^1$ is trained on σ^K. Training is performed keeping the length of the weight vectors fixed:

$$\boldsymbol{w}^\nu(t+1) = \frac{\boldsymbol{w}^\nu(t) + (\eta/N)\sigma^{\nu-1}\boldsymbol{S}}{|\boldsymbol{w}^\nu(t) + (\eta/N)\sigma^{\nu-1}\boldsymbol{S}|} . \tag{6.19}$$

The learning rate η is a parameter controlling the speed of learning.

This problem has been solved analytically in the limit $N \to \infty$ [20,21] for random inputs. The system relaxes to a stationary state, where the angles $\theta_{\nu\mu}$ (or overlaps) between different agents take a fixed value. For small learning rate η, all of these angles are small, i.e., there is good agreement between the agents. But more surprisingly, the state of the system is completely symmetric, that is, there is a single common angle $\theta = \theta_{\nu\mu}$ between all pairs of networks. The agents do not recognize the clockwise flow of information.

Increasing the learning rate η, the common angle θ increases, too. With larger learning steps each agent tends to have an opinion opposite to all of its colleagues. However, due to the symmetry, there is maximal possible angle given by

$$\cos\theta = -\frac{1}{K-1} . \tag{6.20}$$

In fact, increasing η, the system arrives at this maximal angle for some critical value η_c. For larger values $\eta > \eta_c$, the system undergoes a phase transition in which the complete symmetry is broken, but the symmetry of the ring is still conserved:

$$\theta_1 = \theta_{\nu+1,\nu} , \quad \theta_2 = \theta_{\nu+2,\nu} , \quad \dots .$$

For K agents there are $(K-1)/2$ possible values of θ_i if K is odd, and $K/2-1$ possible values if K is even.

This is a simple – but analytically solvable – example of a system of interacting neural networks. We observe a symmetry-breaking transition when increasing the learning rate. However, this system does not solve any problem. In the following section we will extend this scenario to a case where neural networks do indeed interact to solve a special problem, the minority game.

6.7   Competing in the Minority Game

Recently, a mathematical model of economics has received a lot of attention in the statistical physics community [22]. It is a simple model of a closed market. There are K agents who have to make a binary decision $\sigma^\nu \in \{+1, -1\}$ at each time step. All of the agents who belong to the minority gain one point, whilst the majority has to pay one point (to a cashier who always wins). The global loss is given by

$$G = \left| \sum_{\nu=1}^{K} \sigma^\nu \right| . \tag{6.21}$$

If the agents come to an agreement before they make a new decision, it is easy to minimize G: $(K-1)/2$ agents have to choose $+1$, then $G = 1$. However, this is not the rule of the game, the agents are not allowed to cooperate. Each agent knows only the history of the minority decision, $S_1, S_2, S_3, \ldots$, but otherwise he/she has no information. Can the agent find an algorithm to maximize his/her profit?

If each agent makes a random decision, then G is of the order of $\sqrt{K}$. It is not easy to find algorithms which perform better than random [23,24].

Here we use a perceptron for each agent to make a decision based on the past N steps $\boldsymbol{S} = (S_{t-N}, \ldots, S_{t-1})$ of the minority decision. The decision of agent $\boldsymbol{w}^\nu$ is given by

$$\sigma^\nu = \text{sign}(\boldsymbol{w}^\nu \cdot \boldsymbol{S}) . \tag{6.22}$$

After the bit S_t of the minority has been determined, each perceptron is trained on this new example $(\boldsymbol{S}, S_t)$,

$$\Delta \boldsymbol{w}^\nu = \frac{\eta}{N} S_t \boldsymbol{S} . \tag{6.23}$$

This problem can be solved analytically [20,21]. The average global loss for $\eta \to 0$ is given by

$$\langle G^2 \rangle = \left(1 - \frac{2}{\pi} \right) K \simeq 0.363K . \tag{6.24}$$

Hence, for small enough learning rates, the system of interacting neural networks performs better than random decisions. A pool of adaptive perceptrons can organize itself to yield a successful cooperation.

6.8 Predicting Human Beings

As a final example of a perceptron predicting a bit sequence, we discuss a real application. Assume that the bit sequence $S_0, S_1, S_2, \ldots$ is produced by a human being. Now a simple perceptron (6.1) with on-line learning (6.13) takes the last N bits and makes a prediction for the next bit. Then the network is trained on the new true bit, which appears afterwards as part of the input for the following prediction.

Equation (6.13) is a simple deterministic equation describing the change of weights according to the new bit and the past N bits. Can such a simple equation foresee the reaction of a human being? If a person can calculate or estimate the outcome of (6.13), then he/she can just do the opposite, and the network completely fails to predict.

To answer this question we have written a little C program which receives the two bits 0 and 1 from the keyboard [25]. The program needs two fields **neuron** and **weight** which contain the variables S_i and w_i, respectively. Here are the main steps:

1. Repeat:
   ```
   while (1) {
   ```
2. Calculate the vector product $\boldsymbol{w}\,\boldsymbol{S}$:
   ```
   for (h=0; i=0; i<N; i++) h+=weight[i]*neuron[i];
   ```
3. Read a new bit:
   ```
   if(getchar()=='1')   input=1; else input =-1;
   ```
4. Calculate the prediction error:
   ```
   if(h*input<0) {error ++;
   ```
5. Train:
   ```
   for(i=0; i<N; i++) weight[i]+=input* neuron[i]/(double)N;}
   ```
6. Shift the input window:
   ```
   for(i=N-1; i>0; i--) neuron[i]=neuron[i-1]; neuron[0] =input;}
   ```

A graphical version of this program can be accessed over the internet at the address:

http://theorie.physik.uni-wuerzburg.de/~kinzel

or see

http://www.tu-chemnitz.de/physik/HERAEUS/2000/Springer.html

Now we ask a person to generate a bit sequence for which the prediction error of the network is high. From Sect. 6.2, we already know what happens if the candidate produces a rhythm: if its length is smaller than $1.7N$, the perceptron can learn it perfectly, without errors. Hence the candidate should either produce random numbers which give 50% errors or try to calculate the prediction of the perceptron, in which case an error higher than 50% is possible.

We have tested this program on the students in our class. Each student had to send a file with one thousand bits, generated by hand. It turns out that on average the network predicts with an error of about 35%. The distribution of errors is broad with a range between 20% and 50%. Apparently, a human being is not a good random number generator. The simple perceptron (6.1) and (6.13) succeeds in predicting human behaviour!

Some students submitted sequences with 50% error. It was obvious – and later confessed – that they used random number generators, digits of π, the logistic map, etc., instead of their own fingers. One student submitted a sequence with 100% error. He was the supervisor of our computer system, knew the program and submitted the sequence described in Sect. 6.5.

6.9   Summary

The theory of time series generation and prediction is a new field of statistical physics. The properties of perceptrons, simple single-layer neural networks being trained on sequences which were produced by other perceptrons, have been studied. A random bit sequence is more difficult to learn perfectly than random uncorrelated patterns. An analytic solution of this capacity problem is still lacking.

A multilayer network can be used to generate time series. For the continuous transfer function, an analytic solution of the stationary part of the sequence has been found. The sequence has a dimension which is bounded by the number of hidden units. It is not yet completely clear how to extend this solution to the case of a Boolean perceptron generating a bit sequence. For nonmonotonic transfer functions, the network generates a chaotic sequence with a large fractal dimension.

A perceptron which is trained on a quasiperiodic sequence can predict it very well, but it does not obtain much information on the rule generating the sequence. On the other hand, for a chaotic sequence, the overlap between student and teacher is almost perfect, but prediction of the sequence is not possible.

For any prediction algorithm there is a sequence for which it completely fails. For a simple perceptron, such a sequence is rather complex, with huge cycles and low autocorrelations. Another perceptron which is trained on such a sequence reduces the prediction error from 100% to 78% and obtains overlap with the generating network.

When perceptrons learn from each other, the system relaxes to a symmetric state. Above a critical learning rate, there is a phase transition to a state with lower symmetry.

A system of interacting neural networks can develop algorithms for the minority game, a model of a closed economy of competing agents.

Finally it has been demonstrated that human beings are not good random number generators. Even a simple perceptron can predict the bits typed by hand with an error of less than 50%.

Acknowledgements. The author would like to thank Ido Kanter, Richard Metzler, Ansgar Freking and Michael Biehl for their comments.

References

1. J. Hertz, A. Krogh, R.G. Palmer: *Introduction to the Theory of Neural Computation* (Addison Wesley, Redwood City 1991)
2. A. Engel, C. Van den Broeck: *Statistical Mechanics of Learning* (Cambridge University Press, Cambridge 2000)
3. W. Kinzel: Comp. Phys. Comm. **121**, 86–93 (1999)
4. M. Biehl, N. Caticha: 'Statistical Mechanics of On-line Learning and Generalization'. In: *The Handbook of Brain Theory and Neural Networks*, ed. by M.A. Arbib (MIT Press, Berlin 2001)
5. M. Opper, W. Kinzel: 'Statistical Mechanics of Generalization'. In: *Models of Neural Networks III*, ed. by E. Domany, J.L. van Hemmen, K. Schulten (Springer Verlag, Heidelberg 1995) p. 151
6. A. Weigand, N.S. Gershenfeld: *Time Series Prediction* (Addison Wesley, Santa Fe 1994)
7. E. Eisenstein, I. Kanter, D.A. Kessler, W. Kinzel: Phys. Rev. Letters **74**, 6–9 (1995)
8. I. Kanter, D.A. Kessler, A. Priel, E. Eisenstein: Phys. Rev. Lett. **75**, 2614–2617 (1995)
9. L. Schlaefli: *Theorie der Vielfachen Kontinuität*, (Birkhaeuser, Basel 1857)
10. E. Gardner: J. Phys. A **21**, 257 (1988)
11. M. Schröder, W. Kinzel, I. Kanter: J. Phys. A **29**, 7965 (1996)
12. L. Ein-Dor, I. Kanter: Phys. Rev. E **57**, 6564 (1998)
13. M. Schröder, W. Kinzel: J. Phys. A **31**, 9131–9147 (1998)
14. A. Priel, I. Kanter: Phys. Rev. E, in press
15. A. Priel, I. Kanter: Phys. Rev. E **59**, 3368–3375 (1999)
16. A. Priel, I. Kanter: Europhys. Lett. **51**, 244–250 (2000)
17. A. Freking, W. Kinzel, I. Kanter: unpublished
18. C.M. Bishop: *Neural Networks for Pattern Recognition* (Oxford University Press, New York 1995)
19. H. Zhu, W. Kinzel: Neural Computation **10**, 2219–2230 (1998)
20. W. Kinzel, R. Metzler, I. Kanter: J. Phys. A **33**, L141–L147 (2000)
21. R. Metzler, W. Kinzel, I. Kanter: Phys. Rev. E **62**, 2555 (2000)
22. Econophysics homepage: http://www.unifr.ch/econophysics/
23. D. Challet, M. Marsili, R. Zecchina: Phys. Rev. Lett. **84**, 1824–1827 (2000)
24. G. Reents, R. Metzler, W. Kinzel: Cond-mat/0007351 (2000)
25. W. Kinzel, G. Reents: *Physics by Computer* (Springer-Verlag, Heidelberg 1998)

7 Statistical Physics of Cellular Automata Models for Traffic Flow

Michael Schreckenberg, Robert Barlović, Wolfgang Knospe, and Hubert Klüpfel

Summary. We discuss various aspects of statistical physics in the context of traffic flow with the help of cellular automata (CA) models. CA models are, in general, idealizations of physical systems in which space and time are assumed to be discrete. In the context of vehicular traffic, CA models belong to the so-called microscopic approaches, where attention is paid explicitly to each individual vehicle and interactions among these particles are determined by the way the vehicles influence each other. Therefore, vehicular traffic, modeled as a system of interacting particles driven far from equilibrium, offers the possibility of studying various aspects of nonequilibrium systems which are of current interest in statistical physics. We give a brief overview of the Nagel–Schreckenberg (NaSch) model, considered to be one of the simplest CA models for traffic flow, and of developments based on this.

7.1 Introduction

Experimental and theoretical investigations of traffic flow have been the focus of extensive research interest during the past decades. Various theoretical concepts have been developed and numerous empirical observations have been reported. Despite these enormous scientific efforts, both theoretical concepts and experimental findings are still the subject of lively debate. Our main aim is to show how these attempts, particularly the development of CA models for traffic flow, have led to a deeper insight into the typical dynamical phenomena of traffic flow.

A few years ago, a probabilistic cellular automaton for the description of single-lane highway traffic [1] was proposed. Since then, various CA models have been introduced, which allow for a multitude of new applications (for reviews, see for example [2–7]). One of the most important facts concerning CA models is their suitability for large scale computer simulations. For example, it is possible to simulate realistic traffic networks faster than real time [8,9].

The NaSch model itself already leads to a quite realistic flow–density relation (fundamental diagram) and reproduces basic phenomena encountered in real traffic, e.g., the occurrence of phantom traffic jams ('jams out of the blue'). The NaSch model is in fact 'minimal' in the sense that any simplification of the rules no longer produces realistic results. On the other hand, for a description of more complex situations or for a proper modeling of the 'fine-structure' of traffic flow, the basic set of rules has to be improved. Consequently, extended CA models [10] have been proposed in the past few years which are able to reproduce even effects like metastable states of highway traffic or, in a very recent approach [11], a correct representation of the microscopic details observed in experimental data sets.

Even from a theoretical point of view, these kinds of models are of special interest. They belong to the class of one-dimensional driven lattice gases. Driven lattice gases allow study of generic nonequilibrium phenomena, e.g., boundary-induced phase transitions [12].

We proceed as follows. We begin with a short introduction to the basic measurements of traffic flow simulations with CA models. The Nagel–Schreckenberg cellular automaton model for one-dimensional highway traffic is then introduced and its fundamental properties illustrated. Further, a simple extension of the NaSch model is presented, the so-called VDR (Velocity Dependent Randomization) model, and additional features such as phase separation and metastability are discussed. Finally, a high fidelity cellular automaton model for highway traffic (proposed by Knospe et al. [11]) is presented, which can reproduce empirical single-vehicle data [13].

7.2   Basic Measurements of Traffic Flow Simulations with CA Models

Since CA models describe individual vehicles, a lot of information concerning the movement and environment of every car is accessible. The most common measures are the 'global' quantities *flow J*, *mean speed V*, and *density ρ*. The term 'global' is somewhat misleading: on highways, it is only possible to record traffic data for short sections of the road, depending on how long detectors are or how many detectors are installed in a row within a sufficiently short distance, whereas in computer simulations it is easy to determine such measurements. Suppose a simulation is performed on a ring of cells (periodic boundary conditions) then the density is simply given by the ratio of vehicles to cells N/L. If v_i denotes the actual speed of the i-th car, then the main quantities are related via:

$$J = \rho V , \tag{7.1}$$

with

$$\rho = \frac{N}{L} , \tag{7.2}$$

and

$$V = \frac{1}{N} \sum_{i=1}^{N} v_i . \tag{7.3}$$

The relations between these quantities are summarized in the fundamental diagram. As an example, a typical flow–density relation obtained from a simulation with the NaSch model is shown in Fig. 7.3.

Obviously, to compare simulated data with empirical data on a more detailed level, local measurements have to be performed. Simulation data must therefore be evaluated analogously to the empirical setup by virtual inductive loops. For example, as a first step, the speed and the time headway of single vehicles at a given link of the lattice are recorded. Then the data are aggregated over a certain time interval to obtain the *flow* and the *mean speed*. Finally, the density is calculated via the relation $\rho = J/V$.

7.3   Fundamentals of the NaSch Model

In the NaSch model the road is divided into cells of length 7.5 m. Each cell can either be empty or occupied by just one car. The speed v_n of each vehicle $n = 1, 2, \ldots, N$ can take one of the $v_{\max}+1$ allowed integer values $v_n = 0, 1, \ldots, v_{\max}$. The state of the road at time $t+1$ can be obtained from that at time t by applying the following rules to all cars at the same time (parallel dynamics):

- Step 1: *Acceleration*
  $v_n \rightarrow \min(v_n + 1, v_{\max})$.
- Step 2: *Braking*
  $v_n \rightarrow \min(v_n, g_n)$.
- Step 3: *Randomization*
  $v_n \rightarrow \max(v_n - 1, 0)$ with probability p.
- Step 4: *Driving*
  $x_n \rightarrow x_n + v_n$.

Here x_n denotes the position of the nth car and $g_n = x_{n+1} - x_n - 1$ the distance to the next car ahead. The density of cars is given by $\rho = N/L$, where L is the length of the system, i.e., the number of cells. One time step corresponds to approximately 1 s real time.

The first rule reflects the driver's impulse to accelerate until the maximum speed $v_{\max}$ is reached. To avoid accidents, which are explicitly forbidden in the model, the driver has to brake if the speed exceeds the gap in front. This braking event is implemented by the second update rule. In the third, a stochastic

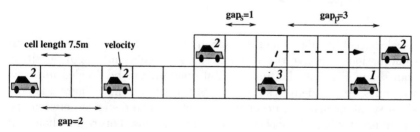

Fig. 7.1. The road in the NaSch model is subdivided into cells. Each car has a discrete speed which is bounded by the headway g to the car ahead and by its own maximum speed. A driver who wants to change lane has to take into account the gaps g_s and g_p on the neighboring lane in order to prevent crashes

element is introduced into the model. This randomization takes into account the different behavioral patterns of the individual drivers, especially nondeterministic acceleration as well as overreaction while slowing down. As mentioned before, the NaSch model is minimal, e.g., without the randomization parameter p, no spontaneous formation of traffic jams is observed. Even a change of order in the update procedure will lead to completely different behavior. Note that the deterministic limit of the NaSch model with $v_{\max} = 1$ is equivalent to CA rule 184 in Wolfram's notation [14].

For a proper mapping of realistic highway networks to computer simulations, the idealized single-lane NaSch model has to be generalized to multilane traffic. A driver who wants to change lane has to take into account the gaps g_s and g_p (see Fig. 7.1) on the neighboring lane in order to prevent crashes. Several lane-changing rules have been suggested so far [15–18]. Lane-changing rules can be symmetric or asymmetric with respect to lanes. If symmetric lane-changing rules are applied, the rules do not depend on the direction of the lane-changing maneuver. In general, the update in two-lane models is divided into two sub-steps: in one sub-step, vehicles may change lanes in parallel, following the lane-changing rules, and in the other sub-step, each vehicle may move forward according to the NaSch rules.

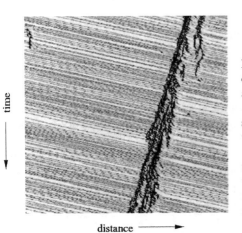

time

distance ⟶

Fig. 7.2. Typical space–time diagram of the NaSch model with $v_{\max} = 5$ and $p = 0.5$. Each *horizontal row of dots* represents the instantaneous positions of the vehicles moving towards the right, while the *successive row of dots* represents the positions of the same vehicles at successive time steps. Note that, while free-flowing vehicles are moving forward (positive x direction), jams move backwards (negative x direction)

A typical space–time diagram of the NaSch model is shown in Fig. 7.2. Spontaneous fluctuations can give rise to traffic jams at sufficiently high densities. As one can see in Fig. 7.3, the fundamental diagram can be divided into two different regimes. At low densities the slope of the fundamental diagram is positive and no spontaneous jams occur in the system. The distance between the vehicles is sufficiently large to absorb velocity fluctuations. This regime holds up until a jamming transition to a congested phase takes place at densities slightly below the point of maximum flow. A detailed analysis shows [19] that only in the deterministic limit $p = 0$ of the model does the transition correspond to a critical point with diverging correlation length, while in the presence of noise, an absence of critical behavior is suggested. The transition point, and therewith the maximum flow, is largely determined by the fluctuation parameter p, as one can see in Fig. 7.3. The simple shape of the fundamental diagram is very profitable when calibrating the model on realistic traffic data. The slope of the free-flow branch is given directly by the mean speed of freely flowing vehicles:

$$v_f = v_{max} - p \,, \tag{7.4}$$

and the maximum flow is restricted by the randomization parameter p.

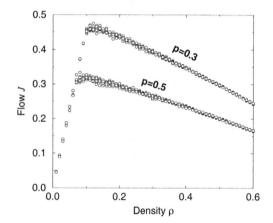

Fig. 7.3. Fundamental diagram of the NaSch model ($v_{max} = 5$) for different p. It is clear that the fundamental diagram can be divided into a free-flow regime with positive slope and a jammed regime with negative slope. The strong dependence of the capacity (maximum possible flow) on the randomization parameter p is remarkable

In the congested regime, velocity fluctuations of freely flowing vehicles cannot be absorbed at all by the distance between the cars, and jams occur in the system. The spatial microscopic behavior of jams is dominated by a branching structure due to spontaneous stops of cars even in the outflow region of jams. Consequently, jammed states in the NaSch model contain clusters with an exponential size distribution [20]. In contrast to this, one finds a clear phase separation between freely flowing and jammed vehicles in the VDR model. This will be introduced in the following section.

7.4   The VDR Model

Measurements on real traffic have revealed the existence of metastable states with very high flow [21]. Such states cannot be observed in the NaSch model. A simple generalization of the model was therefore proposed [10] which incorporates slow-to-start behavior in order to model the restart process of a vehicle in a more realistic fashion. As will be demonstrated below, this rule is an important ingredient for the occurrence of metastable states in CA models for traffic flow.

In the VDR model a velocity-dependent randomization (VDR) parameter $p = p(v(t))$ is introduced, in contrast to the constant randomization in the original formulation. This parameter has to be determined before the acceleration

Step 1. One finds that the dynamics of the model strongly depends on the randomization. For simplicity, we demonstrate the case

$$p(v) = \begin{cases} p_0 & \text{for } v = 0 \text{ ,} \\ p & \text{for } v > 0 \text{ ,} \end{cases} \tag{7.5}$$

with two stochastic parameters p_0 and p, which already contains the expected features. In this case the randomization p_0 for cars at rest is much larger than the randomization p for moving cars, i.e., $p_0 \gg p$. An interesting fact is that an alternative choice of $p(v)$, e.g., $p_0 \ll p$, leads to completely different behavior. Note that the original NaSch model is recovered for $p_0 = p$.

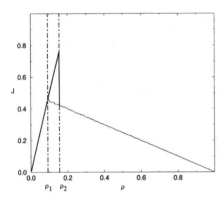

Fig. 7.4. Typical fundamental diagram for the VDR model ($v_{\max} = 5$, $p_0 = 0.5$, $p = 0.01$, $L = 10\,000$) obtained by starting from two different initial conditions, a completely jammed state (*negative slope*) and a homogeneous state (*positive slope*). The metastable region between the densities ρ_1 and ρ_2 can be clearly identified

Typical fundamental diagrams for the VDR model (see Fig. 7.4) show a density regime $\rho_1 < \rho < \rho_2$ where the flow can take two different states depending on the initial conditions. The homogeneous branch is metastable with an extremely long lifetime and the jammed branch shows phase separation between jammed and freely flowing cars. Neglecting interactions among freely flowing vehicles, the fundamental diagram of the model can be derived on the basis of heuristic arguments in good agreement with numerical results.

Obviously the flow in the homogeneous branch is given by

$$J_{\mathrm{h}} = \rho(v_{\max} - p) = \rho v_{\mathrm{f}} \text{ ,} \tag{7.6}$$

because every car can move with its desired speed v_{f}. Assuming that the high density states are phase separated, we can obtain the second branch of the fundamental diagram. The phase-separated states consist of a large jam and a free-flow regime, where each car moves with velocity v_{f}. The density in the free-flow regime ρ_{f} is determined by v_{f} and the average waiting time

$$T_{\mathrm{w}} = \frac{1}{1 - p_0} \text{ ,} \tag{7.7}$$

to accelerate the first car in the jam. This is because, neglecting interactions between cars, the average distance between two consecutive cars is given by

$$\Delta x = T_{\mathrm{w}} v_{\mathrm{f}} + 1 = \rho_{\mathrm{f}}^{-1} \text{ .} \tag{7.8}$$

Using the normalization

$$L = N_J + N_F \Delta x, \tag{7.9}$$

where $N_{F(J)}$ is the number of cars in the free flow regime (jam), we find that the flow is given by

$$J_s(\rho) = (1 - \rho)(1 - p_0) . \tag{7.10}$$

Apparently ρ_f is precisely the branching density ρ_1, because for densities below ρ_f, the jam length is zero. It should be noted that this approach is only valid if $(p_0 \gg p)$ and $v_{max} > 1$ holds. Small values of p have to be considered in order to avoid interactions between cars due to the velocity fluctuations, and $v_{max} > 1$ because otherwise cars can stop spontaneously in the free-flow regime and therefore initiate a jam. Note that the simple linear dependency of the fundamental diagram branches on the randomization parameters p_0 and p is very useful when calibrating the model on empirical traffic data.

The microscopic structure of the jammed states in the VDR model differs from those found in the NaSch model. As already mentioned in the last section, jammed states in the NaSch model contain clusters with an exponential size distribution [20], while one finds phase separation in the VDR model. The reason for this different behavior is the reduction in the outflow of a jam compared with the maximal possible global flow. A large stable jam can only exist if the outflow from the jam is equal to the inflow. If the outflow is the maximal possible flow, a stable jam can only exist at the corresponding density ρ_{max}, but will easily dissolve due to fluctuations. If the outflow of a jam is reduced, the density in the free-flow regime is smaller than the density ρ_{max} of maximal flow where interactions between vehicles lead to jams, so that cars can propagate freely (for $p \ll p_0$) through the low density part of the road. Therefore, no spontaneous jam formation is observed. It is obvious that no phase separation can occur in the NaSch model, due to the fact that the outflow of a jam is maximal.

Figure 7.5 shows the typical structure of a spontaneously emerging jam in the VDR model. The origin of the jam in the initially homogeneous state is

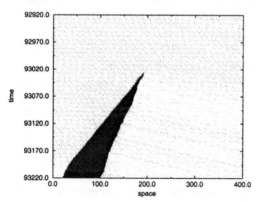

Fig. 7.5. Space–time diagram of the VDR model for $\rho = 0.15$, $L = 400$, $p = 0.01$ and $p_0 = 0.5$. The homogeneous initial state is not destroyed immediately, but after approximately 93 000 lattice updates. In the outflow regime of the jam, the density is reduced compared with the average density

a local velocity fluctuation that can finally lead to a stopped car. As one can see, the density in the outflow regime of the jam is reduced compared with the average density. Consequently, the jam length grows in an approximately linear way until outflow and inflow are equal due to the periodic boundary conditions. The average jam length then strongly fluctuates. This can also lead to a complete dissolution of the jam when these fluctuations are of the order of the jam length.

The phase separation of the jammed state in the VDR model can be directly identified using the jam-gap distribution [22]. It can be shown that for $p = 0$ the jam is compact in the sense that no empty sites appear between the jammed cars. For $p > 0$, holes form due to velocity fluctuations of vehicles entering the jam. Nevertheless, it can be assumed that for $p_0 \gg p$ the jammed states are phase separated, i.e., the size of the jam is of the order of the system size. We stress that this behavior contrasts with the situation in the NaSch model, where jam clusters show an exponential size distribution.

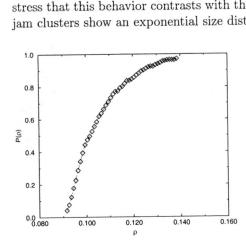

Fig. 7.6. The sensitivity (probability that a finite jam lead to a wide jam) is plotted against density. The simulation results are obtained using a 'damage scenario' where a finite jam is initialized in an undisturbed system

In Fig. 7.6, the sensitivity of the metastable homogeneous branch of the system is plotted vs. density. Here a 'damage scenario' is used, where a finite jam is initialized in an undisturbed system. The starting point is a homogeneous initialization in the density regime $\rho_1 < \rho < \rho_2$. Once the system has relaxed into its steady state, a randomly chosen car is forced to stop until its successor stops as well. The system is updated according to the update rules without any further outside influences. The probability that the induced 'minijam' evolves into a wide jam is referred to as the sensitivity of the system. At this point it should be stressed, however, that although the inflow is larger than the outflow, these quantities are stochastic and therefore, even in the considered regime, a complete dissolution of the emerged jam is possible through fluctuations. It is obvious that the sensitivity shows a fast convergence to one for higher densities. This is due to the fact that the higher the density, the lower the mean gaps between the cars so that it will be harder for the system to absorb the induced minijam. In [23] an analytical description is presented based on random walk arguments, which is able to give a proper description of the jam dynamics in the VDR model.

7.5   Model with Anticipation

Although the NaSch model and its generalization, the VDR model, lead to a quite realistic flow–density relation, the comparison of simulation results with empirical data [13] is not satisfactory at all. So far, the presented models fail to reproduce the microscopic structure observed in measurements of real traffic. Due to the existing discrepancy between simulation and reality, a new CA model generalizing the NaSch model has recently been proposed [11].

In order to allow for a more realistic modeling of car characteristics, such as different acceleration capabilities and car lengths, the standard NaSch cell length is reduced from $l = 7.5$ m to $l = 1.5$ m. Additionally, an anticipation term is introduced into the new model. The update rules for the new model are combined with the original NaSch model. Note that the slow-to-start behavior realized through the VDR approach is also implemented in the new model. The slow-to-start behavior is important for directly tuning the velocity of the upstream end of a traffic jam.

It turns out that, for a realistic choice of the parameters, the outflow of a jam does not achieve the capacity of the road. This fact is known to lead to the existence of metastable states, as explained in the previous section. The next step towards a more realistic description of highway traffic is to introduce anticipation effects, i.e., speed adjustments also take into account the expected behavior of the leading vehicle. The main problem with existing discrete models, such as the NaSch model or the VDR model, is the inability to reproduce platoons of slowly moving vehicles. These patterns are not as stable as in real traffic, i.e., the models overestimate the probability of forming large compact jams.

To improve the platoon behavior, the interaction horizon was prolonged for the case where a braking maneuver of the leading vehicle occurs or, more figuratively, brake lights were installed at the back of the vehicles. This event-driven interaction leads to a timely adjustment of speeds and therefore to a more coherent movement of the vehicles in dense traffic. The reaction to a brake light is simply implemented by an increased randomization parameter $p_b > p$. For a detailed description of the update rules see [11].

The first goal of a traffic model is to reproduce in detail the empirical fundamental diagram. This is obviously the case for the new CA model (see Fig. 7.7). The next parameter which can be directly related to an empirically observable quantity is the braking parameter p_0. This parameter determines the outflow of a jam and therefore the upstream velocity of the jam end. The value $p_0 = 0.5$ is used, which leads to an upstream velocity of the end of a compact jam of approximately 12.75 km/h. This velocity is also in accordance with empirical results [21]. Moreover, empirical work has revealed that metastable states exist in highway traffic due to the fact that the outflow of a jam is smaller than the maximal possible flow under free-flow conditions [21]. The inset of Fig. 7.7 shows that this is well reproduced by the simulations. A more detailed statistical analysis [13] of the time series of flux, velocity and density allows for the identification of three different traffic states. Following the arguments of [13], free flow, congested traffic states, and wide jams can be identified by means of

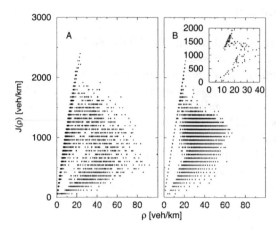

Fig. 7.7. Comparison of the local fundamental diagram obtained by a simulation (B) with the corresponding empirical fundamental diagram (A) of [13]. The *inset* shows the outflow of a megajam

the cross-covariance function. In Fig. 7.5, the cross-covariance cc(J, ρ) of the flow and the locally measured density for different traffic states is shown. It is obvious that this method can identify the different traffic states.

The results show that the new CA approach leads to realistic results for the fundamental diagram and that the model is able to reproduce the three different traffic states. For the correct description of the car–car interaction, the distance headway (Optimal Velocity (OV) curve) gives the most important information for adjustment of speeds. Due to anticipation effects, smaller distances occur, so that in contrast to the NaSch model, it is possible for cars to drive with maximum speed even within very small headways. This strong anticipation becomes weaker with increasing density and cars tend to have smaller speeds than the headway allows, so that the OV curve saturates for large densities. The simulation results for this approach (see Fig. 7.9) show that the empirical OV curve is reproduced in great detail.

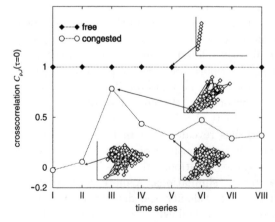

Fig. 7.8. Cross-covariance of the flow and the density for different global densities and homogeneous initializations. Simulations are performed on a ring with a length of 50 000 cells, each corresponding to 1.5 m in reality

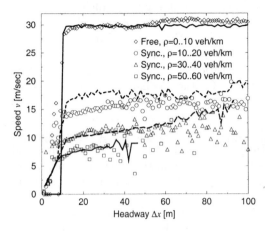

Fig. 7.9. The mean speed chosen by the driver as a function of the gap to his predecessor. Excellent agreement is obtained from comparison between simulations (*lines*) and empirical data (*symbols*)

7.6   Discussion and Conclusion

In this paper various aspects of traffic flow simulations by cellular automata models are discussed. The Nagel–Schreckenberg model is represented and basic simulation results are shown. In the NaSch model, considered to be one of the simplest CA models, spontaneous jams are formed due to the included stochastic elements if the density exceeds a critical value. The shape of the fundamental diagram is rather simple and can be divided into two linear regimes. Obviously, interactions between vehicles are negligible for low densities and therefore all vehicles are able to move with their desired free-flow speed in this so-called free-flow branch. On the other hand, the jammed branch of the fundamental diagram has a negative slope. The higher the density, the more cars are forced to stop due to interactions, until at last the system collapses to a completely blocked state at the density $\rho = 1$. With the two variables of the model, namely the maximum speed and the fluctuation parameter, the fundamental diagram can be adjusted to achieve an acceptable agreement with empirical data.

Unfortunately, metastable states with very high flow observed in real traffic cannot be reproduced with the NaSch model. A simple extension of the NaSch model is thus investigated, introducing velocity-dependent randomization (VDR model). The metastable states in the model are related to a separation between freely flowing and jammed vehicles. On the one hand, fluctuations between freely flowing vehicles are reduced, so that the homogeneous branch is stabilized. On the other, the reason for the phase separation in the jammed branch is a reduced outflow from jams, which destabilizes clusters forming in the outflow region. Therefore, the NaSch model with velocity dependent randomization shows the coexistence of phase-separated and homogeneous states in a certain density interval near the density ρ_2 of maximum flow. Near ρ_2, interactions between cars become important and one finds the spontaneous formation of jams. The fundamental diagram of the VDR model can be confirmed by a phenomenological approach, which gives very accurate results.

Further, a brief discussion of a new CA approach is given, which is able to reproduce microscopic features such as density-dependent optimal velocity curves and characteristic time headways. First of all, the original NaSch cell length must be reduced for a more realistic acceleration behavior which is especially important on highways. For a more realistic car–car interaction, anticipation terms seem to play a very important role. On the one hand, anticipation of the predecessor's movement in the next time step allows smaller time headways and therefore higher flows. On the other hand, braking anticipation by means of brake lights enables a driver to anticipate any imminent speed reductions due to a jam. It is just this braking anticipation, leading to synchronized states and increased time headways, which results in plateaus in the local fundamental diagram. The simulation results from this approach reproduce empirical data in great detail.

These considerations demonstrate the flexibility of the CA approach to traffic flow problems. The formation of metastable states in the fundamental diagram can be described by a rather simple and natural extension of the rules of the NaSch model. With additional anticipation terms, it is even possible to reproduce traffic data in detail. If one is willing to give up an overly realistic description of the interactions between the vehicles, one can obtain rather simple CA models, capable of describing even the fine-structure of traffic flow in a satisfactory way.

7.7   Appendix: Simulating Pedestrian Dynamics and Evacuation Processes

Analogously to vehicular traffic, pedestrian movement can be modelled on different scales, taking either a macroscopic or microscopic approach (for an overview see [3] and the references therein). Macroscopic models are based on the similarities between pedestrian flows and liquids or gases [24–26]. The basis of such models is the continuity equation which has to be supplemented by data about the relation between density and flow. Empirical data are used to calibrate the parameters, like the viscosity or the Reynolds number. Unfortunately, these parameters are hard to identify in real-world systems.

In microscopic models, pedestrians are identified as basic entities [27]. In contrast to macroscopic models, the movement of single persons is modelled. Although this increases the computational effort, it allows us to take individual behaviour into consideration. Every person is described by a set of parameters, e.g., reaction time or walking speed. For a realistic simulation, these are drawn from distributions. The space is either continuous (social force model by Helbing [3]) or discrete. The latter belong to the class of lattice gas or cellular automaton models for crowd motion [28–30].

CA models for road traffic are basically one-dimensional. A similar approach for pedestrian movement, where overtaking or passing is introduced via lane changing, can be found in [30]. Pedestrian motion is generally different, however: the movement is truly two-dimensional, acceleration occurs instantly and

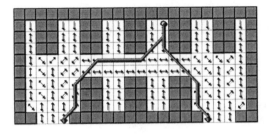

Fig. 7.10. Schematic view of a room. Information about possible walking directions is given by the *small arrows*. Possible routes emerging through orientation along the signposts are shown by the *long thick arrows*

change of direction is possible in very short times. Nevertheless, several cellular automata have been developed for pedestrian dynamics [28–30].

The basic principles of the cellular automaton model for pedestrian flow described here are as follows [31]:

- The floor plan is divided into quadratic cells. The length of one side is 0.4 m (see Fig. 7.10).
- Each cell can be either empty or occupied by only one person.
- Individual persons may differ in their characteristics or abilities. This is reflected by a set of parameters. The values of these parameters vary individually.
- The motion of the people is described by their direction and walking speed and obeys universal laws that hold for everyone.
- Walking speed and direction may be altered non-deterministically with certain probabilities. This accounts for various psychological and social factors not directly represented in the model (stochasticity).

The walking speed is at most 5 cells/s corresponding to 2 m/s, where each person has an individual upper limit v_{max}^i due to his or her physical abilities. For the actual speed, $v^i \leq v_{max}^i$ holds.

Positions are updated sequentially, with the current person being chosen at random. The route choice is supported by looking ahead as far as possible and orientation is with respect to signposts included in the cells. The maximal distance covered is limited by an individual parameter d_{vis}^i; walls, bends, and other people reduce its value (the visibility). If the destination cell is occupied, then alternative cells are chosen. There is a random probability p_{dir}^i for swaying (e.g.,

Fig. 7.11. A sample output from the simulation of a high speed craft. Walls and inaccessible areas are *shaded* a very *light grey*. People are represented by *grey scale dots*, where *black* means zero and *light grey* maximal velocity. Initially, all people are on the *right*, and the exit is on the *left*

126   Michael Schreckenberg et al.

abruptly changing direction) and orientation or indecision p_{vel}^i (e.g., stopping). This accounts for further influences that have not been included yet, such as psychological factors.

References

1. K. Nagel, M. Schreckenberg: J. Physique I **2**, 2221 (1992)
2. D. Chowdhury, L. Santen, A. Schadschneider: Phys. Rep. **329**, 199 (2000) and Curr. Sci. **77**, 411 (1999)
3. D. Helbing: *Verkehrsdynamik: Neue Physikalische Modellierungskonzepte* (in German) (Springer, 1997)
4. D.E. Wolf, M. Schreckenberg, A. Bachem: *Traffic and Granular Flow*, ed. by A. Bachem (World Scientific, 1996)
5. M. Schreckenberg, D.E. Wolf: *Traffic and Granular Flow '97*, ed. by D.E. Wolf (Springer, 1998)
6. D. Helbing, H.J. Herrmann, M. Schreckenberg, D.E. Wolf: *Traffic and Granular Flow '99*, ed. by D.E. Wolf (Springer, 2000)
7. B.S. Kerner: Phys. World **8**, 25 (1999)
8. J. Esser, M. Schreckenberg: Int. J. Mod. Phys. C **8**, 1025 (1997)
9. K. Nagel, J. Esser, M. Rickert: *Annual Review of Computational Physics* **7**, ed. by D. Stauffer (World Scientific, 2000), p. 151
10. R. Barlovic, L. Santen, A. Schadschneider, M. Schreckenberg: Eur. Phys. J. B **5**, 793 (1998)
11. W. Knospe, L. Santen, A. Schadschneider, M. Schreckenberg: J. Phys. A **33**, 477 (2000)
12. J. Krug: Phys. Rev. Lett. **67**, 1882 (1991)
13. L. Neubert, L. Santen, A. Schadschneider, M. Schreckenberg: Phys. Rev. E **60**, 6480 (1999)
14. S. Wolfram: *Theory and Applications of Cellular Automata* (World Scientific, 1986)
15. M. Rickert, K. Nagel, M. Schreckenberg, A. Latour: Physica A **29**, 6531 (1996)
16. P. Wagner, K. Nagel, D.E. Wolf: Physica A **234**, 687 (1997)
17. D. Chowdhury, D.E. Wolf, M. Schreckenberg: Physica A **235**, 417 (1997)
18. K. Nagel, D.E. Wolf, P. Wagner, P. Simon: Phys. Rev. E **58**, 1425 (1998)
19. B. Eisenblätter, L. Santen, A. Schadschneider, M. Schreckenberg: Phys. Rev. E **57**, 1309 (1998)
20. K. Nagel, M. Paczuski: Phys. Rev. E **51**, 2909 (1995)
21. B.S. Kerner, H. Rehborn: Phys. Rev. E **53**, R4275 (1996)
22. G. Diedrich: Diploma thesis, Universität Köln (1999)
23. R. Barlovic, A. Schadschneider, M. Schreckenberg: Physica A **294**, 525 (2001)
24. L.F. Henderson: Transpn. Res. **8**, 509 (1974)
25. D. Helbing: Complex Systems **6**, 391 (1992)
26. S.P. Hoogendoorn, P.H.L. Bovy: In: *Estimators of travel time for road networks*, ed. by P.H.L. Bovy, R. Thijs (Delft University Press, 2000), p. 107
27. P.G. Gipps, B. Marksjö: Mathematics and Computers in Simulation **27**, 95 (1985)
28. M. Muramatsu, T. Irie, T. Nagatani: Physica A **267**, 487 (1999)
29. M. Fukui, Y. Ishibashi: J. Phys. Soc. Japan **8**, 2861 (1999)
30. V.J. Blue: *Bi-directional emergent fundamental pedestrian flows from cellular automata microsimulation*, http://www.ulster.net/~vjblue
31. H. Klüpfel, T. Meyer-König, J. Wahle, M. Schreckenberg: In: *Proceedings of the Fourth International Conference on Cellular Automata for Research and Industry.* (Karlsruhe 2000), accepted

8 Self-Organized Criticality in Forest-Fire Models

Klaus Schenk, Barbara Drossel, and Franz Schwabl

Summary. We review properties of the self-organized critical (SOC) forest-fire model (FFM). Self-organized critical systems drive themselves into a critical state without fine-tuning of parameters. After an introduction, the rules of the model, and the conditions for spiral shaped and SOC large scale structures are given. For the SOC state, critical exponents and scaling relations are introduced. The existence of an upper critical dimension and the universal behavior of the model are discussed. The relations and differences between FFM and percolation systems are outlined, considering an extension of the FFM into the regions beyond the critical point. Phase transitions and the various structures found in these regions are illustrated.

8.1 Introduction

Nature is full of structures with fractal character, like mountains, clouds and coastlines. Fractals [1,2] are self-similar, i.e., they look the same on different scales of observation, since they have no intrinsic length scale. The spatial correlation functions of these patterns are power laws. The temporal counterpart of fractals is $1/f$-noise [3], which can be found in many natural phenomena, (e.g., light intensity of quasars) or in economics (e.g., stock market prices). The name $1/f$-noise indicates that the Fourier transform of the temporal correlation function is of the form $1/f^\alpha$, where α is in many cases close to one.

The ubiquity of fractals and $1/f$-noise has led to the paradigm of self-organized criticality introduced by Bak, Tang, and Wiesenfeld [4]. Self-organized criticality (SOC) refers to the tendency of large dissipative systems to drive themselves into a scale-invariant, i.e., critical state without any special tuning of the parameters. Examples are the sandpile model [4], several earthquake models, and the forest-fire model [5]. Common features of these models are slow driving or energy input and fast relaxation events (avalanches, earthquakes, fires).

In this chapter, we focus on the forest-fire model (FFM). We shall illustrate the rich behavior of this model, which shows many features unknown in equilibrium systems. Although clusters of trees in the FFM bear some resemblance to clusters in percolation theory, critical exponents, scaling relations, and supercritical behavior are different for the two systems. In the following, we will first discuss the spiral state and the SOC state, which occur for certain ranges of the parameters f (lightning probability) and p (tree growth probability). We then change perspective and choose the tree density ρ_t as parameter. Depending on details of the rules, the patterns that occur beyond a critical tree density are either spiral-like or critical, with further phase transitions to disordered or striped states for higher values of the density.

8.2   The Spiral State and the SOC State

We consider an FFM [5,6] on a d-dimensional hypercubic lattice with L^d sites. Each site is either empty, or occupied by a tree or a burning tree. The following rules are simultaneously applied to update the model:

- burning tree $\rightarrow$ empty site,
- tree $\rightarrow$ burning tree if at least one nearest neighbor is burning,
- tree $\rightarrow$ burning tree with probability f if no nearest neighbor is burning,
- empty site $\rightarrow$ tree with probability p.

Variants of these rules introduce an immunity g (see Sect. 8.5.3) against ignition, or clusterwise removal of trees. For a critical value of g, the fire just percolates through the system, which is then in the same universality class as directed percolation in $d+1$ dimensions [7,8]. Many aspects of this model can be described in the terminology of excitable media [9], opening up a wide area of applications (e.g., diseases, oscillating chemical reactions). In the stationary state, the tree density ρ_t, the density of empty sites ρ_e, and the density of burning trees ρ_f are related via $\rho_f = p\rho_e$ and $\rho_e + \rho_f + \rho_t = 1$. In the limit of small ρ_f, trees survive for a long time, and large scale structures can occur.

In the limit of vanishing lightning rate f and small but non-vanishing growth rate p, one finds spiral-shaped structures (see Fig. 8.2). The spatial and temporal correlations in this state have a characteristic length proportional to $1/p$. The structures and the existence of the characteristic length scale become more and more distinct with decreasing p [10,11]. This indicates the increasing determinism of this non-critical pattern. Spiral waves can be found in nature, for example, in epidemics, or in biology (contraction of the heart muscle).

In the limit $p \ll T(s_{max}) \propto (f/p)^{-\nu'}$, tree clusters that are struck by lightning burn down rapidly, before new trees grow in their neighborhood. Here, $T(s_{max})$ is the time the largest cluster needs to burn down, and ν' is an exponent that takes the value ≈ 0.58 in two dimensions. If additionally $f \ll p$, the size of the largest cluster is very large, and the system is close to a critical point. These two inequalities represent a double separation of time scales, a phenomenon that may occur naturally, thus justifying the term 'self-organized critical state' for the observed scale-invariant state. Some properties of the SOC state are summarized in the next subsection [5,10,12–14].

8.2.1   Properties of the SOC State

In the steady state, the number of burning trees equals the number of growing trees, and the mean number of trees destroyed per fire is

$$\bar{s} = \frac{p}{f} \frac{1 - \rho_t}{\rho_t} \, . \tag{8.1}$$

Let $n(s)$ be the mean number of tree clusters consisting of s trees. Computer simulations show that it scales to good approximation as

$$n(s) = s^{-\tau} C(s/s_{\max}) \,, \tag{8.2}$$

where $C(x)$ is a cutoff function which is constant for small x and decays exponentially for large x. The cutoff cluster size scales as $s_{\max} \propto (f/p)^{-\lambda}$. For the parameter range studied in computer simulations, the exponents τ and λ satisfy the scaling relation

$$\lambda = \frac{1}{3 - \tau} \,. \tag{8.3}$$

The cluster radius R is related to the cluster size s via $s \propto R(s)^{\mu}$. Defining the correlation length ξ by

$$\xi^2 = \frac{2 \sum_{s=1}^{\infty} sn(s) \cdot sR^2(s)}{\sum_{s=1}^{\infty} sn(s) \cdot s} \propto (f/p)^{-2\lambda/\mu} \,, \tag{8.4}$$

and noting that the exponent μ seems to be independent of R, we conclude that

$$\xi \propto (f/p)^{-\nu} \text{ with } \nu = \lambda/\mu \,. \tag{8.5}$$

In contrast to percolation theory, the hyperscaling relation $d = \mu(\tau - 1)$ is not satisfied [12], because not all large subsystems contain a spanning cluster. This violated hyperscaling may be related to the observation in [15] that the system has several diverging length scales.

Due to the influence of lightning strokes, the 2-dimensional SOC FFM self-organizes into a patchy structure (see the upper left image of Fig. 8.2). Since the structure of the fires that burn down a whole patch is different from the structure of fires that only burn down a small section of a low density patch, more than one length scale [16] has to be taken into account to describe more details of the SOC FFM.

Finally, we mention that the critical tree density $\rho_t^c = \lim_{f/p \to 0} \rho_t$ is approached as

$$\rho_t^c - \rho_t \propto (f/p)^{1/\delta} \,. \tag{8.6}$$

Table 8.1 shows the values of the exponents τ and μ in $d = 1$ to 8 dimensions [17], and compares them to the corresponding exponents in percolation theory.

The critical exponents for dimensions equal and higher than $d = 6$ [10] are indistinguishable from those of mean field theory (see Sect. 8.5.2), and this is identical to the mean field theory of percolation. This suggests that the SOC forest-fire model has an upper critical dimension $d_c = 6$ [14].

In one dimension, an exact solution [18] has been found, yielding the cluster size distribution (see Sect. 8.5.1)

$$n(s) = \frac{1 - \rho_t}{(s + 1)(s + 2)} \simeq (1 - \rho_t)s^{-2} \,, \tag{8.7}$$

and the critical exponents. τ is exactly $\tau = 2$.

Table 8.1. Comparison of critical exponents

d	1	2	3	4	5	6	7	8
τ	2	2.14(3)	2.23(3)	2.36(3)	2.45(3)	2.50(3)	2.50(3)	2.50(3)
τperc	2	2.05	2.18	2.31	2.41	2.5	2.5	2.5
μ	1	1.96(1)	2.51(3)	3.0	3.2(2)	–	–	–
μperc	1	1.90	2.53	3.06	3.54	4	4	4

In two dimensions, a real-space renormalization group that assumes a uniform density [19] gives approximate values for the exponents ν and τ. The method can be extended to include the inhomogeneous structure of the forest-fire model.

To conclude this section, let us note that the critical behavior of the FFM seems to be universal with respect to several changes in the microscopic details [10], such as the lattice type, the range over which a burning tree can ignite another one, or a (not too large) 'immunity' against fire. However, a change in the rule for tree growth may change the critical exponents and even destroy the critical behavior completely.

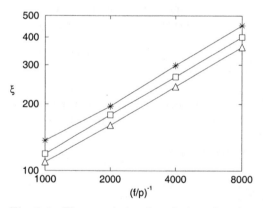

Fig. 8.1. The correlation length ξ as function of $(f/p)^{-1}$ [10]. The slope yields the critical exponent ν. □ square lattice, △ triangular lattice, * next-nearest neighbor interaction

Universal behavior under certain changes of the model rules was checked in [17], where the 2-dimensional simulations were carried out for a triangular lattice ($\rho_t^c \approx 34\%$) and for a square lattice with next-nearest neighbor interaction ($\rho_t^c \approx 28\%$). The critical exponents for these variants of the model were found to be the same. See, for instance, Fig. 8.1, where the correlation length for the three models is plotted. The critical densities are smaller than for the square lattice, since the fire has more possible paths to spread along, due to the larger number of neighbors.

8.3   FFM with the Tree Density as Parameter

For the SOC state discussed in the previous section, the tree density ρ_t cannot be larger than $\rho_t^c \approx 0.41$ in two dimensions. In this section, we modify the rules of the FFM in such a way that the tree density is the parameter of the model, allowing us to study the region $\rho_t > \rho_t^c$ beyond the critical point. In the following, we study two different versions of the model which both have a fixed tree density.

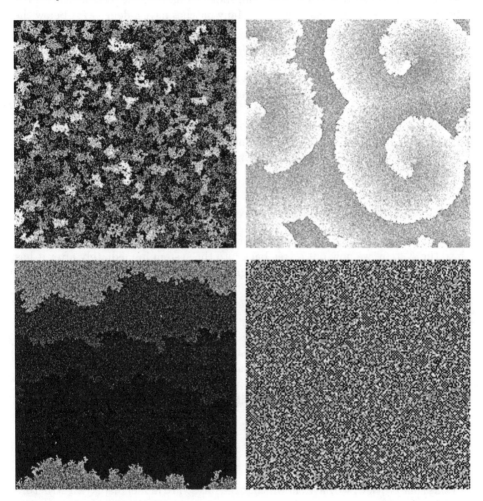

Fig. 8.2. Snapshots of FFM systems with the following densities (from *top left to bottom right*): $\rho_t = 0.393$, 0.41, 0.45 and 0.60. From [20–22]

8.3.1   Removal of Tree Clusters

In this version of the FFM, which is studied in [20], a randomly chosen tree is ignited at each time step, and the cluster connected to this tree is immediately removed (burnt). Then exactly the same number of trees as have been removed are added at randomly chosen empty sites. For $\rho_t < \rho_t^c$, there occur only finite clusters and fires. As ρ_t^c is approached, the size s_{max} of the largest cluster diverges, and all the relations involving f/p for the SOC state can be translated into relations involving ρ_t using (8.6). Although these relations now look similar to those in percolation theory, the values of the critical exponents are different. As shown in the first snapshot of Fig. 8.2, the system consists of different patches of different densities. Many patches have a density far below the percolation threshold, and by the time a patch is struck by lightning, its density is usually far above the percolation threshold, leading to a fractal dimension μ very close to 2.

Since the exponents do not satisfy a hyperscaling relation, the system has no spanning cluster for densities immediately above ρ_t^c. Instead, the system remains critical with a correlation length ξ that diverges as L^{Φ_2}, where the exponent Φ_2 depends continuously on the density. For densities above $\rho^* \approx 0.435$, the system displays synchronized stripes of different densities, as shown in the bottom left part of Fig. 8.2. When the density is increased further, the number of stripes decreases.

8.3.2   One-by-One Removal of Trees

Instead of first burning a whole cluster and then putting the removed trees back into the system, we now put a tree back into the system immediately after it has been burnt, and simultaneously with igniting its neighbors. This version of the model was studied in [21]. For $\rho_t < \rho_t^c$, this variant behaves in the same way as when entire clusters are removed. However, above ρ_t^c, the fire can be sustained even in the absence of lightning strokes, and we find the same spiral structure as in Sect. 8.2 (see Fig. 8.2, top right). This spiral structure can survive for densities down to $\rho_t \approx 0.39$, i.e., the phase transition between the critical and the spiral state shows hysteresis. On the other hand, when ρ_t is increased further, there is another phase transition to a mixed phase with only short-range correlations (see Fig. 8.2, bottom right). This transition also shows hysteresis, as illustrated in Fig. 8.3.

8.4   Conclusion

In this small review, we have shown that the forest-fire model displays a structure which is far richer than in equilibrium systems. Although there are some parallels to percolation, the values of the critical exponents are different, and due to the violation of a hyperscaling relation there exists no spanning cluster for densities immediately above the critical density. Also, different versions of the model show a completely different behavior above the critical density. Equally surprising is the fact that the transition at the critical density shows hysteresis.

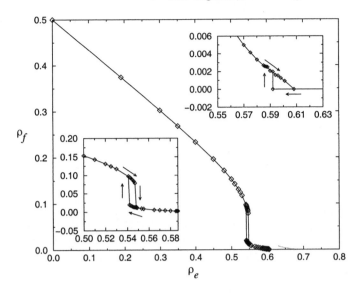

Fig. 8.3. Behavior of ρ_f in a 2-dimensional FFM model with control parameter ρ_e [21]

8.5   Appendixes

8.5.1   Exact Results in One Dimension

In one dimension, exact results have been obtained in [18,23,24], thus proving analytically that non-conservative models can indeed show SOC. Here we give an intuitive derivation of the results. Consider a string of $k \ll p/f$ sites. This string is too short for two trees to grow during the same time step. Lightning does not strike this string before all its trees have grown. Since we are always interested in the limit $f/p \to 0$, the following considerations remain valid even for strings of very large size. Starting with a completely empty string, it passes through the cycle illustrated in Fig. 8.4. During one time step, a tree grows with probability p on any site. After some time, the string is completely occupied by trees. Then the forest in the neighborhood of the string will also be quite dense. The trees on the string are part of a forest cluster which is much larger than k. Eventually that cluster becomes so large that it is struck by lightning with a non-vanishing probability. Then the forest cluster burns down, and the string again becomes completely empty.

This consideration allows us to write down rate equations for the states of the string. In the steady state, each configuration of trees is generated as often as it is destroyed. Let $P_k(m)$ be the probability that the string is occupied by m trees. Each configuration which contains the same number of trees has the same probability. A configuration of m trees is destroyed when a tree grows at one of the empty sites, and generated when a tree grows in a state consisting of $m-1$ trees.

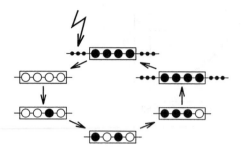

Fig. 8.4. Dynamics on a string of $k = 4$ sites. Trees are *black*, empty sites are *white*. From [10]

The completely empty state is generated when a dense forest burns down. Since all trees on our string burn down simultaneously, this happens each time a given site on the string is set on fire. This in turn happens as often as a new tree grows at this given site, i.e., with probability $p(1 - \rho_t)$ per time step. We therefore have the following equations:

$$pkP_k(0) = p(1 - \rho_t) , \tag{8.8}$$

$$p(k - m)P_k(m) = p(k - m + 1)P_k(m - 1) \quad \text{for} \quad m \neq 0, k . \tag{8.9}$$

We conclude that

$$P_k(m) = \frac{1 - \rho_t}{k - m} \quad \text{for} \quad m < k , \tag{8.10}$$

$$P_k(k) = 1 - (1 - \rho_t) \sum_{m=0}^{k-1} \frac{1}{k - m}$$

$$= 1 - (1 - \rho_t) \sum_{m=1}^{k} \frac{1}{m} . \tag{8.11}$$

These last two equations contain a wealth of information: cluster size distribution, hole distribution, growth velocity, and so on.

A forest cluster of size s is a configuration of s neighboring trees with an empty site at each end. The size distribution of forest clusters is consequently given by

$$n(s) = \frac{P_{s+2}(s)}{\binom{s+2}{s}} = \frac{1 - \rho_t}{(s + 1)(s + 2)} \simeq (1 - \rho_t)s^{-2} . \tag{8.12}$$

This is a power law with critical exponent $\tau = 2$. The size distribution of fires is $\propto sn(s) \propto s^{-1}$. Figure 8.5 shows the numerical result for the fire distribution $sn(s)$. It agrees perfectly with (8.12) in the region $s < s_{\max}$.

$s_{\max}$ is the characteristic size (in one length dimension) where the power law $n(s) \propto s^{-2}$ breaks down. We calculate $s_{\max}$ from the condition that a string of

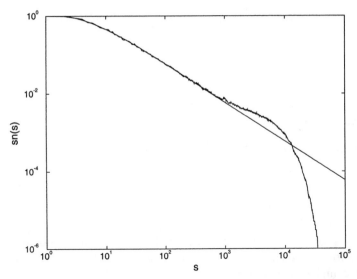

Fig. 8.5. Size distribution of fires for $d = 1$, $f/p = 1/25000$ and $L = 2^{20}$. The size distribution of fires is proportional to the size distribution of forest clusters $n(s)$. The *smooth line* is the theoretical result, valid for cluster sizes $< s_{max}$. From [10]

size $k \leq s_{max}$ is not struck by lightning until all trees have grown. When a string of size k is completely empty at time $t = 0$, it will be occupied by k trees after

$$T(k) = \frac{1}{p} \sum_{m=1}^{k} \frac{1}{m} \simeq \frac{\ln k}{p} \tag{8.13}$$

time steps on average. The mean number of trees after t time steps is

$$m(t) = k[1 - \exp(-pt)] . \tag{8.14}$$

The probability that lightning strikes a string of size k before all trees have grown is

$$f \sum_{t=1}^{T(k)} m(t) \simeq \frac{f}{p} k(\ln k - 1) \simeq \frac{f}{p} k \ln k . \tag{8.15}$$

We conclude that

$$s_{max} \ln(s_{max}) \propto \frac{p}{f} \quad \text{for large } p/f ,$$

leading to $\lambda = 1$.

Next we determine the relation between the mean forest density ρ_t and the parameter f/p. The mean forest density is given by

$$\rho_t \simeq \sum_{s=1}^{s_{max}} sn(s)$$

$$= (1 - \rho_t) \sum_{s=1}^{s_{max}} \frac{s}{(s+1)(s+2)}$$

$$\simeq (1 - \rho_t) \ln(s_{max}) .$$

Thus,

$$\frac{\rho_t}{1 - \rho_t} \simeq \ln(s_{max}) \simeq \ln \frac{p}{f} \quad \text{for large} \quad p/f .$$

The forest density approaches the value 1 at the critical point. This is not surprising since no infinitely large cluster exists in a one-dimensional system as long as the forest is not completely dense.

The exponents characterizing the size distribution of forest clusters and fires remain the same when the fire is allowed to jump over holes of several empty sites [25], proving the universality of the critical exponents.

8.5.2   Mean Field Theory

One of the simplest possible analytical approaches to a model is a mean field theory (MFT), which neglects all correlations in the system and describes it entirely in terms of densities. The neglect of correlations can be modelled in simulations by constructing a random-neighbor model, where the nearest-neighbor connections are chosen randomly at each time step [14]. The probability that a randomly chosen neighbor of a given burning tree is occupied by a green tree is obviously given by the tree density. In the stationary state, the mean field equations are [14,23,26]:

$$\rho_f = p\rho_e , \tag{8.16}$$
$$\rho_f = \rho_t \left(f + (1 - f) \left[1 - (1 - \rho_f)^{2d} \right] \right) , \tag{8.17}$$
$$\rho_e + \rho_t + \rho_f = 1 . \tag{8.18}$$

The first and last equations are exact, since they involve no nearest-neighbor interactions. The second equation includes a mean field approximation, since the probability that one or more neighbors of a given site are burning is given by $[1 - (1 - \rho_f)^{2d}]$, without taking into account any correlations.

From these equations, we find an implicit equation for the fire density alone

$$\rho_f = \left[1 - \rho_f \left(1 + \frac{1}{p} \right) \right] \left[1 - (1 - f)(1 - \rho_f)^{2d} \right] . \tag{8.19}$$

In the SOC limit, p, f, and ρ_f are very small compared to 1, and we find, to leading order,

$$\rho_f = p(1 - 1/2d) + f/2d + O(p^2, pf, \ldots), \qquad (8.20)$$

$$\rho_e = 1 - 1/2d + f/2dp + O(p, f, \ldots), \qquad (8.21)$$

$$\rho_t = 1/2d - f/2dp + O(p, f, \ldots). \qquad (8.22)$$

The tree density approaches its critical value linearly in f/p, leading to $\delta = 1$. The critical tree density is $\rho_t(f/p = 0) = 1/2d$, which means that a burning tree ignites on average just one other tree. This situation is identical to the MFT of percolation or, equivalently, percolation on a Cayley tree, where percolation proceeds to each available neighbor with the same probability and to one neighbor on average. The cluster size distribution and the fractal dimension of clusters for this problem are well known [27]. We thus obtain $\tau = 2.5$ and $\mu = 4$. Scaling relations then give $\lambda = 2$ and $\nu = 0.5$. Percolation has the upper critical dimension 6, and one might expect the SOC forest-fire model to have the same upper critical dimension, as conjectured in [13,14,17]. This is supported by simulation results.

When the limit $p \to 0$ with $f = 0$ is considered instead of the limit $f \to 0$ with $p \ll f/p$, the MFT of the forest-fire model also gives a critical tree density $1/2d$, leading again to the same exponents as the MFT of percolation. In MFT, the structural information is lost and MFT cannot see the qualitative difference between the SOC state and the quasi-deterministic state with spiral-shaped fire fronts.

Fig. 8.6. Snapshot of the forest-fire model near the critical immunity for $L = 200$, $p = 0.05$, $g = 0.48$. From [23]

8.5.3 Immunity

A modification of the model rules is obtained by including a non zero immunity $g \leq 1$, which means that a tree is ignited with the probability $1-g$ when a nearest neighbor is burning or when the tree is struck by lightning. The simulations were performed with immune bonds instead of immune sites. Not all trees that are neighbors of a burning tree catch fire, and consequently the fire no longer burns forest clusters, but rather clusters of trees that are connected by non-immune bonds. A small value of immunity corresponds essentially to a change in the lattice symmetry, since the effective coordination number is decreased. As we have seen, a change in the lattice symmetry does not affect the critical exponents, and this is once again confirmed by simulations. With increasing immunity, the forest density increases (see Fig. 8.6) until the critical forest density becomes $\rho_t^c = 1$ at a critical immunity g_c. Here a new scenario occurs. The forest is completely dense in the limit $f/p \to 0$, and clusters that are destroyed by fire are percolation clusters of bond percolation. Consequently, the exponents $\tau_c = \tau(g = g_c)$ and μ_c are given by percolation theory, and the threshold is $g_c = 1/2$, which is 1 minus the bond percolation threshold. Since the critical forest density is 1 at $g = g_c$, the scaling relation (8.3) now reads

$$\lambda_c = \frac{\gamma_c}{3 - \tau_c} \, , \tag{8.23}$$

with a new exponent γ_c.

For finite f/p, the mean forest density is no longer 1, and one can also determine the other critical exponents λ_c, δ_c, ν_c, and γ_c. The scaling relations (8.23) and (8.5) are confirmed by simulations.

When the immunity is just below its critical value g_c, a crossover from percolation-like to SOC behavior is observed. On length scales smaller than the percolation correlation length $\xi_{\text{perc}} \propto (g_c - g)^{\nu_{\text{perc}}}$, a system close to the percolation threshold cannot be distinguished from a system exactly at the percolation threshold. As long as f/p is so large that fires do not spread further than ξ_{perc}, the exponents are identical to those at $g = g_c$. When f/p becomes very small, there are fires which spread further than the percolation correlation length. These fires are stopped by empty sites that were created by earlier fires. This is the same mechanism as for small g: fires that would spread indefinitely if there were no empty sites are stopped by empty sites. We conclude that these large fires lead to the critical exponents λ, ν, and δ that have been observed for $g = 0$. We make the following scaling ansatz for the correlation length:

$$\xi = (f/p)^{-\nu_c} F \left(\frac{g_c - g}{[f/p]^\phi} \right) \, . \tag{8.24}$$

It is plausible that the crossover from percolation-like to SOC behavior takes place when f/p becomes so small that the correlation length exceeds the percolation correlation length. This suggests that the crossover exponent ϕ is given by $\phi = \nu_c/\nu_{\text{perc}}$. The scaling function $F(x)$ is constant for small x and is proportional to $x^{(\nu - \nu_c)/\phi}$ for large x. Analogous scaling laws hold for $s_{\max}$ and $\rho_t^c - \rho_t$.

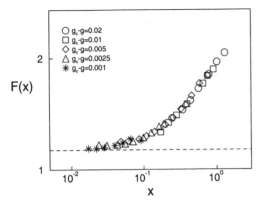

Fig. 8.7. Crossover scaling function $F(x)$ for the correlation length, for different values of the immunity. The *dashed line* represents $F(0)$, as obtained at $g = g_c$. From [10]

Figure 8.7 shows the simulation results for the scaling function of the correlation length $F(x)$, for different values of $g_c - g$. The scaling ansatz (8.24) is well confirmed since all curves coincide. The dashed line represents $F(0)$ as obtained from the simulations at g_c. This crossover is similar to crossover phenomena at equilibrium phase transitions, and it should also be observed in higher dimensions. Although the crossover in τ and μ vanishes in dimensions larger than 6, there is still a crossover in λ and ν, due to the modified scaling relation at g_c. In $d = 1$, the critical immunity is $g_c = 0$, and no crossover can take place.

References

1. B.B. Mandelbrot: *The Fractal Geometry of Nature* (Freeman, New York 1983)
2. A. Bunde, S. Havlin: *Fractals and Disordered Systems*, ed. by A. Bunde, S. Havlin (Springer, New York 1991)
3. P. Dutta, P.M. Horn: Rev. Mod. Phys. **53**, 497 (1981)
4. P. Bak, C. Tang, K. Wiesenfeld: Phys. Rev. Lett. **59**, 381 (1987)
5. B. Drossel, F. Schwabl: Phys. Rev. Lett. **69**, 1629 (1992)
6. P. Bak, K. Chen, C. Tang: Phys. Lett. A **147**, 297 (1990)
7. B. Drossel, F. Schwabl: Physica A **199**, 183 (1993)
8. E.V. Albano: Physica A **216**, 213 (1995)
9. J.J. Tyson, J.P. Keener: Physica D **32**, 327 (1988)
10. S. Clar, B. Drossel, F. Schwabl: J. Phys. Condens. Matter **8**, 6803 (1996)
11. P. Grassberger, H. Kantz: J. Stat. Phys. **63**, 685 (1991)
12. C.L. Henley: Phys. Rev. Lett. **71**, 2741 (1993)
13. P. Grassberger: J. Phys. A **26**, 2081 (1993)
14. K. Christensen, H. Flyvberg, Z. Olami: Phys. Rev. Lett. **71**, 2737 (1993)
15. A. Honecker, I. Peschl: Physica A **239**, 509 (1997)
16. K. Schenk , B. Drossel , S. Clar , F. Schwabl: Eur. Phys. J. B **15**, 177 (2000)
17. S. Clar, B. Drossel, F. Schwabl: Phys. Rev. E **50**, 1009 (1994)
18. B. Drossel, S. Clar, F. Schwabl: Phys. Rev. Lett. **71**, 3739 (1993)
19. V. Loreto, L. Pietronero, A. Vespignani, S. Zapperi: Phys. Rev. Lett. **75**, 465 (1995)

20. S. Clar, B. Drossel, F. Schwabl: Phys. Rev. Lett. **75**, 2722 (1995)
21. S. Clar, K. Schenk, F. Schwabl: Phys. Rev. E **55**, 2174 (1997)
22. S. Clar, B. Drossel, K. Schenk, F. Schwabl: Physica A **266**, 153 (1999)
23. B. Drossel: Ph.D. Thesis, TU München (1994)
24. M. Paczuski, P. Bak: Phys. Rev. E **48**, R3214 (1993)
25. B. Drossel, S. Clar, F. Schwabl: Z. Naturforsch. **49**, 856 (1994)
26. B. Drossel, F. Schwabl, Physica A **204**, 212 (1994)
27. D. Stauffer, A. Aharony: *Introduction to Percolation Theory* (Taylor and Francis, London 1992)

9 Nonlinear Dynamics of Active Brownian Particles

Werner Ebeling

Summary. We consider finite systems of interacting Brownian particles including active friction in the framework of nonlinear dynamics and statistical/stochastic theory. First we study the statistical properties for one-dimensional systems of N masses connected by Toda springs which are imbedded in a heat bath. Including negative friction, we find $N + 1$ attractors of motion, including an attractor describing dissipative solitons. Noise leads to transition between the deterministic attractors. In the case of two-dimensional motion of interacting particles, angular momenta are generated and left/right rotations of pairs and swarms are found.

9.1 Introduction

The main purpose of this work is to study the dynamics of active Brownian particles including interactions. Friction is modelled by a velocity-dependent function derived from an energy supply model [1–5]. Interaction between the particles is modelled by Toda potentials in one-dimensional (1D) systems or Morse potentials in two-dimensional (2D) systems.

Since the classical work of Toda [6], the development of the nonlinear dynamics and statistical thermodynamics of Toda systems has remained a central topic of research. Toda was able to find exact solutions for a special 1D system with an exponential interaction. In particular, Toda proved the existence of soliton solutions and calculated the exact partition function. Since then, solitons have proved to be of remarkable interest as excitations of nonlinear chains of masses. Several interesting results have been obtained for the statistical thermodynamics of Toda systems [7–9]. We consider here systems with active friction and noise by means of analytical tools and simulations. It has been shown that there exists a special temperature regime around a transition temperature T_{tr} between the phonon and the soliton regime, called the localization temperature T_{loc} in another context, where the interaction of solitons and phonons is strong and has a significant effect on several physical phenomena. One such is energy localization at special sites [13] and another is the excitation of a broad-band coloured noise spectrum with a $1/f$ region at low frequencies [14]. Our main interest here is the influence of negative friction on the properties of Brownian motion. Driving the system of N particles by negative friction to far-from-equilibrium states, we find $N + 1$ attractors of deterministic motion including an attractor describing dissipative solitons. Noise leads to transition between the deterministic attractors. In the case of two-dimensional motion of interacting particles, positive or negative angular momenta are generated with equal probability. This leads to left/right rotations of pairs, clusters and swarms. We will show that the collective motion of large clusters of driven Brownian particles very much resembles the typical modes of parallel motions in swarms of living entities.

9.2   Equations of Motion, Friction and Forces

Let us consider a system of N point masses m, numbered $1, 2, \ldots, i, \ldots, N$. We assume that each mass m is connected to its next nearest neighbours on both sides by Toda forces. The distance between the mass i and the mass $i + 1$ is denoted by R_i, the equilibrium distance is assumed to be σ, so that the spring elongation reads $r_i = R_i - \sigma$. In the following, we take σ as the length unit.

The dynamics of the system is given by the following equation of motion for the elongations:

$$\frac{\mathrm{d}^2}{\mathrm{d}t^2} r_i = [V'(r_{i+1}) - 2V'(r_i) + V'(r_{i-1})] - \gamma(v_i)v_i + \mathcal{F}_i(t) , \tag{9.1}$$

where $\mathcal{F}_i(t)$ is a stochastic force with strength D and a δ-correlated time dependence:

$$\langle \mathcal{F}_i(t) \rangle = 0 , \quad \langle \mathcal{F}_i(t)\mathcal{F}_j(t') \rangle = 2D\,\delta(t - t')\delta_{ij} . \tag{9.2}$$

In the case of thermal equilibrium systems, we have $\gamma(v) = \gamma(0) = \text{const}$. In the general case where the friction is velocity dependent, we will assume that it is monotonically increasing with the velocity v and that the limit for large velocities is a well defined constant:

$$\gamma(v) \to \gamma(\infty) = \gamma_0 = \text{const}. \tag{9.3}$$

Our basic assumption is that, in this limit, the loss of energy resulting from friction and the gain in energy resulting from the stochastic force are balanced on the average. From this postulate, the non-equilibrium fluctuation–dissipation theorem follows:

$$D = k_B T \gamma_0 / m , \tag{9.4}$$

where T is the temperature of the heat bath, k_B is the Boltzmann constant, and D is the strength of the stochastic force. For the case of a passive thermal heat bath $\gamma = \gamma_0$, (9.4) agrees with the conventional Einstein relation. The validity of a fluctuation–dissipation relation between noise strength and damping strength is assumed in order to guarantee the existence of a stationary or thermal equilibrium, independent of the limit of the friction parameter $\gamma_0 \geq 0$. We note that $\gamma_0 = 0$, $T = 0$ corresponds to the conservative case.

In the following, we will begin by studying the simplest case of passive friction, i.e., we will assume that the friction function is constant $\gamma(v) \equiv \gamma_0$. More complicated friction functions will then be studied. Historically, velocity-dependent friction forces were first studied by Rayleigh and Helmholtz [15].

Extensions of these models were investigated in many papers on driven Brownian dynamics [16]. A characteristic property of these friction functions is the existence of a zero of friction for a finite velocity v_0 which defines a kind of attractor in the velocity space. The model we use here for active friction, with an attractor for the velocities v_0, was introduced in [1,2]. Detailed studies of this so-called energy depot model may be found in [3,5]. The case of Rayleigh friction was analyzed in detail in [4]. The depot model of the friction function is based on a concrete model of Brownian motion with energy supply, storage in a tank and conversion into motion [1]. We assume that the Brownian particle itself is capable of taking up external energy, and storing some of this additional energy in an internal energy depot $e(t)$. In the energy depot model, the depot may be altered by three different processes.

- Take-up of energy from the environment, where q is the energy pump rate.
- Internal dissipation, assumed to be proportional to the internal energy, with rate of energy loss denoted by c.
- Conversion of internal energy into motion, where d is the rate of conversion of internal to kinetic degrees of freedom. The depot energy is used to accelerate the motion.

Our model of energy supply is motivated by investigations of active biological motion, which relies on the supply of energy. The supplied energy is in part dissipated by metabolic processes, but it can also be converted into kinetic energy. The energy depot model leads in an adiabatic approximation to the friction function

$$\gamma(v) = \gamma_0 - \frac{q}{1 + dv^2} . \tag{9.5}$$

For the energy depot model of active friction, an attracting velocity exists. This is defined by the zero of the friction function $\gamma(v_0) = 0$, which has the value

$$v_0^2 = \frac{\alpha}{d}, \quad \alpha = \frac{q}{\gamma_0} - 1 . \tag{9.6}$$

Here α plays the role of the bifurcation parameter. For $\alpha < 0$, the friction is purely passive, i.e., on average, no energy is supplied. For $\alpha > 0$, the friction function has a negative part near $v = 0$ and a zero point which acts as an attracting set in the velocity space.

The interaction between the masses i and $i + 1$ in the 1D case is modelled by the Toda force

$$f_i = -V'(r_i) = \frac{\omega^2}{b} \left(e^{-br_i} - 1 \right) , \tag{9.7}$$

where b is the stiffness of the springs and ω is the linear oscillation frequency around the equilibrium position.

In the harmonic limit, the force is given by $f_i = \omega r_i$. The spring energy is described by the Toda potential

$$V(r_i; \omega, b) = \frac{m\omega^2}{b^2} \left(e^{-br_i} - 1 + br_i \right) . \tag{9.8}$$

We will use $m\omega^2\sigma^2$ as the energy unit.

In general, the calculation of the distribution functions of interacting particles is not a trivial task since it is connected with the solution of a multi-dimensional Fokker–Planck equation. A full solution is available only for the passive case $\gamma \equiv \gamma_0$. Then the only attractor of the dynamics is the rest state of the particles and the statistical properties in equilibrium are described by a canonical Toda ensemble under pressure P and temperature T [10–12]. The stationary solution of the Fokker–Planck equation, which corresponds to our Langevin equation, reads

$$P_0(p_n, r_n) = Z_1^{-1} \exp\left[-\frac{p_n^2/2m + V_{\text{eff}}}{k_{\text{B}T}} \right], \quad Z_1^{-1} = \frac{bX^{X+Y}}{\sqrt{2\pi m k_{\text{B}}T}e^X \Gamma(X+Y)}, \tag{9.9}$$

with the effective potential

$$V_{\text{eff}}(r_n) = V(r_n) + Pr_n . \tag{9.10}$$

Here $V(r)$ is the Toda potential (9.8) and we have used the abbreviations

$$X = \frac{m\omega^2}{b^2 k_{\text{B}}T}, \quad Y = \frac{P}{bk_{\text{B}}T} . \tag{9.11}$$

Elementary excitations in passive Toda rings, including the noise spectrum and the structure factor, were investigated in detail in [14,17]. The investigation of the dynamics of Toda rings with energy supply by active friction is more difficult and only partial solutions are available [4,5]. We discuss here the special case where the supply of energy is given by the depot model in the approximation of a velocity-dependent friction [1–3]. Due to the driving, slow particles are accelerated and fast particles are damped. For certain conditions, our active friction function has a zero. The deterministic trajectory of our system is then attracted by a cylinder in the 4D space given by

$$v_1^2 + v_2^2 = v_0^2 , \tag{9.12}$$

where v_0 is the value of the stationary velocity (9.6).

Figure 9.1 shows the histogram calculated by J. Dunkel for $N = 8$ particles with Toda interactions [5]. Here the result of 1000 runs with stochastic initial conditions is represented as a function of the average velocity of the particles. The shape of the histogram demonstrates the existence of $N + 1$ attractors of motion. We must stress, however, that the histogram shown in Fig. 9.1 is the result of finite time runs, so it represents the shape of a finite-time distribution function. In the limit of long runs, the distribution may change. For example, at long times, we expect the statistical weights to be shifted from the center of the distribution to the wings.

9.3   Two-Dimensional Dynamics

The effects demonstrated in the previous section are not restricted to 1D Toda lattices but persist at least qualitatively in more realistic 2D and 3D models of

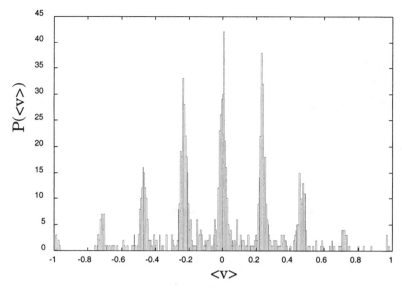

Fig. 9.1. Probability distribution for 8 active Brownian particles as a function of the average velocity in units of v_0

dense fluids consisting of solvent and solute molecules with Morse or Lennard–Jones interactions. Superposition of solitons corresponds to multiple collisions in these systems. In higher dimensions, a weak localisation of potential energy was also observed at the bindings of the bath molecules and was connected to a transition between different lattice configurations [18–20].

We introduce interactions described by the potential $U(r_1, \dots, r_N)$. The dynamics of Brownian particles is then determined by the Langevin equation

$$\dot{r}_i = v_i , \quad m\dot{v}_i = -\gamma(v_i)v_i - \nabla U(r_1, \dots, r_N) + \mathcal{F}(t) , \qquad (9.13)$$

where $\mathcal{F}(t)$ is a stochastic force with strength D and a δ-correlated time dependence, as defined above.

We will now discuss the motion of active particles in 2D space, $r = \{x_1, x_2\}$. The case of constant external forces has been treated by Schienbein and Gruler [21]. Symmetric parabolic external forces were studied in [2,3,5] and the non-symmetric case is investigated in [22]. Here we will study 2D systems of $N \geq 2$ particles (see Fig. 9.2).

We will begin with the study of a 2-particle problem [22]. Let us imagine two Brownian particles which are pairwise bound by a pair potential $U(r_1 - r_2)$. The two molecules will form dumbbell-like configurations. The motion then consists of two independent parts: the free motion of the center of mass with coordinates $X_1 = 0.5(x_{11} + x_{21})$ and $X_2 = 0.5(x_{12} + x_{22})$, and the relative motion under the influence of the potential, described by coordinates $\tilde{x}_1 = (x_{11} - x_{12})$ and $\tilde{x}_2 = (x_{12} - x_{21})$. The motion of the center of mass M is described by the

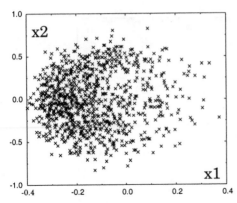

Fig. 9.2. Snapshot of 1000 Brownian particles rotating in a parabolic self-consistent field

equations

$$\dot{X}_1 = V_1 , \qquad M\dot{V}_1 = -\gamma \left(V_1, V_2\right) V_1 + \mathcal{F}_1(t) ,$$
$$\dot{X}_2 = V_2 , \qquad M\dot{V}_2 = -\gamma \left(V_1, V_2\right) V_2 + \mathcal{F}_2(t) . \qquad (9.14)$$

The stationary solution of the corresponding Fokker–Planck equation [3]

$$P_0(\boldsymbol{V}) = C \left(1 + dV^2\right)^{q/2D} \exp\left(-\frac{\gamma_0}{2D} V^2\right) \qquad (9.15)$$

corresponds to the driven motion of a free particle located at the center of mass. The relative motion is described by the equations

$$\dot{\tilde{x}}_1 = \tilde{v}_1 , \qquad \mu\dot{\tilde{v}}_1 = -\gamma \left(\tilde{v}_1, \tilde{v}_2\right) \tilde{v}_1 - \partial_1 U + \tilde{\mathcal{F}}_1(t) ,$$
$$\dot{\tilde{x}}_2 = \tilde{v}_2 , \qquad \mu\dot{\tilde{v}}_2 = -\gamma \left(\tilde{v}_1, \tilde{v}_2\right) \tilde{v}_2 - \partial_2 U + \tilde{\mathcal{F}}_2(t) , \qquad (9.16)$$

with $\mu = m/2$. To simplify, we now specify the potential $U(\boldsymbol{r})$ as a symmetric parabolic potential [5]:

$$U(x_1, x_2) = \frac{1}{2}a \left(x_1^2 + x_2^2\right) . \qquad (9.17)$$

First, we restrict the discussion to a deterministic relative motion, which is described by four coupled first-order differential equations. The relative motion of 2 particles corresponds to the absolute motion of 1 particle in an external field. Consequently, we will now simplify notation by omitting the tilde denoting the relative character of the motion and denoting the mass by m again, rather than μ:

$$\dot{x}_1 = v_1 , \qquad m\dot{v}_1 = -\gamma \left(v_1, v_2\right) v_1 - ax_1 ,$$
$$\dot{x}_2 = v_2 , \qquad m\dot{v}_2 = -\gamma \left(v_1, v_2\right) v_2 - ax_2 . \qquad (9.18)$$

For the 1D case, it is well known that this system possesses a limit cycle corresponding to sustained oscillations [16]. For the 2D case, we have shown in [2] that a limit cycle is developed in the 4D space, which corresponds to left/right rotations with frequency ω_0. The projection of this periodic motion onto the $\{v_1, v_2\}$ plane is the circle

$$v_1^2 + v_2^2 = v_0^2 \ . \tag{9.19}$$

The projection onto the $\{x_1, x_2\}$ plane also corresponds to a circle

$$x_1^2 + x_2^2 = r_0^2 = \frac{v_0^2}{\omega_0^2} \ . \tag{9.20}$$

The energy for motions on the limit cycle is

$$E_0 = \frac{m}{2}(v_1^2 + v_2^2) + \frac{a}{2}(x_1^2 + x_2^2) = \frac{m}{2}v_0^2 + \frac{a}{2}r_0^2 \ . \tag{9.21}$$

We have shown in [2] that any initial value of the energy converges (at least in the limit of strong driving) to

$$H \longrightarrow E_0 = mv_0^2 \ . \tag{9.22}$$

This corresponds to an equal distribution between kinetic and potential energy. As for the 1D harmonic oscillator, both parts contribute the same amount to the total energy. This result was obtained in [2], based on the assumption that the energy is a slow (adiabatic) variable which allows a phase average with respect to the phases of the rotation. In explicit form, we may represent the motion on the limit cycle in the 4D space by the four equations [3]

$$\begin{aligned}
x_1 &= r_0 \sin(\omega_0 t) \ , & v_1 &= r_0 \omega_0 \cos(\omega_0 t) \ , \\
x_2 &= r_0 \cos(\omega_0 t) \ , & v_2 &= -r_0 \omega_0 \sin(\omega t) \ .
\end{aligned} \tag{9.23}$$

The frequency is given by the time the particles need for one period, moving on the circle with radius r_0 with speed v_0. This leads to the relation

$$\omega_0 = \frac{r_0}{v_0} = \left(\frac{m}{a}\right)^{1/2} = \omega \ . \tag{9.24}$$

This means, that the particle oscillates (at least in our approximation) with the frequency given by the linear oscillator frequency ω.

The trajectory on the limit cycle, i.e., the attractor, defined by the above 4 equations is like a hula hoop in the 4D space. The projections onto the x_1–x_2 space are circles, as are the projections onto the v_1–v_2 space. The projections onto the subspaces x_1–v_2 and x_2–v_1 are like a rod. A second limit cycle is obtained by reversal of the velocity. This also forms a hula hoop, different from the first one, although both limit cycles have the same projections onto the $\{x_1, x_2\}$ and $\{v_1, v_2\}$ planes. The motion in the $\{x_1, x_2\}$ plane has the opposite sense of rotation to the first limit cycle. The two limit cycles therefore correspond to opposite angular momenta, $L_3 = +\mu r_0 v_0$ and $L_3 = -\mu r_0 v_0$. Applying similar arguments to the stochastic problem, we find that the two hoop rings are converted into a distribution looking like two embracing hoops with finite size, which converts for strong noise into two embracing tyres in the 4D space. In order to get the explicit form of the distribution, we may introduce the amplitude–phase representation [3]

$$x_1 = \rho \sin(\omega_0 t + \phi) , \quad v_1 = \rho \omega_0 \cos(\omega_0 t + \phi) ,$$
$$x_2 = \rho \cos(\omega_0 t + \phi) , \quad v_2 = -\rho \omega_0 \sin(\omega_0 t + \phi) , \tag{9.25}$$

where the radius ρ is now a slow stochastic variable whilst the phase ϕ is a fast one. By using the standard procedure of averaging with respect to the fast phases, we get the distribution of the radii [3]:

$$P_0(\rho) = C \left(1 + d\rho\omega_0{}^2\right)^{q/2D} \exp\left(-\frac{\gamma_0}{2D}\rho\omega_0{}^2\right) . \tag{9.26}$$

The probability crater is located above the two deterministic limit cycles on the sphere $r_0^2 = v_0^2/\omega_0^2$. Strictly speaking, the whole spherical set is not filled with probability, but only two circle-shaped subsets on it, which correspond to a narrow region around the limit sets. The full stationary probability has the form of two hula hoop distributions in the 4D space. This has been confirmed by simulations [3].

The projections of the distribution onto the $\{x_1, x_2\}$ and $\{v_1, v_2\}$ planes are smoothed 2D rings. The distributions intersect the $\{x_1, v_2\}$ and $\{x_2, v_1\}$ planes perpendicularly. Due to noise, the Brownian particles may switch between the two limit cycles, implying inversion of the angular momentum (direction of rotation) [3,22].

We may summarize our findings for a 2-particle system forming a dumbbell system in the following way. The center of mass of the dumbbell will perform a driven Brownian motion corresponding to a free motion of the center of mass. In addition, the dumbbell is driven in a rotation around the center of mass. What we observe then is a system of rotating Brownian molecules. Internal degrees of freedom are excited and we observe driven rotations. In this way we have shown that the mechanisms described here may also be used to excite the internal degrees of freedom of Brownian molecules.

An extension of the theory of pairs leads to a theory for the motion of clusters of active molecules [22–25]. Figure 9.3 shows a snapshot of simulations of the stochastic dynamics of a cluster of 1000 active interacting Brownian particles.

In order to simplify the simulations, we assume that the interaction of the molecules in the cluster is given by a relatively long range van der Waals-type interaction. For example, we may use the interaction model proposed by Morse, viz.,

$$\phi(r) = A \left(e^{-ar} - 1\right)^2 - A . \tag{9.27}$$

The attracting tail causes the molecules to form clusters. Individual molecules then move in the collective (self-consistent) field of the other molecules, which

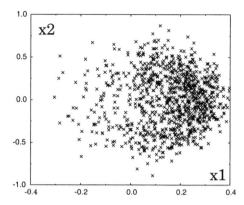

Fig. 9.3. Snapshot of the same swarm after one half period of rotation

can be represented by a mean field approximation

$$V(\tilde{r}) = \int d\mathbf{r}' \phi(\tilde{r} - \mathbf{r}')\rho(\mathbf{r}') \,, \qquad (9.28)$$

where $\tilde{r} = (\tilde{x}_1, \tilde{x}_2)$ is the radius vector from the center of mass and $\rho(\mathbf{r}')$ is a mean density in the cluster. Approximating V by quadratic terms only, we get

$$V(\tilde{x}_1, \tilde{x}_2) = V_0 + \frac{1}{2}\left(a_1\tilde{x}_1^2 + a_2\tilde{x}_2^2\right) + \dots \,. \qquad (9.29)$$

In this way, we arrive once more at the harmonic problem studied above. We may conclude that the individual molecules in the cluster move, at least in certain approximations, in a parabolic potential. From this, it follows that they will perform rotations in the field. We have carried out simulations with 1000 particles moving in a self-consistent potential of parabolic shape. The driving function has a zero at $v_0^2 = 1$. Figures 9.2 and 9.3 show two snapshots of the moving cluster formed by the molecules.

Since the individual particles move in an effective parabolic potential, angular momenta are generated and the swarm starts to rotate. Similarly to the case of the dumbbells, the clusters are driven into spontaneous rotations. A stationary state will eventually be reached, corresponding to a rotating cluster with nearly constant angular momentum (see Fig. 9.4). Under the influence of noise, the cluster may switch to the opposite angular momentum, i.e., to the opposite sense of rotation. The system is thus bistable.

Figure 9.4 shows that, in noisy systems of active Brownian particles, the two values $L_3 = \pm m r_0 v_0$ have the maximal probability. Strong coupling of the particles leads to synchronization of the angular momenta, whilst for weak coupling, the cluster may be decomposed into groups with different angular momenta [22,25]. The rotating swarms simulated in our numerical experiments very much resemble the dynamics of swarms studied in papers by Viscek et al. [23,24] and in other recent work [22,25].

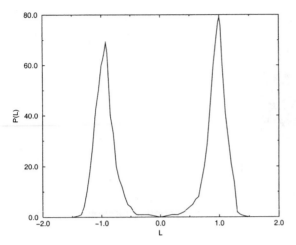

Fig. 9.4. Probability distribution of the angular momentum for active Brownian particles with $v_0^2 = r_0^2 = 1$

9.4   Conclusion

We have studied here the active Brownian dynamics of a finite number of particles including interactions modelled by Toda or Morse potentials and velocity-dependent friction. Besides analytical investigations, we have made a numerical study of relatively small systems. For this we have been investigating 1D model systems with nonlinear Toda interactions and 2D models with parabolic confinement. At low enough temperatures, we observe only harmonic excitations and the models reduce to systems of coupled harmonic oscillators. However, at higher temperatures, the dynamics is completely changed and nonlinear excitations of soliton type come into play. At high temperatures, the behaviour of 1D interacting systems is dominated by soliton-type excitations. Including active friction, we observe in the 1D case $N + 1$ attractors of nonlinear excitations. For interacting Brownian particles in the 2D case, angular momenta are generated (see Fig. 9.4) and we find rotations of the particles around the center of attraction and collective rotational excitations of swarms.

Acknowledgements. The author is grateful to A. Chetverikov (Saratov), J. Dunkel and U. Erdmann (Berlin), M. Jenssen (Eberswalde), Yu.L. Klimontovich (Moscow), F. Schweitzer (Birlinghoven), B. Tilch (Stuttgart), V. Makarov and M. Velarde (Madrid) for collaboration and support of this work. The FORTRAN program used for generating Figs. 9.2–9.4 was written by A. Neiman (St. Louis).

References

1. F. Schweitzer, W. Ebeling, B. Tilch: Phys. Rev. Lett. **80**, 5044 (1998)
2. W. Ebeling, F. Schweitzer, B. Tilch: BioSystems **49**, 17 (1999)
3. U. Erdmann, W. Ebeling, F. Schweitzer, L. Schimansky-Geier: Eur. Phys. J. B **15**, 105 (2000)
4. V. Makarov, W. Ebeling, M. Velarde: Int. J. Bifurc. & Chaos **10**, 1075 (2000)
5. W. Ebeling, U. Erdmann, J. Dunkel, M. Jenssen: J. Stat. Phys. **101**, 442 (2000)
6. M. Toda: *Theory of Nonlinear Lattices* (Springer, Berlin 1981); M. Toda: *Nonlinear Waves and Solitons* (Kluwer, Dordrecht 1983)
7. J.A. Krumhansl, J.R. Schrieffer: Phys. Rev. B **11**, 3535 (1975)
8. J. Bernasconi, T. Schneider (eds.): *Physics in One Dimension* (Springer, Berlin 1981)
9. S.E. Trullinger, V.E. Zakharov, V.L. Pokrovsky (eds.): *Solitons* (North Holland, Amsterdam 1986)
10. H. Bolterauer, M. Opper: Z. Phys. B **42**, 155 (1981)
11. M. Toda, N. Saitoh: J. Phys. Soc. Japan **52**, 3703 (1983)
12. M. Jenssen: Phys. Lett. A **159**, 6 (1991)
13. W. Ebeling, M. Jenssen: Ber. Bunsenges, Phys. Chem. **95**, 1356 (1991); Physica A **188**, 350 (1992)
14. M. Jenssen, W. Ebeling: Physica D **141**, 117 (2000)
15. J.M. Rayleigh: The Theory of Sound, Vol. 1 (MacMillan, London & New York 1984)
16. Yu.L. Klimontovich: *Statistical Physics of Open Systems* (Kluwer, Dordrecht 1995)
17. W. Ebeling, A. Chetverikov, M. Jenssen: Ukrainian J. Physics **45**, 479 (2000)
18. W. Ebeling, V.J. Podlipchuk, A.A. Valuev: Physica A **217**, 22 (1995); J. Mol. Liq. **73,74**, 445 (1997)
19. W. Ebeling, M.V.J. Podlipchuk: Z. physik. Chem. **193**, 207 (1996)
20. W. Ebeling, M. Sapeshinsky, A. Valuev: Int. J. Bifurc. & Chaos **8**, 755 (1998)
21. M. Schienbein, H. Gruler: Bull. Math. Biol. **55**, 585 (1993)
22. U. Erdmann, W. Ebeling, V. Anishchenko: Preprint Sfb-555-5-2001 Humboldt University Berlin (2001)
23. T. Viscek, A. Czirok, E. Ben-Jacob, I. Cohen, O. Shochet: Phys. Rev. Lett. **75**, 1226 (1995)
24. I. Derenyi, T. Viscek: Phys. Rev. Lett. **75**, 294 (1995)
25. F. Schweitzer, W. Ebeling, B. Tilch: Phys. Rev. E, in press (2001); cond-mat/0103360

10 Financial Time Series and Statistical Mechanics

Marcel Ausloos

Summary. A few characteristic exponents describing power law behavior of roughness, coherence and persistence in stochastic time series are compared to each other. Relevant techniques for analyzing such time series are described to show how the various exponents are measured, and what basic differences exist between them. Financial time series, like the JPY/DEM and USD/DEM exchange rates, are used for illustration, but mathematical ones, like Brownian walks (fractional or otherwise) can also be used as indicated.

10.1 Introduction

A great challenge in modern times is the construction of predictive theories for nonlinear dynamical systems for which the evolution equations are barely known, if known at all. General or so-called universal laws are aimed at from very noisy data. A universal law should hold for different systems characterized by different models, but leading to similar basic parameters, like the critical exponents, depending only on the dimensionality of the system and the number of components of the order parameter. In short, this leads to a predictive value or power of the universal laws.

In order to obtain universal laws in stochastic systems, one has to distinguish true noise from chaotic behavior, and sort out coherent sequences from random ones in experimentally obtained signals [1]. The stochastic aspects are not only found in the statistical distribution of underlying frequencies characterizing the Fourier transform of the signal, but also in the amplitude fluctuation distribution and high moments or correlation functions.

Following the scaling hypothesis idea, neither time nor length scales need be considered [2–4]. Henceforth, the fractal geometry is a perfect framework for studies of stochastic systems which do not appear at first sight to have underlying scales. A universal law can be a so-called *scaling law* if a $\log[y(x)]$ vs. $\log(x)$ plot gives a straight line (over several decades if possible) leading to a slope measurement and the exponent characterizing the power law.

When examining such phenomena, it is often recognized that some *coherent factor* is implied. Yet there are states which cannot be reached without going through intricate evolutions, implying concepts like *transience* and *persistence* – well known if one recalls the turbulence phenomenon and its basic theoretical explanation [5]. Finally, the *apparent roughness* of the signal can be expressed in mathematical terms. These concepts are briefly elaborated in Sect. 10.2.

In the function $y(x)$, y and x can be many things. However, the relevant concepts outlined here are well illustrated when x is the time variable t. Time series thus serve as a fundamental testing ground. Several series can be found in the literature. Financial time series and mathematical ones based on the Weierstrass–Mandelbrot function [6], describing fractional (or otherwise) Brownian walks can be used for illustration. The number of points should be large enough to obtain small error bars. A few useful references among many others, discussing tests and other basic or technical considerations on non-linear time series analysis, can be found in [7–12].

Mathematical series, such as Brownian motions (fractional or otherwise) and practical ones, like financial time series, are thus of interest for discussion or illustration, as in Sect. 10.3. There exist several papers and books for the reader interested in financial time series analysis and forecasting. Again, not all can be mentioned, but see [13–15]. On a more general basis, an introduction to financial market analysis per se can be found in [16].

Nevertheless, any scaling exponent should be robust in a statistical sense, with respect to small changes in the data or in the data analysis technique. If this is so, some physical considerations and modeling can be pursued. One question often raised for statistical purposes is whether the data is *stationary* or not, i.e., whether the analyzed raw signal, or any of its combinations depends on the (time) origin of the series. This theoretical question seems somewhat irrelevant practically speaking in financial, meteorological, and several other sciences because the data is obviously never stationary. In fact, in such exotic new applications of physics, a restricted criterion for stationarity is thought to be sufficient: if the statistical mean and extracted parameters do not change too much (up to some statistical significance [7]), the data is called *quasi-stationary*. If this happens, it can be offered for fundamental investigations. Thereafter, the prefix 'quasi' is immediately dropped.

Several characteristic plots lead to *universality considerations*. Hence, fractal-like exponents are described first. They are obtained from different techniques which are briefly reviewed for completeness either in Sect. 10.3 or in appendix. Some useful technical materials can be found in [17–19]. This should serve to distinguish how the various exponents are measured, and what basic differences exist between them. The numerical values pertaining to the words (i) persistence, (ii) coherence, and (iii) roughness will be given and related to each other. For illustration purposes, two cases of foreign currency exchange rates are used, i.e., DEM/USD and DEM/JPY. They are shown in Figs. 10.1 and 10.2 for a time interval ranging from 1 January, 1993 to 30 June, 2000.

For an adequate perspective, let it be recalled that such modern concepts of statistical physics have recently been applied to analyze time series outside finance, in particular those arising from biology [20,21], medicine [22], meteorology [23], electronics [24], image recognition [25], and others. Again, the intention is not to list all references of interest as should be done in a (longer) review paper. Many examples can also be found in this book through contributions from world specialists in computer simulations.

Fig. 10.1. Evolution of the DEM/USD exchange rate from 1 January, 1993 to 30 June, 2000

10.2   Phase, Amplitude and Frequency Revisited

10.2.1   Coherence

When mentioning the word coherence to any student in physics, he/she immediately thinks of lasers – that is where the word has been most striking in scientific memory. It is recalled that the laser is such an interesting instrument because all photons are emitted in phase. More exactly, the difference in phases between emitted waves is a constant in time. Thus there is a so-called *coherence* in the light beam. There are other cases in which phase coherence occurs. For example, light bugs emit coherently [26]; young girls in dormitories have their period in a coherent way; driving conditions are best if some coherence is imposed, as on Belgian highways to and from the seashore at peak hours during summer time; crystals have a better shape and properties if they are grown in a coherent way; sand piles and stock markets would also appear to have coherent properties [27,28].

10.2.2   Roughness

A wave is clearly characterized by its *amplitude*. Indeed this also has some importance in measurements, often as the measure of an intensity (the square of the amplitude). More interestingly, one can define the *roughness* of a profile by observing how the signal amplitude varies in time (and space, if necessary), and in particular the correlation between the various amplitude fluctuations.

Fig. 10.2. Evolution of the exchange rate DEM/JPY from 1 January, 1993 to 30 June, 2000

10.2.3 Persistence

On the other hand, it is easy to show that a periodic signal can be decomposed into a series of sin or cos functions for which the frequencies are in arithmetic order. Thus, the third 'parameter' describing the wave is its *frequency*. Some (regular) frequency effect is apparent in all cycling phenomena, starting with biology, climatology, meteorology, astronomy, and including stock markets, foreign currency exchange markets, tectonic events, traffic and turbulence, and politics. For non-periodic signals, the Fourier transform has been introduced in order to sort out the distribution of frequencies of interest, i.e., the 'density of modes'. The distribution defines the type of *persistence* displayed by a phenomenon. We may also ask whether the frequencies are distributed in a geometrical progression, rather than following an ordinary or typical arithmetic progression, i.e., whether the phenomena might be log-periodic, as they are for antennas, earthquakes and stock market crashes [29,30].

10.3 Power Law Exponents

10.3.1 Persistence and Spectral Density

Data from a (usually discrete) time series $y(t)$ are one-dimensional sets and are simpler to analyze than spatial ones [31,32]. For simplicity, we assume that measurements are taken at equal time intervals. Thus for financial time series, there is no holiday or weekend. A more general situation is hardly necessary here. Two classic examples of mathematically univariate stochastic time series are those resulting from Brownian motion and Levy walks [33]. In both cases, the power spectral density $S_1(f)$ of the time series $y(t)$ (assumed self-affine) has a single power law dependence on the frequency f,

$$S_1(f) \sim f^{-\beta} , \tag{10.1}$$

following from the Fourier transform

$$S_1(f) = \int dt\, e^{ift} y(t) . \tag{10.2}$$

For $y(t)$, one can use the Weierstrass–Mandelbrot (fractal) function [6]

$$\mathcal{W}(t) = \sum_m \gamma^{(2-D)m} \left[1 - e^{i\gamma^n t}\right] e^{i\phi_n t} , \tag{10.3}$$

with $\gamma > 1$, and $1 < D < 2$. The phase ϕ_n can be stochastic or deterministic. For illustration, Berry and Lewis [6] took $\phi_n = n\mu$, with $\mu = 0$ or π. The function obeys

$$\mathcal{W}(\gamma t) = \gamma^{(2-D)} e^{-i\gamma^{\mu}} \mathcal{W}(t) . \tag{10.4}$$

It is stationary and the trend of the real and imaginary parts is given in the deterministic cases by

$$\operatorname{Re} \mathcal{W}(t) \sim \operatorname{Im} \mathcal{W}(t) \sim t^{(2-D)}/\ln\gamma . \tag{10.5}$$

To lowest order, the power spectrum is easily calculated [6] to be

$$S_1(f) \sim (1/\ln\gamma)\, f^{2D-5} . \tag{10.6}$$

One could ask whether the moments of the function obey power laws with characterizing exponents. Equation (10.1) allows one to put the phenomena into the *self-affine* class of persistent phenomena, characterized by the value of β. The range over which β is well defined in (10.1) indicates the range of persistence in the time series. A Brownian motion is characterized by $\beta = 2$, and white noise by $\beta = 0$. (Notice that the differences between adjacent values of a Brownian motion amplitude result in white noise.) A very interesting set of such mathematical signals and the corresponding power spectrum is given in [12].

The Fourier transforms, or power spectra, of the financial signals used for illustration here are shown in Figs. 10.3 and 10.4. Notice the large error bars, allowing us to estimate that β is roughly equal to 2, as for the trivial Brownian motion case. However, the coherence and/or roughness aspect are masked in this one-shot analysis. Only the persistence behavior is touched upon.

If the distribution of fluctuations is not a power law, or if marked deviations exist, say the *statistical correlation coefficient* is less than 0.99, indicating that a mere power law for $S(f)$ is doubtful, a more thorough search of the basic frequencies is in order. A crucial step is to extract deterministic or stochastic components [1], e.g., the stochastic aspects found in the statistical distribution of

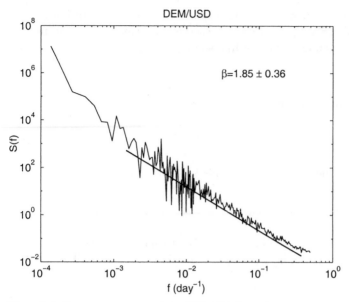

Fig. 10.3. Power spectrum of the DEM/USD exchange rate for the data in Fig. 10.1

values show persistence to be either nonexistent (white noise case) or existent, i.e., $\beta \neq 2$. If so, the persistence can be qualitatively thought of as strong or weak. The Fourier transform can sometimes indicate the presence of specific frequencies, much more abundant than others, especially if cycles exist, as in meteorology and climatology.

The range of persistence is obtained from correlations between events. A short or long range is checked through the autocorrelation function, usually c_1. This function is a particular case of the so-called qth order structure function [34] or qth order height–height correlation function of the (normalized) time-dependent signal $y(t_i)$,

$$c_q(\tau) = \frac{\langle |y(t_{i+r}) - y(t_i)|^q \rangle_\tau}{\langle |y(t_i)|^q \rangle_\tau} , \qquad (10.7)$$

where only non-zero terms are considered in the average $\langle \ \rangle_\tau$ taken over all pairs (t_{i+r}, t_i) such that $\tau = |t_{i+r} - t_i|$. In so doing, one can obtain a set of exponents β_q.

If the autocorrelation is larger than unity for some long time one can talk about strong persistence, otherwise it is weak. This criterion defines the time regime in $y(t)$ for which there is long or short (time) range persistence. Notice that the lower limit of the time regime is due to the discretization step, and this sets the highest frequency as the inverse of twice the discretization interval. The upper limit is obvious.

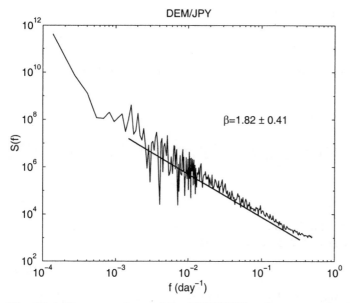

Fig. 10.4. Power spectrum of the DEM/JPY exchange rate for the data in Fig. 10.2

10.3.2   Roughness, Fractal Dimension, Hurst Exponent, and Detrended Fluctuation Analysis

The fractal dimension [2–4,35] D is often used to characterize the roughness of profiles [35]. Several methods are used for measuring D, like the box-counting method, but are not all efficient; many others are found in the literature, as in [2–4,33] and below. For topologically one-dimensional systems, the fractal dimension D is related to the exponent β by

$$\beta = 5 - 2D . \tag{10.8}$$

Brownian motion is characterized by $D = 3/2$, and white noise by $D = 2$ [6,12].

Another measure of signal roughness is sometimes given by the Hurst exponent H, first defined in Hurst's rescale range theory [36,37] for Nile flooding and drought amplitudes. The Hurst method consists in listing the differences between the observed value at a discrete time t over an interval with size N on which the mean has been taken. The upper (y_M) and lower (y_m) values in that interval define the range $R_N = y_M - y_m$. The root mean square deviation S_N is also calculated, and the rescaled range R_N/S_N is expected to behave like N^H. This means that for a (discrete) self-affine signal $y(t)$, the neighborhood of a particular point on the signal can be rescaled by a factor b using the roughness (or Hurst [3,4]) exponent H and defining the new signal $b^{-H}y(bt)$. For the exponent value H, the frequency dependence of the signal so obtained should be indistinguishable from the original one, i.e., $y(t)$.

The exponent H can be calculated as

$$H = 1 + H_1 , \tag{10.9}$$

from the height–height correlation function $c_1(\tau)$ assumed to behave according to

$$c_1(\tau) = \langle |y(n_{i+r}) - y(n_i)| \rangle_\tau \sim \tau^{H_1} , \tag{10.10}$$

rather than from the box-counting method. For a *persistent* signal, $H_1 > 1/2$, whilst for an *anti-persistent* signal, $H_1 < 1/2$. Flandrin has proved theoretically [38] that

$$\beta = 2H - 1 , \tag{10.11}$$

so that $\beta = 1 + 2H_1$. This implies that the classical random walk (Brownian motion) is such that $H = 3/2$. It is clear that

$$D = 3 - H . \tag{10.12}$$

Fractional Brownian motion values are found to lie between 1 and 2 [17,18,27]. Since white noise is a truly random process, it can be concluded that $H = 1.5$ implies an uncorrelated time series [33].

Thus $D > 1.5$ or $H < 1.5$ implies antipersistence, whilst $D < 1.5$ or $H > 1.5$ implies persistence. From preimposed H values of a fractional Brownian motion series, it is found that the above equality usually holds true in a very limited range and β converges only slowly towards the value H [12].

The inertia axes of the *2-variability diagram* [39–41] seem to be related to these values and could be used for fast measurements as well.

A generalized Hurst exponent H_q is defined through the relation

$$c_q(\tau) \propto \tau^{qH_q} , \quad q \geq 0 , \tag{10.13}$$

where $c_q(\tau)$ has been defined above.

The above results can be compared with those obtained from the detrended fluctuation analysis (DFA) [27,42] method. DFA [20] consists in dividing a random variable sequence $y(n)$ over N points into N/τ boxes, each containing τ points. The best linear trend $z(n) = an + b$ in each box is defined. The fluctuation function $F(\tau)$ is then calculated from

$$F^2(\tau) = \frac{1}{\tau} \sum_{n=(k-1)\tau+1}^{k\tau} |y(n) - z(n)|^2 , \quad k = 1, 2, \ldots, N/\tau . \tag{10.14}$$

Averaging $F(\tau)^2$ over the N/τ intervals gives the fluctuations $\langle F(\tau)^2 \rangle$ as a function of τ. If the $y(n)$ data are random uncorrelated variables or short range correlated variables, the behavior is expected to be a power law

$$\langle F^2 \rangle^{1/2} \sim \tau^h , \tag{10.15}$$

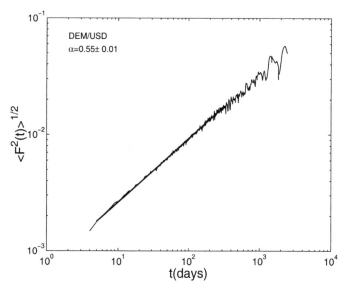

Fig. 10.5. The DFA result for the DEM/USD exchange rate data in Fig. 10.1

with the so-called Hausdorff [3,33] exponent h different from 0.5. It is expected, although not always proved, as emphasized by [6,43], that

$$h = 2 - D , \tag{10.16}$$

where D is the self-affine fractal dimension [2,33]. It is immediately seen that

$$\beta = 1 + 2h . \tag{10.17}$$

For Brownian motion, $h = 0.5$, while for white noise $h = 0$ and $D = 2$.

The DFA log–log plots of the DEM/JPY and DEM/USD exchange rates are given in Figs. 10.5 and 10.6. It is seen that the value of h fulfills the above relations for the DEM/JPY and DEM/USD data, since $h = 0.55$. Both h and D readily measure the roughness and persistence strength. Fractional Brownian motion h values are found to lie between 0 and 1. The effect of a trend is supposedly eliminated here. However, only linear or cubic detrending have been studied to my knowledge [42]. Other trends emphasizing some characteristic frequency, like seasonal cycles, could be studied further.

The *intermittency* of the signal can be studied through the so-called singular measure analysis of the small-scale gradient field, obtained from the data through

$$\varepsilon(r; l) = \frac{r^{-1} \sum_{i=l}^{l+r-1} |y(t_{i+r}) - y(t_i)|}{\langle |y(t_{i+r}) - y(t_i)| \rangle} , \tag{10.18}$$

where

$$l = 0, \ldots, \Lambda - r \tag{10.19}$$

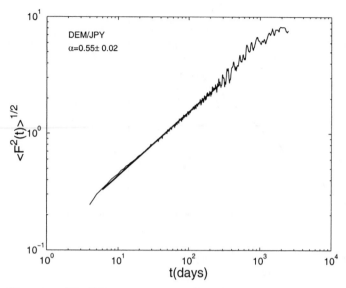

Fig. 10.6. The DFA result for the DEM/JPY exchange rate data in Fig. 10.2

and

$$r = 1, 2, \ldots, \Lambda = 2^m \, , \tag{10.20}$$

and m is an integer. The scaling properties of the generating function are then sought using

$$\chi_q(\tau) = \langle \varepsilon(r; l)^q \rangle \sim \tau^{-K_q} \, , \quad q \geq 0 \, , \tag{10.21}$$

with τ as defined above.

The exponent K_q is closely related to the generalized dimensions $D_q = 1 - K_q/(q-1)$ [44]. The nonlinearity of both characteristic exponents, qH_q and K_q, describes the multifractality of the signal. If a linear dependence is obtained, then the signal is monofractal or in other words, the data follows a simple scaling law for these values of q. Thus the exponent [34,45]

$$C_1 = - \left| \frac{dK_q}{dq} \right|_{q=1} \tag{10.22}$$

is a measure of the intermittency in the signal $y(n)$ and can be numerically estimated by measuring K_q around $q = 1$. Some conjectures on the role/meaning of H_1 can be found in [27]. From some financial and political data analysis, it seems that H_1 is a measure of the information entropy of the system.

10.4   Conclusion

It has been emphasized that the analysis of stochastic time series, such as those describing fractional Brownian motion and foreign currency exchange rates, reduces to examining the distribution of signal amplitude variations, together with correlations between amplitudes, frequencies and phases of harmonic-like components of the signal. Due to some scaling hypothesis, characteristic exponents can be obtained to describe power laws. These exponents serve to determine universality classes, and in short to build physical or algorithmic models. In the case of financial time series, the exponents can even serve to devise an investment strategy [27]. Notice that, due to the non-stationarity of the data, such exponents vary with time, and multifractal concepts must be brought in at a refining stage – even when devising an investment strategy.

In this spirit, we should emphasize the analogy between a H_1, C_1 diagram and the ω, k diagram of dynamical second-order phase transitions [46]. In the latter, the frequency and phase of a time signal are considered on the same footing, and encompass the critical and hydrodynamical regions. In the present cases, an analog diagram relates the roughness and intermittency. This has already been examined in [47,48].

Among various other physical data analysis techniques which have been presented recently in order to obtain some information on the deterministic and/or chaotic content of univariate data, we should mention the wavelet technique. This has been widely used (several references exist, see below) for DNA and meteorology studies [45,49–52]. The H_1, C_1 technique used in turbulence and meteorology [53] is also relevant.

Many other techniques have not been mentioned here, such as the weighted fixed point [54], and the time-delay embedding [55]. Several others have also been used, but only those relevant to the present purpose have been discussed above. As a summary, Table 10.1 indicates the range of values found for different signals and their relationship to stationarity, persistence and coherence. In the years to come, it seems relevant to ask for more data on multivariate functions, thus extending the above considerations to other real mathematical and physical cases in higher dimensions.

Table 10.1. Values of the most relevant exponents in various regimes (i.e., stationary, persistent, antipersistent) of univariate stochastic series: D fractal dimension, h Hausdorff measure, H Hurst exponent, α from the DFA technique, β power spectrum exponent

Signal name	D	h	H	α	β	
–	–	–	0	–	-1	–
–	–	–	–	Antipersistent	–	Stationary
–	–	–	0.5	–	0	Uncorrelated
–	–	–	–	Persistent	–	Stationary
White noise	2	0	1	–	1	–
Fractional Brownian motion	–	–	–	Superpersistent	–	Non-stationary
Brownian motion	3/2	0.5	3/2	–	2	–
Flat spectrum	1	1	2	Superpersistent	3	–

Finally, two appendixes follow. Another time series analysis technique often used by experts for prediction purposes, viz., the *moving average technique* is discussed briefly in Sect. 10.5.1. In addition, these notes would be incomplete without mentioning the intrinsic *discrete scale invariance* implication in time series. Such a substructure leads to log-periodic oscillations in the time series, and whence to fascinating effects and surprises in predicting crash-like events. This is mentioned in Sect. 10.5.2.

10.5   Appendixes

10.5.1   Moving Averages

It can be shown that it is irrelevant to take into account the trend for short-range correlation events. In any case, the trend is quite ill-defined since it is a statistical mean, and thus depends on the size of the interval, i.e., the number of data points which is taken into account. Nevertheless, many technical analyses, like the moving average method [56], rely on signal averages over various time intervals. These methods should be examined from a physical point of view. An interesting observation has resulted from checking the density of intersections of such mean values over different time interval windows, which are continuously shifted. This corresponds to obtaining a spectrum of the so-called moving averages [56], used by analysts in order to point to 'gold' or 'death' crosses in a market. The density ρ of crossing points between any two moving averages is obviously a measure of long-range power law correlations in the signal. It has been found that ρ is a symmetric function of the difference between the interval sizes on which the averages are taken, and it has a simple power law form [57,58]. This leads to a very fast and rather reliable measure of the fractal dimension of the signal. The method is easily implemented to obtain the time evolution of D, and hence applied to elementary investment strategies.

10.5.2   Discrete Scale Invariance

It has been proposed that an economic index $y(t)$ follows a complex power law [59,60], i.e.,

$$y(t) = A + B(t_c - t)^{-m}\{1 + C\cos[\omega \ln(1 - t/t_c) + \phi]\}, \qquad (10.23)$$

for $t < t_c$, where t_c is the crash time or rupture point, and A, B, m, C, ω, ϕ are parameters. This index evolution is a power law divergence (for $m > 0$) on which log-periodic oscillations are taking place. The law for $y(t)$ diverges at $t = t_c$ with an exponent m (for $m > 0$), while the period of the oscillations converges to the rupture point at $t = t_c$. This law is similar to that of critical points at second-order phase transitions [46], but generalizes the scaleless situation for cases in which discrete scale invariance is presupposed [61]. This relationship has already been proposed to fit experimental measurements of sound wave rate emissions prior to the rupture of heterogeneous composites stressed up to failure [62]. The same type of complex power law behavior has been observed as a precursor of the Kobe earthquake in Japan [63].

Fits using (10.23) were performed on the S&P500 data [59,60] for the period preceding the 1987 October crash. The parameter values have not been found to be robust against small perturbations, like a change in the phase of the signal. It is known that a nonlinear seven parameter fit is highly unstable from a numerical point of view. Indeed, eliminating the contribution of the oscillations in (10.1), i.e., setting $C = 0$, implies that the best fit leads to an exponent $m = 0.7$, significantly larger than $m = 0.33$ for $C \neq 0$ [59]. Feigenbaum and Freund [60] also reported various values of m ranging from 0.53 to 0.06 for different indexes and events (upsurges and crashes).

Universality in this case means that the value of m should be the same for any crash and for any index. So a single model should describe the phase transition, and the exponent would define the model and be the only parameter. A limiting case of power law behavior is logarithmic behavior, corresponding to $m = 0$, i.e., the divergence of the index y for t close to t_c should be

$$y(t) = A + B \ln(1 - t/t_c)\{1 + C \cos[\omega \ln(1 - t/t_c) + \phi]\} . \qquad (10.24)$$

This logarithmic behavior is known in physics to characterize the specific heat (four-point correlation function) of the Ising model, and the Kosterlitz–Thouless phase transition [64] in two spatial dimensions. It is thus specific to systems with a low number of components in the order parameter. It is nevertheless a smooth transition. The mean value of the order parameter [46] is not defined over long-range scales, but a phase transition nevertheless exists because there is some ordered state on small scales. In addition to the physical interpretation of the latter relationship, the advantages are that

- the number of parameters is reduced by one,
- the log-divergence seems to be close to reality.

In order to test the validity of (10.24) in the vicinity of crashes, we have separated the problems of the divergence itself and the oscillation convergences, in order to extract two values for the rupture point t_c:

- t_c^{div} for the power (or logarithmic) divergence,
- t_c^{osc} for the oscillation convergence.

In so doing, the long-range and short-range fluctuation scales are examined on an equal footing. The final t_c is obtained at the intersection of two straight lines, by successive iteration fits. The results of the fit, as well as the correlation fitting factor, are given in [65,66].

A technical point is in order. The rupture point t_c^{osc} is estimated by selecting the maxima and minima of the oscillations through a double envelope technique [66]. Finally, notice that the basic oscillation frequency depends on the connectivity of the underlying space [18,19]. The log-periodic behavior also corresponds to a complex fractal dimension [61,67].

Acknowledgements. The author would like to thank K. Ivanova for very useful discussions, and for preparing the data and drawings. A. Pękalski kindly reviewed the manuscript and has much improved it due to his typical and well known pedagogical insight. Warm thanks also go to the organizers of the WE-Heraeus-Ferienkurs für Physik 2000 in Chemnitz, for inviting me to present these ideas and techniques, and to the students for their questions, comments and output during the exercises.

References

1. C.J. Cellucci, A.M. Albano, P.E. Rapp, R.A. Pittenger, R.C. Josiassen: Chaos **7**, 414 (1997)
2. M. Schroeder: *Fractals, Chaos and Power Laws* (W.H. Freeman and Co., New York 1991)
3. P.S. Addison: *Fractals and Chaos* (Inst. of Phys., Bristol 1997)
4. K.J. Falconer: *The Geometry of Fractal Sets* (Cambridge University Press, Cambridge 1985)
5. P. Bergé, Y. Pomeau, Ch. Vidal: *L'ordre dans le chaos* (Hermann, Paris 1984)
6. M.V. Berry, Z.V. Lewis: Proc. R. Soc. Lond. A **370**, 459 (1980)
7. S.L. Meyer: *Data Analysis for Scientists and Engineers* (Wiley, New York 1975)
8. P.E. Rapp: Integrat. Physiol. Behav. Sci. **29**, 311 (1994)
9. Th. Schreiber: Phys. Rep. **308**, 1 (1999)
10. H. Kantz, Th. Schreiber: *Nonlinear Time Series Analysis* (Cambridge University Press, Cambridge 1997)
11. C. Diks: *Nonlinear Time Series Analysis* (World Scientific, Singapore 1999)
12. B.D. Malamud, D.L. Turcotte: J. Stat. Plann. Infer. **80**, 173 (1999)
13. P.J. Brockwell, R.A. Davis: *Introduction to Time Series and Forecasting*, Springer Text in Statistics (Springer, Berlin 1998)
14. Ph. Franses: *Time Series Models for Business and Economic Forecasting* (Cambridge University Press, Cambridge 1998)
15. Ch. Gourieroux, A. Monfort: *Time Series and Dynamic Models* (Cambridge University Press, Cambridge 1997)
16. D. Blake: *Financial Market Analysis* (Wiley, New York 2000)
17. M. Ausloos, N. Vandewalle, K. Ivanova. In: *Noise of Frequencies in Oscillators and Dynamics of Algebraic Numbers*, ed. by M. Planat (Springer, Berlin 2000) pp. 156–171
18. M. Ausloos, N. Vandewalle, Ph. Boveroux, A. Minguet, K. Ivanova: Physica A **274**, 229 (1999)
19. M. Ausloos: Physica A **285**, 48 (2000)
20. R.N. Mantegna, S.V. Buldyrev, A.L. Goldberger, S. Havlin, C.-K. Peng, M. Simmons, H.E. Stanley: Phys. Rev. Lett. **73**, 3169 (1994)
21. S. Mercik, K. Weron, Z. Siwy: Phys. Rev. E **60**, 743 (1999)
22. B.J. West, R. Zhang, A.W. Sanders, S. Miniyar, J.H. Zuckerman, B.D. Levine: Physica A **270**, 552 (1999)
23. K. Ivanova, M. Ausloos: Physica A **274**, 349 (1999)
24. N. Vandewalle, M. Ausloos, M. Houssa, P.W. Mertens, M.M. Heyns: Appl. Phys. Lett. **74**, 1579 (1999)
25. T. Lundahl, W.J. Ohley, S.M. Kay, R. Siffert: IEEE Trans. Med. Imag. **MI-5**, 152 (1986)

26. J.F. Lennon: Pour la Science, special issue (January 1995) p. 111
27. N. Vandewalle, M. Ausloos: Physica A **246**, 454 (1997)
28. G. Bonanno, N. Vandewalle, R. Mantegna: Phys. Rev. E **62**, R7615 (2000)
29. A. Johansen, D. Sornette, H. Wakita, U. Tsunogai, W.I. Newman, H. Saleur: J. Phys. I France **6**, 1391 (1996)
30. N. Vandewalle, M. Ausloos, Ph. Boveroux, A. Minguet: Eur. Phys. J. B **9**, 355 (1999)
31. S. Prakash, G. Nicolis: J. Stat. Phys. **82**, 297 (1996)
32. J.E. Wesfreid, S. Zaleski: *Cellular Structures in Instabilities*, Lect. Notes Phys. **210** (Springer, Berlin 1984)
33. B.J. West, B. Deering: *The Lure of Modern Science: Fractal Thinking* (World Scientific, Singapore 1995)
34. A.L. Barabási, T. Vicsek: Phys. Rev. A **44**, 2730 (1991)
35. B.B. Mandelbrot, D.E. Passoja, A.J. Paulay: Nature **308**, 721 (1984)
36. H.E. Hurst: Trans. Amer. Soc. Civ. Engin. **116**, 770 (1951)
37. H.E. Hurst, R.P. Black, Y.M. Simaika: *Long Term Storage* (Constable, London 1965)
38. P. Flandrin: IEEE Trans. Inform. Theory, **35** (1989)
39. K. Ivanova, M. Ausloos: Physica A **265**, 279 (1999)
40. A. Babloyantz, P. Maurer: Phys. Lett. A **221**, 43 (1996)
41. K. Ivanova, M. Ausloos: Physica A **274**, 349 (1999)
42. N. Vandewalle, M. Ausloos: Int. J. Comput. Anticipat. Syst. **1**, 342 (1998)
43. B.B. Mandelbrot: Proc. Natn. Acad. Sci. USA **72**, 3825 (1975)
44. H.P.G.E. Hentschel, I. Procaccia: Physica D **8**, 435 (1983)
45. A. Davis, A. Marshak, W. Wiscombe. In: *Wavelets in Geophysics*, ed. by E. Foufoula-Georgiou, P. Kumar (Academic Press, New York 1994) p. 249; A. Marshak, A. Davis, R. Cahalan, W. Wiscombe: Phys. Rev. E **49**, 55 (1994)
46. H.E. Stanley: *Phase Transitions and Critical Phenomena* (Oxford University Press, Oxford 1971)
47. B. Mandelbrot, J.R. Wallis: Water Resour. Res. **5**, 967 (1969)
48. J.B. Bassingthwaighte, G.M. Raymond: Ann. Biomed. Engin. **22**, 432 (1994)
49. E. Bacry, J.F. Muzy, A. Arneodo: J. Stat. Phys. **70**, 635 (1993)
50. A. Arneodo, E. Bacry, J.F. Muzy: Physica A **213**, 232 (1995)
51. Z.R. Struzik. In: *Fractals: Theory and Applications in Engineering*, ed. by M. Dekking, J. Levy-Vehel, E. Lutton, C. Tricot (Springer, Berlin 1999)
52. E. Koscielny-Bunde, A. Bunde, S. Havlin, H.E. Roman, Y. Goldreich, H.J. Schellnhuber: Phys. Rev. Lett. **81**, 729 (1998)
53. K. Ivanova, T. Ackerman: Phys. Rev. E **59**, 2778 (1999)
54. V.I. Yukalov, S. Gluzman: Int. J. Mod. Phys. B **13**, 1463 (1999)
55. L. Cao: Physica A **247**, 473 (1997)
56. A.G. Ellinger: *The Art of Investment* (Bowers & Bowers, London 1971)
57. N. Vandewalle, M. Ausloos: Phys. Rev. E **58**, 6832 (1998)
58. N. Vandewalle, M. Ausloos, Ph. Boveroux: Physica A **269**, 170 (1999)
59. D. Sornette, A. Johansen, J.-P. Bouchaud: J. Phys. I France **6**, 167 (1996)
60. J.A. Feigenbaum, P.G.O. Freund: Int. J. Mod. Phys. B **10**, 3737 (1996)
61. D. Sornette: Phys. Rep. **297**, 239 (1998)
62. J.C. Anifrani, C. Le Floc'h, D. Sornette, B. Souillard: J. Phys. I France **5**, 631 (1995)
63. A. Johansen, D. Sornette, H. Wakita, U. Tsunogai, W.I. Newman, H. Saleur: J. Phys. I France **6**, 1391 (1996)

64. J.M. Kosterlitz, D.J. Thouless: J. Phys. C **6**, 1181 (1973)
65. N. Vandewalle, M. Ausloos: Eur. J. Phys. B **4**, 139 (1998)
66. N. Vandewalle, M. Ausloos, Ph. Boveroux, A. Minguet: Eur. Phys. J. B **9**, 355 (1999)
67. D. Bessis, J.S. Geronimo, P. Moussa: J. Physique Lett. **44**, L-977 (1983)

11 'Go with the Winners' Simulations

Peter Grassberger and Walter Nadler

Summary. We describe a general strategy for sampling configurations from a given distribution (Gibbs–Boltzmann or other). It is not based on the Metropolis concept of establishing a Markov process whose stationary state is the desired distribution. Instead, it builds weighted instances according to a biased distribution. If the bias is optimal, all weights are equal and importance sampling is perfect. If not, 'population control' is applied by cloning/killing configurations whose weight is too high/low. It uses the fact that nontrivial problems in statistical physics are high-dimensional. Consequently, instances are built up in many steps, and the final weight can be guessed at an early stage. In contrast to evolutionary algorithms, the cloning/killing is done in such a way that the desired distribution is strictly observed without simultaneously keeping a large population in computer memory. We apply this method (closely related to diffusion-type quantum Monte Carlo methods) to several problems of polymer statistics, population dynamics, and percolation.

11.1 Introduction

For many statistical physicists, the term 'Monte Carlo' is synonymous with the Metropolis strategy [1] where one sets up an ergodic Markov process which has the desired Gibbs–Boltzmann distribution as its unique asymptotic state. There exist numerous refinements concerned with more efficient transitions in the Markov process (e.g., cluster flips [2] or pivot moves [3]), or with distributions biased in such a way that false minima can be more easily escaped from and autocorrelations are reduced (e.g., multicanonical sampling [4] and simulated tempering [5]). However, most of these schemes remain entirely within the framework of the Metropolis strategy.

On the other hand, stochastic simulations that are not based on the Metropolis strategy have been used from early on. Well known examples are evolutionary (in particular genetic) algorithms [6–8], diffusion-type quantum Monte Carlo simulations [9–11], and several algorithms devised for the simulation of long chain molecules [12–16]. However, these methods were developed independently in different communities and it was not in general recognized that they are realizations of a common strategy. Perhaps the first to point this out clearly were Aldous and Vazirani [17] who also coined the name 'go with the winners'. For later references, which also stress the wide range of possible applications for this strategy, see [18,19]. Reference [19] even points to applications in lattice spin systems and Bayesian inference, fields which will not be treated in the present review.

As we shall see, the main drawbacks of the go-with-the-winners strategy are as follows.

- The method yields correlated samples, just as the Metropolis method does. This makes a priori error estimates difficult [17]. A posteriori errors, estimated from fluctuations of measured observables, are of course always possible. But they can be very misleading when sampling is so incomplete that the really large fluctuations have not yet been seen. However, there is also a positive side compared with the Metropolis case: it is easier to estimate whether this has happened, and hence whether the method gives reliable results or not.
- Efficiency is not guaranteed. The go-with-the-winners strategy allows a lot of freedom with respect to implementation details, and its efficiency depends on making a good choice amongst them. There are thus cases where it has not yet been successful at all, although in other problems its efficiency has proved unmatched by any other method we aware of. On the other hand, the flexibility of the general strategy represents a strong positive point.

Instead of giving a formal definition of the go-with-the-winners strategy, we shall present an example from which the basic concepts will become clear. In later sections, we shall then see how these concepts are implemented in detail and how they are applied to other problems as well.

11.2   An Example: A Lamb in front of a Pride of Lions

The example here is a very idealized problem from population dynamics (or chemical reactions, if you wish) [20,21]. Consider a 'lamb', represented by a random walker on a 1-dimensional lattice $x = \ldots, -1, 0, 1 \ldots$, with discrete time and hopping rate p per time unit, leading to a diffusion constant D_{lamb}. It starts at time $t = 0$ at $x = 0$. Starting at the same time, there are also N 'lions', n_{L} of them at $x_i = -1$ $(i = 1, \ldots, n_{\mathrm{L}})$ and $n_{\mathrm{R}} = N - n_{\mathrm{L}}$ at $x_i = +1$. They likewise perform random walks, but with a diffusion constant D_{lion} which may differ from D_{lamb}. Two lions can jump onto the same site without interacting with each other. But if a lion and the lamb meet at the same site, the lamb is eaten immediately, and the process is finished. Note that both the lamb and the lions are absolutely short-sighted and stupid: there is no question of evasion or pursuit. It is for this reason that the model can also be interpreted as the capture of a diffusing molecule by diffusing adsorbers.

The survival probability $P(t)$ of the lamb up to time t can be estimated easily for a single lion, $N = 1$. In this case the relative distance makes a random walk with diffusion constant $D_{\mathrm{lamb}} + D_{\mathrm{lion}}$ which starts at $\Delta x = 1$, and $P(t)$ is equal to the probability that the walk has not yet hit an absorbing wall at $\Delta x = 0$. This probability is well known to decrease as $t^{-1/2}$, thus

$$P_{N=1} \sim t^{-1/2} \, . \tag{11.1}$$

The problem is less trivial but still solvable for $N = 2$ (see [21] and the literature quoted there). One again finds a power law

$$P_{N=2} \sim t^{-\alpha_2} , \qquad (11.2)$$

but with an exponent which depends on the ratio D_{lamb}/D_{lion} and on whether both lions are on the same side or on different sides of the lamb. For the first case one gets

$$\alpha_2 = \left[2 - \frac{2}{\pi} \arccos \frac{D_{lamb}}{D_{lamb} + D_{lion}} \right]^{-1} , \quad N = n_R = 2 , \quad n_L = 0 . \qquad (11.3)$$

For the case with both lions on opposite sides, one obtains a similar expression. If $D_{lamb} = D_{lion}$, (11.3) reduces to $\alpha_2 = 3/4$.

For any $N > 2$ one can still prove rigorously that, for large t,

$$P_N \sim t^{-\alpha_N} \qquad (11.4)$$

holds asymptotically, but this time one cannot give closed expressions for α_N. Numerical values for the case where all lions are on one side have been obtained for several N by direct simulation ($\alpha_3 \approx 0.91$, $\alpha_4 \approx 1.03$, $\alpha_{10} \approx 1.4$ [20]), but these estimates become more and more difficult for increasing N because of the exceedingly small chance for the lamb to survive sufficiently long to allow precise measurements.

Such numerical estimates would be welcome in order to test an asymptotic estimate for $N \to \infty$ [21]. In this limit, the location of the *outermost* of a group of lions moves nearly deterministically. If the lion at the front at time t lags behind, there will be another lion that overtakes it, so that the front continues to move on with maximal speed. Assuming that the fluctuations in the motion of the front can be neglected, the authors of [21] find

$$\alpha_N \approx \frac{D_{lion}}{4D_{lamb}} \ln N , \quad N = n_R \gg 1 , \quad n_L = 0 . \qquad (11.5)$$

In the same spirit, the optimal strategy of a lamb squeezed between $N/2$ lions to its left and $N/2$ to its right would be to stand still. Assuming that this single path dominates in the limit $N \to \infty$, one finds simply $\alpha_N \approx N/2$.

As we said, straightforward simulations to check these predictions are inefficient. In order to improve efficiency, one can think of two tricks.

Trick 1: Make occasional 'enrichment' steps.

In particular, this might mean that one starts with $M \gg 1$ instances. As soon as the number of surviving instances has decayed to a number $< M/2$, one makes a clone of each instance. (Note that lambs can be cloned also in reality, but on the computer we clone the entire configuration consisting of lamb and lions!) This boosts the number of instances again up to $\approx M$, and one can repeat the game. One just has to remember how often the sample has been enriched when computing survival probabilities, i.e., each instance generated carries a relative statistical weight $w = 1/2^c$, with c the number of cloning steps.

> **Trick 2:** Replace the random walks by biased random walks

Not only should the lamb preferentially run away from the lions, but also the lions should run away from the lamb in order to obtain long-lived samples that contribute to (11.4). If just this were done without compensation, it would of course give incorrect results. But we can correct for this bias by giving *weights* to each instance. For each step of either the lamb or the lion, made according to a biased pair of probabilities $\{p_L, p_R\}$ instead of $\{p, p\}$, we should multiply the weight by a factor p/p_R if the actual step was to the right, and by a factor p/p_L if the step was to the left. In this way, the weights compensate exactly, on average, the fact that not all walks were sampled with the same probability. This trick is indeed very general. In any sampling procedure where some random move should be made with probability $p > 0$ in order to obtain an unbiased sample, one can replace p by any other probability $p' \neq 0$, if at the same time we use weighted samples and multiply the current sample weight by p/p'.

Actually, in view of the second trick, the first one is clearly not optimal. Instead of cloning irrespective of its weight, one would like to clone preferentially those configurations which have high weight. We thus replace the first trick by

> **Trick 1':** Clone only configurations with high weight.

We choose a cloning threshold $W_+(t)$. It can in principle be an arbitrary function of t, and it need not be kept fixed during the simulation; it can thus be optimized on-line. Good choices will be discussed later. If a configuration at some time t has weight $w > W_+(t)$, it is cloned and both clones are given weight $w/2$.

On the other hand, too strong a bias and too frequent cloning could result in configurations which have too little weight. Such configurations are just costly in terms of CPU time, without adding much to the precision of the result. But we are not allowed to kill ('prune') them straight away, since they do still carry some weight. Instead, we use

> **Trick 3:** Kill configurations with low weight in a probabilistic manner.

We choose a pruning threshold $W_-(t)$. The same remarks apply to it as to $W_+(t)$. If $w < W_-(t)$, we call a random number r uniformly in $[0, 1]$. If $r < 1/2$, we prune. Otherwise, if $r > 1/2$, we keep the configuration and double its weight, $w \to 2w$. Again this does not introduce a bias, as far as averages are concerned.

In principle, that is all. One can modify the tricks 1', 2 and 3 by making more than one clone at each enrichment step, by killing with probability $\neq 1/2$, or by letting $W_\pm$ depend also on other variables. Whether such further improvements are helpful will depend on the problem at hand. In the case of lamb & lions, it seems they are not. Indeed, in this problem, even pruning is not needed if the bias is not too strong, although this is a somewhat special case.

Before going on and describing the detailed implementation, let us just see some results. Probabilities $P_N(t)$ for all lions at the same side and the resulting

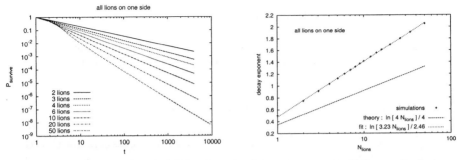

Fig. 11.1. *Left*: survival probabilities for a lamb starting next to N lions, all of whom are on the same side. Lamb and lions both make ordinary random walks with $D_{\text{lion}} = D_{\text{lamb}} = 1/2$. *Right*: corresponding decay exponents. The *lower dashed line* represents the prediction from (11.5)

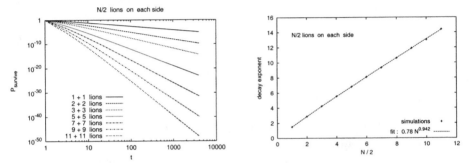

Fig. 11.2. Same as Fig. 11.1 but for $N/2$ lions on each side of the lamb. This time, neglecting fluctuations of the front of the group of lions would give $\alpha_N = N/2$

decay exponents are shown in Fig. 11.1, for N up to 50. We see that (11.5) is qualitatively correct in predicting a logarithmic increase in α_N, but not quantitatively. Obviously, fluctuations of the front of the pride of lions are not negligible. The data on the right show a slight downward curvature. This might be an indication that (11.5) is asymptotically correct, but then asymptotes would set in only at extremely large values of N. The same conclusion is reached when there are $N/2$ lions on either side, as can be seen from Fig. 11.2. In that case the raw data shown on the left clearly demonstrate the power of the algorithm: we are able to obtain reliable estimates of probabilities as small as 10^{-50}, which would have been impossible with straightforward simulation.

11.3 Other Examples

11.3.1 Multiple Spanning Percolation Clusters

Let us now consider percolation [22] on a large but finite rectangular lattice in any dimension $2 \leq d < 6$. We single out one direction as 'spanning direction'. In this direction, boundary conditions are open (surface sites just have no neighbours outside the lattice), while boundary conditions in the other direction(s) may be either open or periodic. Up to about six years ago, there was a general belief, based on a misunderstood theorem, that there is at most one spanning cluster in the limit of large lattice size, keeping the shape of the rectangle fixed ($L_i = x_i L$, $L \to \infty$, $i = 1, \ldots, d$). A 'spanning cluster' is a cluster which touches both boundaries in the spanning direction.

Since there is no spanning cluster for subcritical percolation (with probability decreasing exponentially in the lattice size L), and since there is exactly one in the supercritical case, the relevant case is only critical percolation. For that case, it is now known that the probabilities P_k of having exactly k spanning clusters are all non-zero in the limit $L \to \infty$. In two dimensions, they were calculated exactly by Cardy using conformal invariance [23], but in dimensions ≥ 3 no exact results are known. The only analytical 'result' is a conjecture by Aizenman [24], stating that for a lattice of size $L \times \ldots \times L \times (rL)$ (rL is the length in the spanning direction)

$$P_k \sim e^{-\alpha r} , \tag{11.6}$$

where

$$\alpha \propto k^{d/(d-1)} \qquad \text{for } k \gg 1 . \tag{11.7}$$

Cardy's formula in $d = 2$ agrees with (11.6) and (11.7). For periodic transverse boundary conditions, it is

$$\alpha = \frac{2\pi}{3} \left(k^2 - \frac{1}{4} \right) , \quad k \geq 2 , \quad d = 2 . \tag{11.8}$$

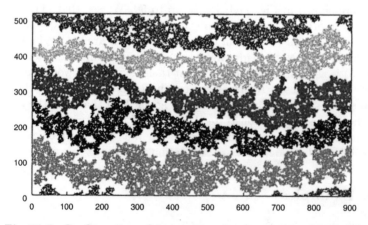

Fig. 11.3. Configuration of 5 spanning site percolation clusters on a lattice of size 500×900. Any two clusters keep a distance of at least 2 lattice units. Lateral boundary conditions are periodic. The probability of finding 5 such spanning clusters in a random disorder configuration is $\approx 10^{-92}$

It has recently been generalized [25] to the case where the clusters are seperated by at least two lattice units (i.e., there are at least two non-intersecting paths on the dual lattice between any two clusters). In that case,

$$\alpha = \frac{2\pi}{3} \left[\left(\frac{3k}{2}\right)^2 - \frac{1}{4} \right] . \tag{11.9}$$

In order to test (11.7)–(11.9) for a wide range of values of k and r, one has to simulate events with tiny probabilities, $\ln P_k \sim -10^2$ to -10^3. It is thus not surprising that previous numerical studies have verified (11.8) only for small values of k, and have been unable to verify or disprove (11.7) [26,27].

In order to demonstrate that such rare events can be simulated efficiently using the go-with-the-winners strategy, we show in Fig. 11.3 a rectangular lattice of size 500×900 with 5 spanning clusters which keep distances ≥ 2. Equation (11.9) predicts $P_k = \exp(-336\pi/5) \approx 10^{-92}$ for this case. The configuration was obtained by letting 5 clusters grow simultaneously, using a standard cluster growth algorithm [28], starting from the left border. Precautions were taken to ensure that they grow with the same speed towards the right, i.e., if one of them lagged behind, the growth of the others was stopped until the lagging cluster had caught up. If one of them died, or if two came closer than two lattice units, the entire configuration was discarded. If not, cloning was done as described in trick 1′. Note that here the growth was made without bias (it is not obvious what this bias should have been), and therefore the weight was just determined by the cloning. Consequently, and since there are no Boltzmann weights, no configuration could get too high a weight, and therefore no pruning was necessary either.

In this way we were able to check (11.8) and (11.9) with high precision. We do not show these data. Essentially they just test the correctness of our algorithm.

More interesting is the test of (11.7) for $d = 3$. Estimated probabilities for up to 16 spanning clusters, on lattices of sizes up to $64 \times 64 \times 2000$, are shown on the left in Fig. 11.4. Note that probabilities are now as small as 10^{-300}. Values of α

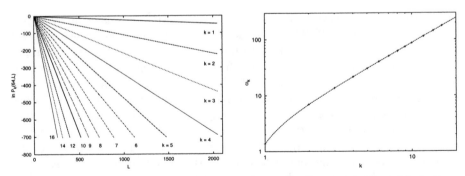

Fig. 11.4. *Left:* probabilities of having k spanning clusters on simple cubic lattices of size $64 \times 64 \times L$, with $k \leq 16$ and $L \leq 2000$. *Right:* decay exponents α_k versus k obtained from the data in the left panel, and from similar data with transverse lattice sizes 16×16, 32×32, and 128×128. The *dashed line* is a fit with $\alpha_k \sim k^{3/2}$ for large k

obtained from these simulations and from similar ones at different lattice sizes (in order to eliminate finite-size corrections) are shown on the right of Fig. 11.4. The dashed line there is a fit [29]

$$\alpha = 2.76(k^2 - 0.61)^{3/4} \, . \tag{11.10}$$

Even if we should not take this fit too seriously, we see clearly that $\alpha \propto k^{3/2}$ for $k \to \infty$, in perfect agreement with (11.7).

11.3.2 Polymers

Another big class of problems where the go-with-the-winners strategy is naturally applied are configurational statistics of long polymer chains. It is well known that linear polymers in good solvents form random coils which differ from random walks by having size

$$R \sim N^\nu \, , \tag{11.11}$$

with $\nu \neq 1/2$. In fact, $\nu = 3/4$ in 2 dimensions, and $\nu \approx 0.588$ in $d = 3$ [30,31]. The canonical model which gives this anomalous scaling is the self-avoiding random walk. Anomalous scaling laws in other universality classes are obtained by attaching polymers to impenetrable boundaries or attractive walls, adding monomer–monomer attraction, etc. Simulating long chain molecules has thus been a vigorous problem since the very early days of electronic computers.

The most straightforward method for simulating a self-avoiding random walk on a regular lattice is to start from one end and to make steps in random directions. As long as no site is visited twice, every configuration should have the same weight. But as soon as a site is visited which has already been visited before, the energy becomes infinite because of hard core repulsion, the weight becomes zero, and the configuration can be discarded. This leads to exponential 'attrition' – the number of generated configurations of length t decreases as $C(t) \sim \exp(-at)$ – and to a very inefficient code.

A first proposal to avoid, or at least reduce this attrition was made by Rosenbluth and Rosenbluth [12]. They proposed to bias the sampling by replacing steps to previously visited sites by steps to unvisited ones, if possible. For example, take a simple cubic lattice. Except for the very first step, there will be at most 5 free neighbours for the next move. If there is no free neighbour at any given moment, the configuration must be discarded. Otherwise, if there are $m \geq 1$ free neighbours, one selects one of them randomly and moves to it. At the same time, in order to compensate for this bias, one multiplies the weight of the configuration by a factor $\propto m$ (the value of the proportionality constant is irrelevant for estimates of averages, and only affects the partition sum in a trivial way).

Although this allows much longer chains to be simulated, the Rosenbluth method is far from perfect because it leads to very large weight fluctuations [32]. As an alternative, enrichment was therefore proposed [13], in the form of trick 1 of Sect. 11.2. But more efficient than either is the full go-with-the-winners strategy with all three steps 1', 2, and 3. Population control (pruning/cloning) is of course done on the basis of full statistical weights, including both Boltzmann and bias correction factors. This was first used in [15] and later with a different implementation in [16]. In the latter, it was called the 'pruned-enriched Rosenbluth method' (PERM).

PERM is particularly efficient near the so-called theta- or coil-globule transition. This transition occurs when we start with a good solvent and make it worse, e.g., by lowering the temperature T. The repulsive interaction between monomers and solvent molecules leads to an effective monomer–monomer attraction which would like to make the polymer collapse into a dense globule. If T is sufficiently high, this is outweighed by the loss of entropy associated with the collapse. But at $T < T_\theta$, the entropy is no longer sufficient to prevent the collapse. According to the generally accepted scenario, the theta point is a tricritical point with upper critical dimension $d = 3$ [30,31].

At the 3-d theta point, bias correction and Boltzmann factors nearly cancel. Hence, long polymers have essentially random walk configurations with very small (logarithmic) corrections. Consequently, an unbiased random walk (with just a non-reversal bias: no 180 degree reversals are allowed) is already sufficient to give good statistics with very few pruning and enrichment events. In [16], chains made of up to $1\,000\,000$ steps were sampled with high statistics within modest CPU time.

Applications of PERM to other polymer problems are treated in [33–44]. We want to discuss here just two applications, namely the 'melting' (denaturation) of DNA [40] and the low energy ('native') states of heteropolymers [35].

DNA Melting. As is well known, in physiological conditions, DNA forms a double helix. Changing the pH value or increasing T can break the hydrogen bonds between the A–T and C–G pairs, and a phase transition occurs to an open coil, with higher energy but also with higher entropy. This transition has been studied experimentally for about 40 years. It seems to be very sharp, and experimental data are consistent with a first order transition [45]. While a second order transition would be easy to explain [46,47], constructing models which give first order transitions has turned out to be much more difficult [48].

The model studied in [40] lives on a simple cubic lattice. A double strand of DNA with length N is described by a diblock copolymer of length $2N$, made of N monomers of type A and N monomers of type B. All monomers have excluded volume interactions, i.e., two monomers cannot occupy the same lattice site, with one exception: the kth A monomer and the kth B monomer can occupy the same site, where $k = 1, \ldots, N$ is their index counted from the center at which both strands join together. If they do so, then they even gain an energy $-\epsilon$. This models the binding of complementary bases.

The surprising result of simulations of chains with N up to 4000 is that the transition is first order, but shows finite scaling behaviour as expected for a second order transition with crossover exponent $\phi = 1$. To demonstrate this, we first show in Fig. 11.5 the specific heat as a function of ϵ at $T = 1$, for

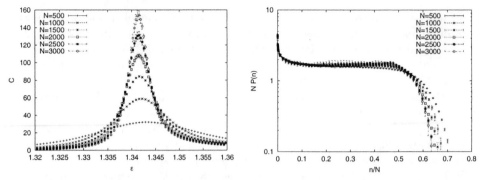

Fig. 11.5. *Left*: specific heat as a function of ϵ at $T = 1$, for single strand length $N = 500, \ldots, 3000$. *Right*: histograms of the number of contacts, for the same chain lengths, at $\epsilon = \epsilon_c$. n/N is plotted on the *horizontal axis*, as appropriate for a first order transition

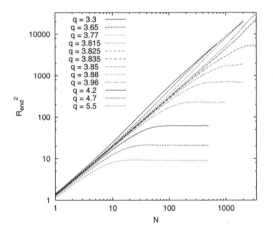

Fig. 11.6. Average squared end-to-end distance R_{end}^2 for various values of $q = e^\epsilon$, plotted against N on a double logarithmic scale. Since all curves are based on independent runs, their typical fluctuations relative to each other indicate the order of magnitude of their statistical errors

several chain lengths. We see a linear increase in the peak height with N, which indicates a first order transition. In the right hand panel of Fig. 11.5, energy histograms are plotted for the same chain lengths. Energy is measured in terms of the number of contacts n, divided by chain length N. One sees two maxima, one at $n = 0$ and the other at $n \approx N/2$, whose distance scales proportionally to N. This again points to a first order transition. But in contrast to typical first order transitions, the minimum between these two maxima does not become deeper with increasing N. This is due to the absence of any analogue of surface tension. Finally, in Fig. 11.6, we show average squared end-to-end distances. They obviously diverge for infinite N, when the transition point is approached from low temperatures. This is typical of a second order transition. A more detailed analysis shows that this divergence is $R_{\mathrm{end}} \sim (\epsilon - \epsilon_c)^\nu$, as one would expect for a transition with $\phi = 1$ [40].

Native Configurations of Toy Protein Models. Predicting the native states of proteins is one of the most challenging problems in mathematical biology [49]. It is not only important for basic science, but could also have enormous technological applications. At present, such predictions are mostly made by analog methods, i.e., by comparing with similar amino acid sequences whose native states are already known. More direct approaches are hampered by two difficulties.

- Molecular force fields are not yet precise enough. Energies between native and misfolded states are usually just a few eV, which is about the typical precision of empirical potentials. Quantum mechanical ab initio calculations of large biomolecules are impossible today.
- Even if perfect force fields were available, present day algorithms for finding ground states are too slow. One should add that the accepted dogma is that native states – at least of not overlarge proteins – are essentially energetic ground states.

In view of the second problem, there exists a vast literature on finding ground states of artificially constructed heteropolymers. Most of these models are formulated on a (square or simple cubic) lattice and use only a few monomer types. The best known example is the HP model of K. Dill [50] which has two types of amino acids: hydrophobic (H) and hydrophilic (polar, P). With most algorithms, one can find ground states, typically for random chains of length up to ~ 50.

In [35], we used PERM to study several sequences, of the HP model and of similar models, which had been discussed previously by other authors. In all cases, we found the same lowest energy states as these authors, but in several cases we found new lowest energy states. A particularly impressive example

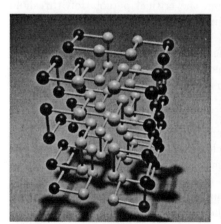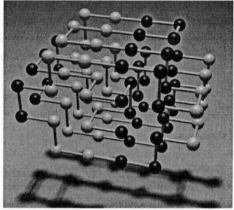

Fig. 11.7. *Left*: putative native state of the 'four helix bundle' sequence of [51]. It has $E = -94$, fits into a rectangular box, and consists of three homogeneous layers. Structurally, it can be interpreted as four helix bundles. *Right*: true ground state with $E = -98$. Its shape is highly symmetric although it does not fit into a rectangular box. It is degenerate with other configurations not discussed here

is a chain of length 80, with two types of monomers and somewhat artificial interactions: two monomers on neighbouring lattice sites contribute an energy -1 if they are of the same type, but do not contribute any energy if they are different [51]. A particular sequence was constructed that should fold into a bundle of four 'helices' with an energy -94 [51]. Even with a specially designed algorithm, the authors of [51] were not able to recover this state. With PERM, we not only found it easily, but we also found several lower states, the lowest having energy -98 and a completely different structure. Instead of being dominated by α helices, it has mostly β sheets (as far as these structures can be identified on a lattice). This is illustrated in Fig. 11.7. Since PERM gives not only the ground state but the full partition sum, we were also able to follow the transition between mostly helical states at finite T and the sheetlike ground state. We found a peak in the specific heat associated with this transition which could have been mistaken as a sign of a transition between a molten globule and the frozen native state.

11.3.3   Lattice Animals (Randomly Branched Polymers)

Consider the set of all connected clusters C_n of n sites on a regular lattice, with the origin being one of these sites, and with a weight defined on each cluster. The (n-site) lattice animal problem is defined by giving the same weight to each cluster. The last requirement distinguishes animal statistics from statistics of percolation clusters. Take site percolation for definiteness, with 'wetting' probability p. Then a cluster of n sites with b boundary sites carries a weight $p^n(1-p)^b$ in the percolation ensemble, while its weight in the animals ensemble is independent of b. In the limit $p \to 0$, this difference obviously disappears, and the two statistics coincide. Due to universality, we do indeed expect the scaling behavior to be the same for any value of p less than the critical percolation threshold p_c. It is generally believed that lattice animals are a good model for randomly branched polymers [52].

While there exists no simple and efficient algorithm for simulating large animals which also gives estimates for the partition sum, there exist very simple and efficient algorithms for percolation clusters. The best known is presumably the Leath algorithm [28] which constructs the cluster in a 'breadth first' way (see next section).

Our PERM strategy now consists in generating subcritical percolation clusters by the Leath method, and in making clones of those growing clusters which contribute more than average to the animal ensemble [53]. Since we work at $p < p_c$, each cluster growth would stop sooner or later if there were no enrichment. Hence we do not need explicit pruning. The threshold W_+ for cloning is chosen such that it depends both on the present animal weight and on the anticipated weight at the end of growth.

Usually, with growth algorithms like the Leath method, cluster statistics is updated only after clusters have stopped growing. But, as outlined below, one can also include contributions of still growing clusters. For percolation, this slightly reduces the statistical fluctuations of the cluster size distributions, but

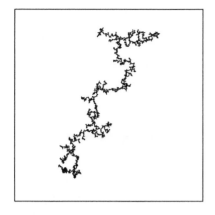

Fig. 11.8. A typical lattice animal with 8000 sites on the square lattice

the improvement is small. On the other hand, this improved strategy is crucial when using PERM to estimate animal statistics.

Consider a growing cluster during Leath growth. It contains n wetted sites, b boundary sites which are already known to be non-wetted, and g boundary sites at which the cluster can still grow since their status has not yet been decided ('growth sites'). This cluster will contribute to the percolation ensemble only if growth actually stops at all growth sites, i.e., with weight $(1-p)^g$. As the relative weights of the percolation and animal ensembles differ by a factor $(1-p)^{(b+g)}$ (since $b+g$ is now the total number of boundary sites), this cluster has weight $w(\mathcal{C}) \propto (1-p)^{-b}$ in the animal ensemble. If we used only this weight as a guide for cloning, we would clone when $w(\mathcal{C})$ is larger than some W_+ which is independent of b and g, and which depends on n in such a way that the sample size becomes independent of n. But clusters with many growth sites will of course have a bigger chance of continuing to grow and will contribute more to the precious statistics of very large clusters. In view of this, it is not clear a priori what the optimal choice for W_+ actually is, but numerically, we obtained the best results for $W_+ \propto (1-p)^g$.

In this way we were able to obtain good statistics for animals of several thousand sites, independently of the dimension of the lattice. A typical 2-d animal with 8000 sites is shown in Fig. 11.8. We were also able to simulate animal collapse (when each nearest neighbor pair contributes $-\epsilon$ to the energy), and animals near an adsorbing surface. Details will be published in [53].

11.4   Implementation Details

In this section, the notation will be appropriate for the lamb-and-lions problem of Sect. 11.2, but all statements also hold mutatis mutandis for the other problems.

11.4.1   Depth First Versus Breadth First

As described in trick 1 of Sect. 11.2, enrichment was originally implemented 'breadth first', i.e., many replicas of the process are kept simultaneously in the computer, and advanced simultaneously. This is also the traditional way of implementing evolutionary algorithms [6,7]. In that case, it is required for two reasons:

- because of crossover moves, where two configurations ('replicas', 'instances', 'individuals', etc.) are combined to give a new configuration;
- because of tournament selection, where the least fit of a randomly selected pair is killed and replaced by the fitter of the two.

In the present case, there are no crossovers, and tournament selection is replaced by comparing the 'fitness' w against thresholds $W_\pm$ which will be determined by some average fitness.

This allows us to use a depth-first implementation where only a single replica is kept in computer memory at any given time, and only when this is pruned or has reached its final time t_{max} is a new replica started. The names 'breadth first' and 'depth first' come from searching rooted trees. The first searches the tree in full breadth before increasing the depth, while the second follows a branch in full depth and only then considers alternative branches [54,55].

The main advantages of depth-first algorithms are reduced storage requirements and elegance of programming. (The former can be important even with today's parallel machines, where breadth first algorithms can be implemented by putting each configuration on its own node). While the 'natural' coding paradigms for breadth first algorithms are iterations and first-in/first-out queues, they are recursions and stacks for depth first approaches. In order to implement the lamb-and-lions problem, we just need a recursive function

$$\mathrm{STEP}(t, w, x, x_1, \ldots, x_N) \,,$$

where the arguments have the obvious meaning. When called, it increases $t \to t+1$, selects new positions from the neighbours of the previous ones, updates w accordingly, and calls itself either twice (cloning), once (normal evolution) or not at all (pruning). A pseudocode for this is given in [16].

It is often said that recursions are inefficient in terms of CPU time [56], and large recursion depths should be avoided. They can be avoided since each recursion can also be re-coded as an iteration (for example, FORTRAN 77 has no recursion and is nevertheless a universal language). On the other hand, the speed-up is negligible on modern compilers (less than 10%, typically), depths of 10^4–10^5 cause no problems, and readability of the code is much worse if recursion is not used. We see only one reason for avoiding recursion, and that is its less efficient use of main memory. In very large problems where stack size limitations can be crucial, recoding in terms of iterations may be needed.

11.4.2   Choosing $W_\pm$

In general, thresholds for pruning/cloning should not be too far apart since otherwise the weights fluctuate too much and most of the total weight is carried by only a few configurations. We obtained best results with $3 < W_+/W_- < 10$ in most applications, but other authors [44] also report good results for $W_+/W_- \approx 100$. Obviously, the precise value is not very important.

More important is W_+ itself. As a rule of thumb, it should be chosen such that the total number of configurations $C(t)$ created at time t is independent of t. If $C(t)$ decreases with t, most of the CPU time is spent on small t and the statistics at large t depend on only a small number of realizations. Conversely, if $C(t)$ increases with t, all configurations at large t are descendants of only a small number of ancestors and are thus strongly correlated.

There is a very simple way to guarantee the approximate constancy of $C(t)$ (up to a factor ~ 1) [16]. Let $Z(t)$ denote the path integral (or partition sum), and $\hat{Z}(t)$ its estimate from the current simulation,

$$\hat{Z}(t) = M^{-1} \sum_j w_j(t) , \tag{11.12}$$

where M is the number of starting configurations which have already been treated, and $w_j(t)$ is the weight of the jth configuration at time t. $C(t)$ will be roughly independent of t if

$$W_\pm = c_\pm \hat{Z}(t) , \tag{11.13}$$

where $c_\pm$ are constants of order unity (typically, $c_+ \approx 1/c_- \approx [W_+/W_-]^{1/2}$). We thus start the simulation with some guess for $W_\pm$. The precise values are largely irrelevant, and any large values would do just as well. We then replace them by (11.13) as soon as there has already been a configuration at t, i.e., as soon as we have $\hat{Z}(t) > 0$.

More sophisticated ways [35,33] of choosing $W_\pm$ are needed only for very hard problems with excessive pruning and cloning. In this case, the above method would occasionally give excessively large 'tours'. (A tour is the set of all configurations which descend from the same ancestor, i.e., which are obtained by cloning from the same starting configuration.) To cut them short, one should make $W_\pm$ larger than given by (11.13), if a tour is already very large. We should, however, warn the reader that in such hard cases the estimates of partition sums are no longer reliable, and results should be treated with some suspicion.

11.4.3   Choosing the Bias

As a general rule, the bias should be such that the bias correction factor exactly cancels the Boltzmann weight (if there is one) and minimizes the number of pruning/cloning events. A bad choice of the bias is immediately seen in an increase of these events, and in a decrease of the number of tours which reach large values of t. In t, a simulation corresponds essentially to a random walk with reflecting boundary at $t = 0$. While normal evolution steps correspond to forward steps in t, pruning events correspond to backward jumps to the previous cloning time. A proper choice of $W_\pm$ eliminates any drift from this random walk, while a good bias maximizes the effective diffusion constant. If W_+ is chosen according to (11.13), the CPU time needed to create an independent configuration at large t increases essentially as t^2, and the prefactor can be substantially decreased by choosing a good bias.

Unfortunately, there is no universal recipe for such a good bias. There is a general prescription in the case of diffusion quantum Monte Carlo (see next section) and for related Markov processes, but even this is not usually easy to implement. In other cases, such as polymers with self avoidance, one only has heuristics. Sometimes even the algorithm without bias is already very efficient, e.g., for multiple spanning percolation clusters. In other cases, as in the lamb-and-lions problem, whilst the qualitative properties of the bias are obvious, we used trial and error for its quantitative implementation.

One possible way of determining a good bias (or 'guiding', as it is sometimes called) is to look k steps ahead. For a polymer, for example, one might try all extensions of the chain by k monomers, and on the success of these extensions decide which single step to take next. This *scanning method* [57] is efficient in guiding the growth, but also very time-consuming. The effort increases exponentially with k. For polymers at low temperatures, where Boltzmann factors can become very large, this may nevertheless be efficient. But for athermal self-avoiding walks (SAWs) and for lattice polymers in the open coil phase, another method is much more efficient.

Markovian Anticipation. In this alternative strategy [37] for guiding polymer growth, the idea is essentially not to look forward, but rather to look backward k steps. We thus remember the last k steps during the growth. On a lattice with coordination number $\mathcal{N}$, this means we label the present configuration by an integer $i = 1, \ldots \mathcal{N}^k$. Assume now that the next step is in direction j, $j = 1, \ldots, \mathcal{N}$. During the initial steps of the simulation (or during an auxiliary run), we build up a histogram $H(i, j)$ of size $\mathcal{N}^{(k+1)}$. There we add up the weights with which all configurations with history (i, j) between the steps $n - k$ and n contribute to the partition sum of chains with length $n + m$, for $m \gg 1$ (we typically use $m \approx 100$–200). The ratio

$$\hat{p}(j|i) = \frac{H(i, j)}{\sum_{j'} H(i, j')} \tag{11.14}$$

is then an estimate of how efficient the extension j was in the long run. After some obvious modifications, taking into account that there is no history yet for the first k steps, and that no anticipation is useful for the last few steps, we use $\hat{p}(j|i)$ (which is properly normalized already!) as the probability with which we make step j, given the history i.

Note that this can also be used for stretched polymers, for example, where the $\hat{p}(j|i)$ are not isotropic, and where one can anticipate that the next monomer should be added preferentially in the direction of stretching.

11.4.4 Error Estimates and Reliability Tests

Errors can in principle be estimated *a priori* and *a posteriori*. In the former case, they are known even before making the simulations. For instance, if one draws n realizations of a random number with variance σ, the average has variance σ/N. A posteriori error estimates, in contrast, are obtained from fluctuations between the different realizations.

A priori error estimates for go-with-the-winners simulations are possible [17] but difficult because the generated sample is correlated. Indeed, making such estimates was the main objective of [17], but compromises regarding efficiency are such that the results obtained there do not seem very practical.

A posteriori error estimates can easily be made by dividing a long run into several bunches, computing averages over each bunch, and studying the fluctuations between them. This is essentially the same as the strategy in standard Metropolis simulations, but here the situation is even simpler. Since each 'tour' (see Sect. 11.4.2) is independent from any other, the break-up into bunches just has to be between tours. No problem due to correlations of uncertain range occurs here as it does in Metropolis simulations.

Nevertheless, the problems of critical slowing down and of being trapped in local free energy minima which plague Metropolis simulations are not absent in go-with-the-winners simulations. They just appear in new guises. Namely, single tours can become extremely large. If that happens, nearly the entire weight accumulated during a long simulation can be carried by a single tour or, what is even worse, the tours of *really* large weight may not be found at all. The latter is the analogy with not yet having reached equilibrium in Metropolis simulations.

Although this is first and foremost a problem of error bars, if one is not very careful, it can easily turn into a source of systematic errors. This is because one is not primarily interested in the partition sum (which is always sampled without a bias), but in its logarithm or in derivatives thereof. Consider for example a situation in which we wish to carry out several independent runs to make sure that everything has been done correctly. From each simulation we estimate a free energy, and then take the average value as our final estimate. If the problem is really hard, fluctuations in the partition sums will be non-Gaussian, with very many small downward fluctuations compensated by few large positive fluctuations. By taking the logarithm, the latter are cut down, and a negative bias results.

There is no foolproof protection against this danger. But there is an easy and straightforward way to check that at least that part of phase space which has been visited has been sufficiently sampled during a single run. For this, we make a histogram of tour weights on a logarithmic scale, $P(\log(w))$, and compare it with the weighted histogram $wP(\log(w))$. If the latter has its maximum for values of $\log(w)$ where the former is already large (i.e., where the sampling is already sufficient), we are presumably safe. However, if $wP(\log(w))$ has its maximum at or near the upper end of the sampled range, we should be skeptical.

As an example, we show results for a self avoiding 2-d walk in a random medium [39]. This medium is an infinite square lattice with (frozen) random energies E_i on each site i. In particular, E_i is either -1 (with probability p) or 0 (with probability $1 - p$). The polymer is free to float in the entire lattice. Previous simulations [58] had suggested that for any finite p there is a phase transition at $T = T_c(p)$, possibly because the polymer becomes localized in an 'optimal' part of the lattice for $T < T_c$.

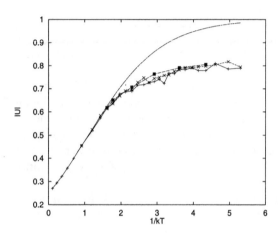

Fig. 11.9. Absolute value of U for $p = 1/4$ and $N = 200$. The *continuous line* is the exact theoretical result. *Plus signs* + have low statistics (ca. 10–20 min CPU time per point). *Crosses* × have medium statistics (roughly a factor of 10 more). *Squares* ■ have roughly a factor 20 more statistics than crosses

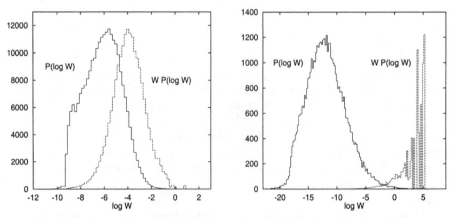

Fig. 11.10. *Full lines* are histograms of logarithms of tour weights for $1/kT = 0.92$ (*left*) and 2.30 (*right*), normalized as tours per bin. *Broken lines* show the corresponding weighted distributions, normalized so as to have the same maximal heights. Weights W are only fixed up to a β-dependent multiplicative constant

However, one can show rigorously that no such transition can exist [39]. In order to understand the source of the problem, we simulated with PERM and monitored the distribution of tour weights. Results are shown in Figs. 11.9 and 11.10. Figure 11.9 shows the average values of the energy,[1] $|U| = -\langle\sum_{i\in\mathcal{C}} E_i\rangle$ for $p = 1/4$ and chain length $N = 200$. Three curves obtained from simulations are shown together with a curve obtained analytically. The fact that the three curves, obtained with vastly different statistics, agree with each other but deviate from the theoretical curve at the same value of T indicates the supposed phase transition. However, histograms obtained in the regions where theory and simulations (dis)agree (see Fig. 11.10) clearly show that the simulations for $1/kT > 2$ are not reliable (right panel), while those for $1/kT < 1.5$ are.

11.5   Diffusion Quantum Monte Carlo

For completeness we sketch here the main idea of diffusion type quantum MC (QMC) simulations [9–11], to show how they are related to the previous calculations and to understand why a perfect bias is in principle possible in QMC but not in classical applications.

We start from the simplest version of a time dependent Schrödinger equation,

$$i\frac{\partial\psi(x,t)}{\partial t} = -(2m)^{-1}\nabla^2\psi(x,t) + V(x)\psi(x,t) . \tag{11.15}$$

We are interested in finding its ground state energy, i.e., we want to solve the time-independent Schrödinger equation

$$\cdot \ \ E_{\min}\psi(x) = -(2m)^{-1}\nabla^2\psi(x) + V(x)\psi(x) , \tag{11.16}$$

for the smallest eigenvalue $E_{\min}$. Replacing t by an imaginary 'time' and $(2m)^{-1}$ by a diffusion coefficient D, we end up with a diffusion equation

$$\frac{\partial\psi(x,t)}{\partial t} = D\nabla^2\psi(x,t) - V(x)\psi(x,t) , \tag{11.17}$$

with an external source/sink $V(x)$ and ψ viewed as a classical density. The ground state energy of the original problem is now transformed into the slowest relaxation rate. If we want to simulate this by diffusing particles, we can take the last term into account either by killing particles (if $V > 0$) and cloning them (if $V < 0$), or by giving them a weight $\exp[\int dt V(x(t))]$. Neither is very efficient. For efficiency, we should rather replace the random walk by a biased ('guided') motion for which neither weighting nor killing/cloning is needed.

For this purpose we choose a 'guiding function' $g(x)$ and write

$$\psi(x,t) = \rho(x,t)/g(x) . \tag{11.18}$$

---

[1] Here, $\mathcal{C}$ denotes the set of sites occupied by the walk, and averaging is carried out over all walks and all disorder realizations

Equation (11.17) then leads to the following equation for ρ:

$$\frac{\partial \rho}{\partial t} = D\nabla^2 \rho - \left[V(x) - D\frac{\nabla^2 g(x)}{g(x)}\right]\rho - \nabla\left(\left[2D\frac{\nabla g(x)}{g(x)}\right]\rho\right). \qquad (11.19)$$

This is now a diffusion equation with drift (last term) and with a modified source/sink term. If the latter is constant, i.e.,

$$D\nabla^2 g(x) - V(x)\,g(x) = \text{const.}\,g(x)\,, \qquad (11.20)$$

then no killing/pruning is needed and the weight increase/decrease is uniform and thus trivial. But (11.20) is just the time-independent Schrödinger equation (11.16) we wanted to solve. It seems that we have gained nothing. For an optimal implementation, we must already know the solution we want to get.

Of course, things are not really so bad since we can proceed iteratively: start with a rough guess for $g(x)$, obtain with it an estimate for $\psi(x)$, and use it as the next guess for $g(x)$, etc.

At this point, we may make the following crucial observation: if $g(x)$ satisfies (11.20), the density $\rho(x)$ of the guided diffusers is just equal to the quantum mechanical density,[2] $|\psi(x)|^2$. Thus random sampling of (11.19) corresponds precisely to random sampling of the quantum-mechanical density. It follows that we really have solved the problem of importance sampling.

Note that if we had *started* instead with (11.17) as a *classical* problem, and were interested in the *classical* density, we would not carry out perfect importance sampling: the particles would not be sampled in the simulation with the density they should have. Although using (11.20) for the guiding function would still be formally correct, it would not lead to minimal statistical fluctuations.

11.6   Conclusion

We have seen that stochastic simulations *not* following the traditional Metropolis scheme can be very efficient. We have illustrated this with a wide range of problems. The Ising model was conspicuous by its absence. The reason is simply that no go-with-the-winners algorithm for the Ising model has been proposed which is more efficient than, say, the Swendsen–Wang [2] algorithm. But there is no reason why such an algorithm should not exist. In principle, the go-with-the-winners strategy has at least as wide a range of applications as the Metropolis scheme. Its only requirement is that instances (configurations, histories, etc.) should be built up in small steps, and that the growth of their weights during the early steps of this build-up should not be too misleading.

The method is not new. It has its roots in algorithms which have been regularly used for several decades. Some of them, like genetic algorithms, are familiar to most scientists, but it is in general not well appreciated that they can be made

---

[2] Note that $\psi(x)$ is real here since we are interested in the ground state. However, a similar derivation is possible when using a time-dependent guiding function $g(x,t)$. Then (11.20) becomes the adjoint time-dependent Schrödinger equation.

into a general purpose tool. It seems even less appreciated how closely methods developed for quantum MC simulations, polymer simulations, and optimization methods are related. We firmly believe that this close relationship can be put to good use in many more applications to come.

Acknowledgements. We are indebted to Drs. Rodrigo Quian Quiroga and Günter Radons for discussions and for carefully reading the manuscript. WN is supported by DFG, SFB 237.

References

1. N. Metropolis et al.: J. Chem. Phys. **21**, 1087 (1953)
2. R.H. Swendsen, J-S. Wang: Phys. Rev. Lett. **58**, 86 (1987)
3. N. Madras, A.D. Sokal: J. Stat. Phys. **50**, 109 (1988)
4. B. Berg, T. Neuhaus: Phys. Lett. B **285**, 391 (1991); Phys. Rev. Lett **69**, 9 (1992)
5. E. Marinari, G. Parisi: Europhys. Lett. **19**, 451 (1992)
6. I. Rechenberg: *Evolutionstrategie* (Fromman-Holzboog Verlag, Stuttgart 1973)
7. J. Holland: *Adaptation in Natural and Artificial Systems* (University of Michigan Press, Ann Arbor 1975)
8. H.P. Schwefel: *Evolution and Optimum Seeking* (Wiley, New York 1995)
9. M.H. Kalos: Phys. Rev. **128**, 1791 (1962)
10. J.B. Anderson: J. Chem. Phys. **63**, 1499 (1975)
11. W. von der Linden: Phys. Rep. **220**, 53 (1992)
12. M.N. Rosenbluth, A.W. Rosenbluth: J. Chem. Phys. **23**, 356 (1955)
13. F.T. Wall, J.J. Erpenbeck: J. Chem. Phys. **30**, 634, 637 (1959)
14. S. Redner, P.J. Reynolds: J. Phys. A **14**, 2679 (1981)
15. T. Garel, H. Orland: J. Phys. A **23**, L621 (1990)
16. P. Grassberger: Phys. Rev. E **56**, 3682 (1997)
17. D. Aldous, U. Vazirani: *'Go with the winners' algorithms*, in Proc. 35th IEEE Sympos. on Foundations of Computer Science (1994)
18. P. Grassberger, H. Frauenkron, W. Nadler: 'PERM: a Monte Carlo Strategy for Simulating Polymers and Other Things', in *Monte Carlo Approach to Biopolymers and Protein Folding*, ed. by P. Grassberger et al. (World Scientific, Singapore 1998)
19. Y. Iba: e-print cond-mat/0008226 (2000)
20. M. Bramson, D. Griffeath: 'Capture problems for coupled random walks', in *Random Walks, Brownian Motion, and Interacting Particle Systems: A Festschrift in Honor of Frank Spitzer*, ed. by R. Durrett, H. Kesten (Birkhäuseauser, Boston 1991)
21. S. Redner, P.L. Krapivsky: Amer. J. Phys. **67**, 1277 (1999)
22. D. Stauffer, A. Aharony: *Introduction to Percolation Theory* (Taylor & Francis, London 1992)
23. J. Cardy: J. Phys. A **31**, L105 (1998)
24. M. Aizenman: Nucl. Phys. B **485**, 551 (1997)
25. M. Aizenman, B. Duplantier, A. Aharony: Phys. Rev. Lett. **8** (2000)
26. P. Sen: Int. J. Mod. Phys. C **8**, 229 (1997); **10**, 747 (1999)
27. L. Shchur: e-print cond-mat/9906013 v2 (1999)
28. P. Leath: Phys. Rev. B **14**, 5046 (1976)
29. P. Grassberger, R. Ziff: to be published

30. P.G. de Gennes: *Scaling Concepts in Polymer Physics* (Cornell Univ. Press, Ithaca 1979)
31. J. des Cloizeaux, G. Jannink: *Polymers in Solution* (Clarendon Press, Oxford 1990)
32. J. Batoulis, K. Kremer: J. Phys. A **21**, 127 (1988)
33. U. Bastolla, P. Grassberger: J. Stat. Phys. **89**, 1061 (1997)
34. H. Frauenkron, P. Grassberger: J. Chem. Phys. **107**, 9599 (1997)
35. H. Frauenkron, U. Bastolla, E. Gerstner, P. Grassberger, W. Nadler: Phys. Rev. Lett. **80**, 3149 (1998); PROTEINS **32**, 52 (1998)
36. G.T. Barkema, U. Bastolla, P. Grassberger: J. Stat. Phys. **90**, 1311 (1998)
37. H. Frauenkron, M.S. Causo, P. Grassberger: Phys. Rev. E **59**, R16 (1999)
38. S. Caracciolo, M.S. Causo, P. Grassberger, A. Pelissetto: J. Phys. A **32**, 2931 (1999)
39. P. Grassberger: J. Chem. Phys. **111**, 440 (1999)
40. M.S. Causo, B. Coluzzi, P. Grassberger: Phys. Rev. E **62**, 3958 (2000)
41. U. Bastolla, P. Grassberger: J. Molec. Liquids **84**, 111 (2000)
42. U. Bastolla, P. Grassberger: e-print cond-mat/0004169 (2000)
43. J. Hager, L. Schäfer: Phys. Rev. E **60**, 2071 (1999)
44. T. Prellberg, A.L. Owczarek: Phys. Rev. E **62**, 3780 (2000)
45. R.M. Wartell, A.S. Benight: Phys. Rep. **126**, 67 (1985)
46. D. Poland, H.A. Sheraga: J. Chem. Phys. **45**, 1456, 1464 (1966)
47. M.E. Fisher: J. Chem Phys. **45**, 1469 (1966)
48. Y. Kafri, D. Mukamel, L. Peliti: e-print cond-mat/0007141 (2000)
49. T.E. Creighton (ed.): *Protein Folding* (Freeman, New York 1992)
50. K.A. Dill: Biochemistry **24**, 1501 (1985)
51. E. O'Toole, A. Panagiotopoulos: J. Chem. Phys. **97**, 8644 (1992)
52. T.C. Lubensky, J. Isaacson: Phys. Rev. Lett. **41**, 829 (1978)
53. P. Grassberger, W. Nadler: to be published
54. R. Tarjan: SIAM J. Comput. **1**, 146 (1972)
55. R. Sedgewick: *Algorithms in C*, 3rd edn., Parts 1–4 (Addison–Wesley, Reading, MA 1998)
56. W.J. Thompson: Computers in Physics **10**, 25 (1996)
57. H. Meirovitch, H.A. Lim: J. Chem. Phys. **92**, 5144 (1990)
58. A. Baumgärtner: J. Chem. Phys. **109**, 10011 (1998)

12 Aperiodicity and Disorder – Do They Play a Role?

Uwe Grimm

Summary. The effects of an aperiodic order or a random disorder on phase transitions in statistical mechanics are discussed. A heuristic relevance criterion based on scaling arguments as well as specific results for Ising models with random disorder or certain kinds of aperiodic order are reviewed. In particular, this includes an exact real-space renormalization treatment of the Ising quantum chain with coupling constants modulated according to substitution sequences, related to a two-dimensional classical Ising model with layered disorder.

12.1 Introduction

Equilibrium statistical mechanics bridges the gap between the fundamental laws of physics at microscopic scales on the one hand and the macroscopic phenomena that we may experience in everyday life on the other hand [1]. Its main achievement is to understand the behaviour of macroscopic systems on the basis of their microscopic constituents and their mutual interactions.

One of the most fascinating topics in this field concerns phase transitions, i.e., a qualitative change observed in some physical properties of a macroscopic system after a slight variation of certain parameters, such as the temperature or an external magnetic field. Here, statistical mechanics reveals its full power by explaining how even short-range interactions at the microscopic scale may lead to such violent cooperative behaviour.

In mathematical terms, a phase transition corresponds to a non-analyticity of the thermodynamic potential that describes the equilibrium state of the system as a function of the external parameters. In essence, one may regard it as a battle between energetic and entropic terms, the first driving the system towards an ordered phase at low temperature, the second favouring a disordered state at higher temperatures. However, this is a somewhat simplistic view of a phenomenon whose complexity should not be underestimated. As an example that teaches caution, one might think of the entropy-driven phase transitions observed in non-interacting hard-core systems [2]. Counter-intuitively, these may lead to an ordered state that has a higher entropy than the disordered state.

While the general picture is rather well understood, it turns out to be much more difficult to calculate the properties of a specific macroscopic system from its microscopic interactions explicitly. In fact, complete solutions are known only for a few simple models [3]. Among these is the paradigm of a ferromagnet, the celebrated Ising model [4,5], after Onsager [6] succeeded in calculating the free energy for the two-dimensional (2D) Ising model without an external magnetic field. However, this extremely simplistic model has, so far, defied all attempts at an analytic solution in higher dimensions, or even in 2D in the presence of an external magnetic field. Recent progress from complexity theory may even support the widespread belief in its general unsolvability.

Moreover, the models that have been solved analytically (or 'exactly', which usually does not mean rigorously in the mathematical sense) are almost exclusively tied to regular periodic structures, as for instance the Ising model on a square lattice or some other regular planar lattices. However, nature is never completely regular, and this immediately spurs the question as to what happens if the regular periodic structure is disturbed by some randomness, mimicking a more realistic crystalline substance, or even given up completely, resulting in either aperiodically ordered, or disordered structures. While the effects of a random disorder have been studied for many decades, aperiodically ordered systems entered the scene after the experimental discovery of aperiodically ordered solids known as *quasicrystals* in the 1980s. The reader is referred to [7] for a recent compilation of introductory lectures on this issue.

The situation is, however, not as hopeless as it may seem at first sight. One of the ideas coming to our rescue is the important concept of *universality* of critical phenomena [8]. According to the universality hypothesis, the critical behaviour of a system does *not* depend on details of the microscopic interactions of the model, but merely on a number of general properties such as the dimension of space and the symmetries of the model. The reader should also refer to Chaps. 8, 13, 16 and 17 in this volume. This means that second-order or 'smooth' phase transitions, which are characterized by an infinite correlation length, can be classified into *universality classes* of systems showing the same physical behaviour at and close to the critical point. In particular, this implies that the critical exponents, which describe the singularities of physical quantities upon approaching the critical point, only depend on the universality class of a model. The origin of universality and the *scaling relations* between different critical exponents can be understood in terms of the so-called *renormalization group* (see Chap. 17 and [9]) because critical points correspond to fixed points under the renormalization flow. Of course, things are not quite as simple in reality, and the situation may, in general, turn out to be more complex. Such arguments should therefore be taken with a grain of salt.

In 2D we have another powerful tool at our disposal. There exists an intimate relation between 2D statistical models at criticality and *conformal field theory* [10] that explains the surprising fact that critical exponents in many cases are found to be rational numbers, and furthermore allows the computation of correlation functions. This yields an alternative, and more substantiated, explanation of the universality classes and a partial classification of the possible types of critical behaviour in 2D. Such strong results, albeit by no means complete, are limited to the 2D case due to the particular properties of the 2D conformal symmetry group.

Nevertheless, because of universality, one may expect a certain robustness of critical properties, i.e., introducing disorder or aperiodic order may not affect the critical behaviour of some systems at all. The obvious question one would like to answer is, given an aperiodic or disordered model, does it show the same critical behaviour as its perfectly ordered counterpart or not? And, if not, how does it change?

In what follows, we shall concentrate on the example of the Ising model, simply because it is the most frequently studied and, besides numerous numerical treatments, it also admits analytic solutions. After a brief introduction to the Ising model and its critical properties in the subsequent section, we discuss a rather general, albeit heuristic, relevance criterion that is based on scaling arguments in Sect. 12.3. Thereafter, in Sect. 12.4, we give a short overview of the various types of disordered Ising models considered in the literature and the main results concerning their critical behaviour. Next we turn our attention to aperiodically ordered Ising models which are the topic of Sect. 12.5. Finally, we conclude with a brief summary in Sect. 12.6.

12.2 The Ising Model

There have been numerous review articles and even entire books [11] devoted to the Ising model, and many of these give an extensive introduction to the history of this simple, yet fascinating model. In addition, most textbooks on statistical mechanics treat at least the 1D Ising model. For a mathematically rigorous approach, see, e.g., [12]. Here, we shall restrict ourselves to the most important facts, the interested reader is referred to the literature for a more complete account of the historical background, for instance in [13]. The recent review [14] on aperiodic Ising models also contains a brief historical introduction, and an extensive list of references.

12.2.1 The Square-Lattice Ising Model

Originally, the Ising model was devised by Lenz as a simplistic toy model of a ferromagnet [4]. The 1D model was later solved by Ising [5], a student of Lenz, who obtained the disappointing result that the model does not show a phase transition at non-zero temperature. He supposed that this should hold in higher dimensions as well, but, fortunately, it was not long until he was proven wrong – otherwise the model carrying his name would not have attracted the interest of physicists for almost a century.

So how is the model defined? The local magnetic moments of the solid are modeled by 'Ising spins' which can only take two values, ± 1, on the sites of a regular lattice. The ferromagnetic interaction between these spins is given by a simple coupling of neighbours on the lattice. Let us concentrate on the square-lattice case for definiteness. A *configuration*

$$\sigma = (\sigma_{1,1}, \sigma_{1,2}, \dots, \sigma_{N,M}) \in \{\pm 1\}^{NM}$$

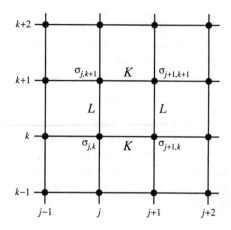

k+2

k+1 $\sigma_{j,k+1}$ K $\sigma_{j+1,k+1}$

        L   L

k $\sigma_{j,k}$ K $\sigma_{j+1,k}$

k−1

j−1   j   j+1   j+2

Fig. 12.1. The square-lattice Ising model

of spins $\sigma_{j,k}$ on an $N \times M$ rectangular section of the lattice is assigned an energy

$$E(\sigma) = -\sum_{j=1}^{N}\sum_{k=1}^{M}\left(K\sigma_{j,k}\sigma_{j+1,k} + L\sigma_{j,k}\sigma_{j,k+1} + H\sigma_{j,k}\right) , \qquad (12.1)$$

where we may assume certain boundary conditions such as, for instance, periodic $(\sigma_{j,M+1} \equiv \sigma_{j,1},\ \sigma_{N+1,k} \equiv \sigma_{1,k})$ or free $(\sigma_{j,M+1} \equiv \sigma_{N+1,k} \equiv 0)$ boundary conditions. Here, $K > 0$ and $L > 0$ denote the ferromagnetic coupling constants in the horizontal and vertical directions, respectively (see Fig. 12.1), and H is the magnetic field.

The *partition function*, a sum of Boltzmann factors over all configurations σ, is given by

$$Z_{N,M}(T,H) = \sum_{\sigma} \exp[-E(\sigma)/k_{\mathrm{B}}T] , \qquad (12.2)$$

where T denotes the temperature and k_{B} is Boltzmann's constant. It can also be expressed in terms of a so-called *transfer matrix*

$$Z_{N,M}(T,H) = \mathrm{Tr}[U_N(T,H)^M] , \qquad (12.3)$$

where U_N is a $2^N \times 2^N$ matrix whose elements are Boltzmann weights of the configuration corresponding to a single row of the lattice, and the trace is due to the periodic boundary conditions that we assumed. The trace thus performs the sum over the configuration $(\sigma_{1,1}, \sigma_{2,1}, \dots, \sigma_{N,1})$ of the first row, whereas the summation over the configurations $(\sigma_{1,k}, \sigma_{2,k}, \dots, \sigma_{N,k})$, $2 \leq k \leq N$, along the other rows is hidden in the matrix product.

12.2.2   Phase Transitions and Critical Exponents

In the thermodynamic limit of an infinite system, the *free energy* per site

$$f(T,H) = \lim_{N,M \to \infty}\left[\frac{-k_{\mathrm{B}}T}{NM}\ln\left[Z_{N,M}(T,H)\right]\right] \qquad (12.4)$$

is an analytic function of the parameters T and H, except at the phase transition. The transition is of first order if a first derivative of the free energy is discontinuous. Otherwise, the transition is termed a second- or higher-order transition. Examples of first-order transitions are the melting of ice or the evaporation of water, where one has to supply energy, the latent heat, to the system without changing the temperature. The Curie point of a ferromagnet, or the critical point of water at the end of the first-order transition line between liquid and gas, are examples of critical points.

Correspondingly, one encounters singularities in thermodynamic quantities that can be obtained as derivatives of the free energy, such as the specific heat

$$c_H(T) = -T\frac{\partial^2 f(T,H)}{\partial T^2}$$

(12.5)

at constant magnetic field H, the spontaneous magnetization

$$m_0(T) = \lim_{H \searrow 0} m(T,H)\,, \qquad m(T,H) = -\frac{\partial f(T,H)}{\partial H}\,,$$

(12.6)

or the magnetic susceptibility

$$\chi_T(H) = \frac{\partial m(T,H)}{\partial H} = -\frac{\partial^2 f(T,H)}{\partial H^2}\,.$$

(12.7)

For a second-order phase transition, on which we shall now concentrate, these singularities are characterized by so-called *critical exponents* describing power-law singularities on approaching the critical point. For magnetic systems such as the Ising model, some zero-field ($H = 0$) critical exponents are

$$c_0(T) \sim |t|^{-\alpha}\,,$$ (12.8)

$$m_0(T) \sim (-t)^{\beta}\,, \qquad \text{for } t < 0,\; m_0(T) \equiv 0 \text{ for } t \ge 0\,,$$ (12.9)

$$\chi_T(0) \sim |t|^{-\gamma}\,,$$ (12.10)

as $t \to 0$, where $t = (T - T_c)/T_c$ is a 'reduced temperature' parametrizing the distance to the critical temperature $T = T_c$ (see also Chap. 13). The power laws are related to a particular behaviour of the correlation $G(r, r')$ at the critical point. Instead of a standard exponential decay with distance, given by $G(r, r') \sim \exp(-|r - r'|/\xi)$, with a characteristic correlation length ξ, correlation functions at a critical point show a power-law decay $G(r, r') \sim |r - r'|^{-x}$. In other words, the correlation length ξ diverges as

$$\xi \sim |t|^{-\nu}\,,$$

(12.11)

as the critical point is approached. Of course, several other exponents can be considered, and there also exist phase transitions where the exponents differ when the critical point is approached from the ordered phase $t < 0$ or from the disordered phase $t > 0$.

12.2.3   The Ising-Model Phase Transition

As mentioned above, it was not clear from the beginning that the Ising model has any phase transition at all. However, as early as 1936, Peierls [15] gave an argument that showed the existence of a phase transition in two or more dimensions. The argument is simple, and indeed also holds for rather general models, including Ising models with ferromagnetic random couplings.

Essentially, the argument goes as follows. At $T = 0$, the system is in one of its magnetized ground states, where all spins are either $+1$ or -1. The states that the system can explore at low temperature differ from the ground state by small islands of turned spins, and their energy is given by the surface of these islands where neighbouring spins differ. Hence, even at small finite temperatures, the system is magnetized, and there must be a transition to a disordered high-temperature state at nonzero temperature. Indeed, this argument breaks down in 1D, because a flip of the local spins along an entire infinite half-line affects only a single bond. Consequently, even an arbitrarily small temperature suffices to destroy the magnetic order (see [16] for a more precise discussion).

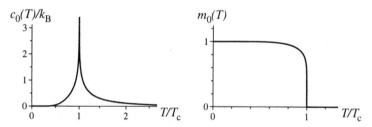

Fig. 12.2. The specific heat $c_0(T)$ and spontaneous magnetization $m_0(T)$ of the square-lattice Ising model

The critical temperature for the zero-field Ising model was first determined by Kramers and Wannier [17] in 1941 under the assumption that there is only a single phase transition. In this case, the transition has to take place at a fixed point of a duality transformation that maps low- and high-temperature phases onto each other. For the square-lattice Ising model, the critical temperature T_c is given by

$$\sinh(2K/k_B T_c) \sinh(2L/k_B T_c) = 1 , \qquad H = 0 . \tag{12.12}$$

Only three years later, Onsager's celebrated solution of the zero-field square-lattice Ising model [6] revealed a logarithmic singularity of the specific heat,

$$c_0(t) \sim \ln|1/t| , \tag{12.13}$$

i.e., the corresponding critical exponent $\alpha = 0$. The correlation length diverges at the critical point with an exponent $\nu = 1$. The magnetization exponent $\beta = 1/8$ was later calculated by Yang [18] by means of a perturbative approach. Figure 12.2 shows the specific heat and spontaneous magnetization for the zero-field Ising model.

12.2.4 The Ising Quantum Chain

Before moving on to discuss disordered models, we need to introduce a 1D quantum version of the Ising model [19] that is closely related to the 2D classical Ising model (see also Chap. 13). Consider the Hamiltonian

$$H = -\frac{1}{2}\sum_{j=1}^{N}\sigma_j^x\sigma_{j+1}^x - \frac{\lambda}{2}\sum_{j=1}^{N}\sigma_j^z \tag{12.14}$$

on the Hilbert space $\bigotimes_{j=1}^{N}\mathbb{C}^2 \cong \mathbb{C}^{2^N}$. Here, the local spin operators are defined as

$$\sigma_j^{x,z} = \underbrace{\mathbb{I}\otimes\mathbb{I}\otimes\ldots\otimes\mathbb{I}}_{j-1 \text{ factors}}\otimes\sigma^{x,z}\otimes\underbrace{\mathbb{I}\otimes\ldots\otimes\mathbb{I}\otimes\mathbb{I}}_{N-j \text{ factors}} \tag{12.15}$$

with the standard Pauli matrices

$$\sigma^x = \begin{pmatrix} 0 & 1 \\ 1 & 0 \end{pmatrix}, \qquad \sigma^z = \begin{pmatrix} 1 & 0 \\ 0 & -1 \end{pmatrix}, \qquad \mathbb{I} = \begin{pmatrix} 1 & 0 \\ 0 & 1 \end{pmatrix}. \tag{12.16}$$

A different basis is often used, where the coupling term involves the products $\sigma_j^z\sigma_{j+1}^z$, and the 'transverse field' term has the form $\sum_{j=1}^{N}\sigma_j^x$ (Chap. 13). For the homogeneous Hamiltonian (12.14), the spectrum is completely known [20].

The Hamiltonian (12.14) can be obtained as an anisotropic limit of the transfer matrix U_N in (12.3) of the zero-field ($H = 0$) square-lattice Ising model (12.1), where the coupling $K \to 0$ along the chain direction and the coupling $L \to \infty$ in the 'time' direction, while $\beta K\exp(2\beta L)$ is kept fixed [21]. Consequently, the 'transverse field' parameter λ in the quantum chain acts analogously to the temperature in the classical model; and the Ising quantum chain at zero temperature displays a quantum phase transition (Chap. 13), i.e., a change in the ground-state properties, at the critical value $\lambda_c = 1$ which belongs to the same universality class as the critical point of the classical square-lattice Ising model. In particular, the ground-state energy per site, $-E_0/N$, of the Hamiltonian (12.14) corresponds to the free energy of the classical Ising model, and the energy gap $E_1 - E_0$ between the ground state and the first excited state corresponds to the inverse $\xi_\parallel^{-1}$ of the correlation length along the chain, i.e., the energy gap vanishes at criticality. According to finite-size scaling [23], the gap vanishes as $E_1 - E_0 \sim N^{-z}$, where $z = \nu_\parallel/\nu_\perp$ denotes the ratio of the correlation length exponents along the chain and in the 'time' direction. In spite of the anisotropic limit, at criticality the homogeneous system (12.14) behaves isotropically, with $\nu_\parallel = \nu_\perp = 1$ as in the classical square-lattice Ising model. However, as we shall see below, this may be different if one considers disordered or aperiodically ordered Ising quantum chains, which then correspond to classical 2D Ising models with a layered disorder or aperiodicity.

12.2.5 Effects of Disorder: Some General Remarks

Phase transitions are cooperative phenomena leading to singularities in physical quantities such as the specific heat or the magnetic susceptibility. From a naïve point of view, any disorder in a system, if it affects the phase transition at all, will tend to *weaken* critical singularities – it is hard to imagine how it could possibly do the opposite. Thus, one might expect a first-order transition in a perfectly ordered system to be weakened to a higher-order phase transition in a disordered or aperiodically ordered system, and indeed there exist examples where such behaviour has been observed [22]. For a second-order transition, disorder may change the critical exponents, or the singularities may be weakened to higher-order singularities, or even be washed out completely so that the phase transition disappears.

So given a specific model, how can one estimate the effect of a certain type of disorder on the system? A quite simple answer to this question is provided by a heuristic relevance criterion based on scaling arguments. Although these arguments rely on several assumptions that may not always be fulfilled, they have proven to be rather successful in predicting the correct critical behaviour.

12.3   Heuristic Scaling Arguments

The relevance criterion that we are now going to discuss was first put forward by Harris in 1974 [24] for disordered Ising models. It was later generalized by Luck [25] to the case of aperiodically ordered Ising models. The argument presented here closely follows the discussion of the Harris–Luck criterion in [23], where one can also find further applications of the criterion.

We restrict ourselves to models of Ising type with purely ferromagnetic couplings; the case of frustration in disordered systems or spin glasses (see Chap. 1) is not discussed here. Furthermore, a certain homogeneity in the distribution of the ferromagnetic coupling constants $\varepsilon_{j,k} \geq 0$ is needed, such that, for instance, the mean coupling $\bar{\varepsilon} = \varepsilon_0$ is well defined. We shall consider this as the coupling of our unperturbed reference system denoted by the index 0. We characterize the distribution of coupling constants by their *fluctuations*, expressed in terms of the deviations from the mean coupling. More precisely, consider an approximately spherical volume V located somewhere in our infinite system (see Fig. 12.3). Within V, the mean coupling is

$$\bar{\varepsilon}_V = \frac{2}{n_0 N_V} \sum_{\langle j,k \rangle \in V} \varepsilon_{j,k} \; , \tag{12.17}$$

where N_V denotes the number of spins in V, and n_0 is the mean coordination number, i.e., the average number of neighbours in the system. The deviation of the accumulated couplings from the mean is given by

$$G_V = \sum_{\langle j,k \rangle \in V} \varepsilon_{j,k} - \frac{\varepsilon_0 n_0 N_V}{2} = \frac{n_0 N_V}{2} \left(\bar{\varepsilon}_V - \varepsilon_0 \right) \; . \tag{12.18}$$

As a measure of the fluctuations, we consider the asymptotic behaviour of the standard deviation Δ_{G_V} for large volumes V,

$$\Delta_{G_V} \sim (N_V)^\omega \; , \tag{12.19}$$

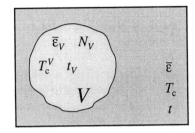

Fig. 12.3. A volume V within the system and the associated local quantities

thus defining the *fluctuation exponent* ω, which we assume to be well defined. Note that we did not specify the type of disorder in the system, we may consider both models with varying coupling constants on a regular lattice, or models that are defined on topologically disordered graphs, for instance random graphs or periodic lattices with randomly distributed vacancies. Another example we shall encounter below concerns Ising models that are defined on aperiodic graphs. These also yield fluctuations, although no randomness is involved.

Now comes the crucial intuitive step. The critical temperature T_c clearly depends on $\bar{\varepsilon}$. Thus, due to the fluctuations in the couplings $\varepsilon_{j,k}$, the system may correspond *locally*, i.e., within V, to a different 'local critical temperature' T_c^V that, for sufficiently small disorder, may be expected to depend linearly on the local average coupling $T_c^V \sim \bar{\varepsilon}_V$. One may think of T_c^V as the critical temperature of a system that essentially looks the same in any ball of volume V. Correspondingly, we may define a 'local reduced temperature' t_V.

If the disorder is *irrelevant* for the critical behaviour, the shift in the local reduced temperature $\delta t = t_V - t$ has to vanish at criticality, i.e., $\delta t \to 0$ as $t \to 0$. The relevance criterion now follows from this consistency requirement by using scaling relations involving the critical exponent ν. In the case of *irrelevant* disorder, the latter should coincide with the exponent ν_0 of the pure system. The argument proceeds as follows. The volume V of spins that are correlated is of order $V \sim \xi^d$, where d denotes the dimension of the system. Then, from (12.19) and our assumption $T_c^V \sim \bar{\varepsilon}_V$, we deduce

$$\delta t \sim \xi^{-d(1-\omega)} \sim |t|^{\nu d(1-\omega)} . \tag{12.20}$$

In the irrelevant scenario $\nu = \nu_0$, this yields

$$\delta t / t \sim |t|^{-\Phi} \qquad \text{with } \Phi = 1 - d\nu_0(1-\omega) . \tag{12.21}$$

Consequently, the disorder is *irrelevant* if the crossover exponent $\Phi < 0$.

We can summarize the Harris–Luck relevance criterion as follows [23–25]. A modulation in $d_m \le d$ space dimensions of a ferromagnetic d-dimensional Ising model with correlation exponent ν_0 is *relevant* for $\omega > \omega_c$, *marginal* for $\omega = \omega_c$, and *irrelevant* for $\omega < \omega_c$, where

$$\omega_c = 1 - \frac{1}{d_m \nu_0} . \tag{12.22}$$

For the 2D Ising model, we have $\nu_0 = 1$. Thus, for a model with layered disorder, i.e., $d_m = 1$, we obtain $\omega_c = 0$. In this case, any divergent fluctuation is relevant. For planar disorder, i.e., $d_m = 2$, the marginal value is $\omega_c = 1/2$. Thus, for randomly distributed couplings, which correspond to $\omega = 1/2$, one is precisely in the marginal situation where the criterion does not give a definite prediction. Generally, one might expect to find logarithmic corrections to scaling in the marginal case, and this is indeed the case in the 2D random-bond Ising model [29–32].

In the original formulation by Harris, the criterion was expressed in terms of the specific heat exponent α instead of the correlation exponent ν. The two exponents are related by the hyperscaling relation $d\nu = 2 - \alpha$ [3]. This holds for the 2D Ising model, which again corresponds to the marginal case $\alpha = 0$ when there is random disorder [24].

12.4   Random Ising Models

Traditionally, randomness was introduced into Ising models to describe the magnetic behaviour of alloys where some magnetic atoms were replaced by non-magnetic atoms. Such models are therefore termed *dilute* Ising models and, in the simplest case, just correspond to a regular Ising model where some spins, or some bonds, have been removed (see the reviews [26–28]). Clearly, there is some relation to percolation (Chap. 17), because one at least needs infinite clusters in order to support a phase transition. Somewhat more general are random-bond Ising models, where the coupling constants are chosen randomly, often from a bimodal distribution.

In these systems, the disorder is regarded as 'frozen' or 'quenched', which means that it is static. One then has to average the free energy over the possible realizations, an extremely difficult task (Chap. 1). Some analytical results have nevertheless been obtained. As shown above, we expect the 2D system to be marginal with respect to random disorder. Using the replica method, Dotsenko and Dotsenko [29] found a double logarithmic singularity in the specific heat

$$c(t) \sim \ln[1 + g \ln(1/|t|)] , \qquad (12.23)$$

instead of the pure Ising behaviour (12.13), and

$$m(t) \sim \exp\{-a[\ln\ln(1/|t|)]^2/2\} , \qquad \chi(t) \sim t^{-2} \exp\{-a[\ln\ln(1/|t|)]^2\} , \qquad (12.24)$$

corresponding to critical exponents $\beta = 0$ and $\gamma = 2$, which differ from the pure Ising values. However, by means of bosonization techniques and conformal field theory, Shalaev, Shankar, and Ludwig [30–32] arrived at different results,

$$m(t) \sim t^{1/8}(\ln|t|)^{-1/16} , \qquad \chi(t) \sim t^{-7/4}(\ln|t|)^{7/8} , \qquad (12.25)$$

for the magnetization and the susceptibility, while the result for the specific heat coincides with (12.23). Here, the critical exponents keep their pure Ising values

$\beta = 1/8$ and $\gamma = 7/4$, but logarithmic corrections show up. Due to the slow variation of the logarithm, it turns out to be very difficult to obtain reliable results from numerical calculations, because the logarithmic dependence on the reduced temperature translates to a logarithmic dependence on the system size, and there is no hope of distinguishing a simple logarithmic dependence from a double logarithmic dependence using finite-size calculations. Still, the majority of the numerous finite-size scaling studies based on Monte Carlo simulations or transfer matrix calculations favour the latter result (12.25) (see [26–28] and references therein), and it has also been substantiated by a recent series expansion investigation [33].

The correlation exponent of the 3D pure Ising model $\nu \approx 0.63$, so that $\alpha > 0$, and disorder should be relevant in that case. However, a recent result indicates that for purely topological disorder introduced by considering an Ising model on a 3D random graph, the critical behaviour stays the same as for the regular cubic lattice [34].

In recent years, there has been an increasing activity in the mathematical literature for stochastic Ising models [35–37], leading to a number of rather general rigorous results. For instance, for the random-field Ising model, it can be shown rigorously that in dimensions $d \leq 2$ there exists a unique Gibbs state. There is therefore *no* phase transition in the 2D random-field model, whereas there is a phase transition in $d \geq 3$. For ferromagnetic random-bond Ising models on certain periodic planar graphs, it has been proven that at most two extremal Gibbs states exist, so there are no more than two phases (see [38] and references therein).

12.5 Aperiodic Ising Models

In the remainder of this article, we shall concentrate on the case of aperiodically ordered Ising models [14]. To this end, we need to introduce aperiodic sequences and tilings (see [39] for a recent review of the mathematical aspects). In the theoretical description of quasicrystalline structures [7], aperiodic tilings constitute the analogues of periodic lattices in conventional crystallography. For the sake of space, we cannot go into much detail here. The interested reader is referred to the MATHEMATICA [40] program packages accompanying this article [53], which give an insight into the construction and properties of 1D and 2D aperiodic tilings. The different methods employed and the corresponding MATHEMATICA routines are described in detail in [41].

12.5.1 Aperiodic Tilings on the Computer

Aperiodic sequences are commonly constructed by means of *substitution rules*. As examples, consider an alphabet of two letters a and b and rules

$$\varrho^{(k)} : \begin{cases} a \to ab \\ b \to a^k , \quad k = 1, 2, \dots , \end{cases} \tag{12.26}$$

that replace the single letters a and b by words $w_a^{(k)} = ab$ and $w_b^{(k)} = a^k$, the word consisting of k letters a, in the two-letter alphabet $\{a, b\}$. We restrict ourselves to *primitive* substitutions, i.e., after a finite number of iterations the words obtained from the basic letters should contain all letters. Corresponding semi-infinite *substitution sequences* $w_\infty^{(k)}$ are then obtained as fixed points of the substitution rules by an iterated application $w_n^{(k)} = \varrho^{(k)}(w_{n-1}^{(k)})$ of the rules (12.26) on some initial word, say $w_0^{(k)} = a$. The substitution rules (12.26) yield some prominent examples of aperiodic sequences, for instance the Fibonacci sequence $abaababa\ldots$ for $k = 1$, the so-called periodic-doubling sequence $abaaabab\ldots$ for $k = 2$, and a sequence known as the binary non-Pisot sequence $abaaaababab\ldots$ for $k = 3$.

Many properties of the substitution sequences can be conveniently calculated from the associated substitution matrices

$$M^{(k)} = \begin{pmatrix} 1 & k \\ 1 & 0 \end{pmatrix} , \qquad \lambda_\pm^{(k)} = \frac{1 \pm \sqrt{4k+1}}{2} , \qquad (12.27)$$

whose elements just count the number of letters a and b in their substitutes $w_a^{(k)}$ and $w_b^{(k)}$, respectively. The largest eigenvalue $\lambda_+^{(k)}$ of $M^{(k)}$ determines the asymptotic growth of the sequence in one iteration step, while the entries of the corresponding eigenvector, suitably normalized, determine the frequencies $p_a = 1 - p_b = 1 - 2/(3 + \sqrt{4k+1})$ of the letters a and b in $w_\infty^{(k)}$. The second eigenvalue measures the *fluctuations* of the letter frequencies

$$g^{(k)}(N) = \#_a[w_\infty^{(k)}|_N] - p_a N , \qquad h^{(k)}(N) = \max_{M \le N} \left| g^{(k)}(M) \right| \sim N^{\omega^{(k)}} , \qquad (12.28)$$

where $\#_a[w_\infty^{(k)}|_N]$ denotes the number of letters a in the first N letters of the limit word $w_\infty^{(k)}$. Thus, $g^{(k)}(N)$ measures the deviation from the mean $p_a N$, and we use the maximal deviation $h^{(k)}(N)$ to define the fluctuation exponent $\omega^{(k)}$ which enters the Harris–Luck criterion [compare (12.19)]. It is given by

$$\omega^{(k)} = \frac{\ln |\lambda_-^{(k)}|}{\ln \lambda_+^{(k)}} , \qquad \omega^{(1)} = -1 , \quad \omega^{(2)} = 0 , \quad \omega^{(3)} \approx 0.317 > 0 . \qquad (12.29)$$

Thus, according to the Harris–Luck criterion, a layered aperiodicity introduced into the Ising model according to these substitution rules is *irrelevant* for $k = 1$, *marginal* for $k = 2$, and *relevant* for $k = 3$. So, contrary to what one might have expected, *deterministic* substitution sequences, albeit non-random, actually provide examples for all interesting classes of 'disorder'.

The MATHEMATICA program FibonacciChain.m that is included in the program package [41] offers the reader the opportunity to gain experience with substitution sequences and their properties. In addition, it also shows how quasiperiodic examples, such as the Fibonacci chain, can be derived from a *projection* of the 2D square lattice. This approach, known as the cut-and-project method,

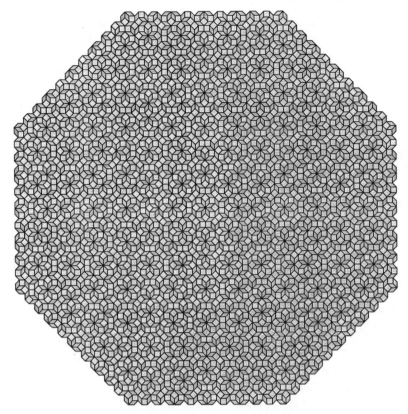

Fig. 12.4. Eightfold patch of the Ammann–Beenker tiling

can be generalized to higher dimensions to obtain quasiperiodic tilings (or in mathematical terminology, model sets [39]) with interesting symmetry properties. Roughly speaking, a certain part of the high-dimensional periodic lattice is projected onto a suitably chosen subspace. As an example, Fig. 12.4 shows a central patch of an eightfold symmetric planar tiling obtained from projection of the hypercubic lattice $\mathbb{Z}^4$. The tiling consists of squares and rhombi and is known as the octagonal or Ammann–Beenker tiling [42]. In addition, this geometric structure also admits an inflation/deflation symmetry that is analogous to the substitution rule for the 1D sequences. It consists of a dissection of the two basic tiles into copies of itself, such that the resulting tiling is, apart from an overall scaling, invariant under the procedure.

Both approaches are used in the *Mathematica* program OctagonalTiling.m [41]. The remaining programs ChairTiling.m and SphinxTiling.m deal with two further planar examples of inflation tilings, while GridMethod.m introduces a variant of the cut-and-project scheme [41]. The package PenrosePuzzle.m employs yet another method to construct the most famous among the zoo of quasiperiodic tilings, Penrose's tiling of the plane. Here, the two basic rhombi are marked by certain arrow decorations on their edges, and the task consists in filling the entire plane with tiles, without holes or overlaps, and without violating the *matching rules* that require arrow decorations on adjacent edges to match. The reader should be warned that this puzzle method, although it seems to be the simplest, is not suitable for constructing a large patch of the tiling. This is because the matching rules do not uniquely specify how to add tiles and, in particular, do not prevent implicit mistakes which will only be revealed when, at a certain stage, no further legal additions of tiles are possible. But if indeed the reader manages to fill the entire plane without violating the rules, then the resulting tiling will be a Penrose tiling with all the other 'magic' properties, such as an inflation/deflation symmetry.

12.5.2   Ising Models on Planar Aperiodic Graphs

In what follows, we shall consider planar Ising models defined on quasiperiodic graphs, and, subsequently, Ising quantum chains with coupling constants varying according to substitution sequences. The latter correspond to 2D classical Ising models with a layered aperiodicity, so that $d_m = 1$ in the Harris–Luck criterion, and $\omega_c = 0$.

It may be surprising, but it is indeed possible to construct planar aperiodic Ising models that are exactly solvable [14,43] in the sense of commuting transfer matrices [3]. For such systems, the coupling constants have to be chosen in a particular way to ensure integrability [43]. As a consequence, they have the rather unusual property that the free energy does *not* depend on the actual distribution of coupling constants, but only on the frequencies of the various couplings. Other evidence of the particular arrangement is the fact that the local magnetization, i.e., the expectation value of a spin at a certain site, does not depend on the site, but is uniform across the entire system. Consequently, the geometric arrangement of couplings does not play a role, and the critical behaviour is always the same as in the pure system. These models are nevertheless interesting counterexamples to the Harris–Luck criterion, because the previous statement also holds for arrangements with strong fluctuations that should have been relevant. However, the solvable examples are certainly very special. They possess some hidden underlying symmetry, and one should expect the criterion to hold for any *generic* distribution of coupling constants. Still, it reveals the limited predictive power of such criteria.

Ising models on quasiperiodic graphs, in particular the Penrose tiling, have been thoroughly investigated by means of Monte Carlo simulations and by approximate renormalization group treatments (see [14] and the literature cited therein). The coupling is usually taken as uniform along the bonds of the graph, so the fluctuations arise solely from the locally differing coordination numbers. This type of planar aperiodicity is irrelevant according to the Harris–Luck criterion, because the fluctuations are small for quasiperiodic cut-and-project sets, such as the Penrose tiling or the Ammann–Beenker tiling shown in Fig. 12.4. Essentially, this is due to the fact that the quasiperiodic tiling is the projection of a slab of a periodic lattice, and thus, despite being aperiodic, these tilings show a pronounced regularity. The numerical results unanimously corroborate the prediction of the Harris–Luck criterion and these models undoubtedly belong to the same universality class as the square-lattice Ising model.

Apart from these predominantly numerical approaches, some analytical techniques have also been employed. This concerns, for instance, high- and low-temperature expansions [44] that can be adapted to quasiperiodic systems. For the Penrose and Ammann–Beenker tiling, the high-temperature expansion of the relevant part $\tilde{f}$ of the free energy,

$$\tilde{f} = \sum_{n=2}^{\infty} g_{2n}\, w^{2n} \, , \qquad w = \tanh(J/k_{\mathrm{B}}T) \, , \tag{12.30}$$

where J denotes the coupling constant, has recently been calculated to 18th order in w [45]. The coefficients g_{2n} can be calculated from the frequencies of certain graphs of circumference $2n$ in the tiling, which can be calculated explicitly for cut-and-project sets. Furthermore, each graph carries a certain weight and these weights can be computed recursively. As an example, the expansion for the Ammann–Beenker tiling is given by

$$\tilde{f} = w^4 + \lambda w^6 + \left(47\tfrac{1}{2}-17\lambda\right) w^8 + (138-50\lambda)\, w^{10}$$
$$+ \left(803\tfrac{1}{3}-310\tfrac{1}{2}\lambda\right) w^{12} + (586\lambda-1220)\, w^{14} + \left(96\tfrac{3}{4}+295\tfrac{1}{2}\lambda\right) w^{16}$$
$$+ \left(46566\tfrac{1}{3}\lambda-108706\right) w^{18} + O(w^{20}) \, ,$$

where $\lambda = 1 + \sqrt{2}$. This may be compared with the square-lattice result [44]

$$\tilde{f} = w^4 + 2w^6 + 4\tfrac{1}{2}w^8 + 12w^{10} + 37\tfrac{1}{3}w^{12} + 130w^{14}$$
$$+ 490\tfrac{1}{4}w^{16} + 1958\tfrac{2}{3}w^{18} + O(w^{20}) \, ,$$

where the coefficients are rational numbers. From the radius of convergence of the series, and the behaviour close to it, one can in principle derive the critical temperature and the thermal critical exponent. However, the information about the critical behaviour that one can extract from the expansion is rather poor, because, in contrast to the square-lattice case, the extrapolated values show extremely strong fluctuations [45,46], and many more terms would be necessary in order to give reliable estimates of the critical temperature and the critical exponents. But this is not feasible; a total of 244 638 different subgraphs of the Ammann–Beenker tiling contribute to the coefficient g_{18}, as compared with a mere 1975 different graphs in the square-lattice case. A series analysis of other quantities such as the magnetic susceptibility may yield more stringent evidence, but this has not yet been calculated because the number of graphs that have to be considered is even larger.

Another method involves computing the zeros of the partition function in the complex temperature variable $z = \exp(2J/k_{\mathrm{B}}T)$. The set of zeros accumulates on certain curves or areas in the complex plane that separate different analytic domains, hence different phases of the system. Thus, a zero on the real positive axis will correspond to a phase transition point of the model, and the

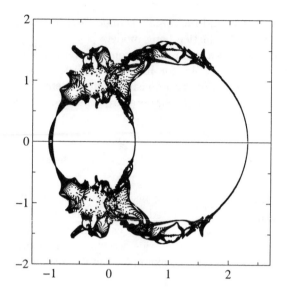

Fig. 12.5. Some partition function zeros in the complex variable $z = \exp(2J/k_BT)$ for a periodic approximant of the Ammann–Beenker tiling with 41 spins per unit cell

zeros close to this point contain information about the critical exponents. For periodic approximants of aperiodic tilings with rather large unit cells, the zero patterns can be calculated explicitly [46]. The result for an approximant of the Ammann–Beenker tiling, shown in Fig. 12.5, is rather involved, whereas for the square lattice case, the zeros are restricted to two circles with radius $\sqrt{2}$, centred at $z = \pm 1$. The two intersections of the zero pattern with the positive real axis correspond to the ferromagnetic ($J > 0$) and antiferromagnetic ($J < 0$) critical point, respectively, which are related to each other because the tiling is bipartite. The corresponding estimates of the critical temperature are given in Table 12.1. Clearly, this method allows a very precise determination of the critical temperature, in good agreement with the less accurate Monte Carlo results [47].

Table 12.1. Critical temperatures $w_c = \tanh(J/k_BT_c)$ for periodic approximants of the Ammann–Beenker tiling, and estimated value for the aperiodic tiling

m	N	w_c
1	7	0.396 850 570
2	41	0.396 003 524
3	239	0.395 985 346
4	1393	0.395 984 811
5	8119	0.395 984 795
6	47321	0.395 984 795
∞	∞	0.395 984 79(1)

12.5.3   Aperiodic Ising Quantum Chains

We now turn our attention to the case of 1D aperiodicity and consider Ising quantum chains with aperiodically modulated coupling constants [14]

$$H = -\frac{1}{2}\sum_{j=1}^{N} \varepsilon_j\, \sigma_j^x \sigma_{j+1}^x - \frac{1}{2}\sum_{j=1}^{N} h_j \sigma_j^z , \tag{12.31}$$

where, in contrast to (12.14), the couplings $\varepsilon_j \in \{\varepsilon_a, \varepsilon_b\}$ and the transverse fields $h_j \in \{h_a, h_b\}$ are chosen according to the jth letter of an aperiodic substitution sequence.

Such models can be treated by an exact real-space renormalization approach, exploiting the recursive structure that is already inherent in the substitution sequence [23,48–51]. The basic idea behind this approach is that a renormalization step *reverses* a substitution step, i.e., the Hamiltonian is transformed into a Hamiltonian at the previous substitution level, which just differs from the original Hamiltonian by renormalized values of the parameters. The renormalization transformation then becomes a mapping in the finite-dimensional parameter space of the Hamiltonian, and the renormalization transformation is exact because no additional parameters have to be introduced. This is very interesting because it provides one of the few examples with an exact renormalization transformation for a non-trivial model, keeping in mind that the usual decimation approach does not work even for the square-lattice Ising model. Furthermore, it works for *arbitrary* substitution sequences [23,50], not just for our examples (12.26), and can even be applied to the case of random substitutions [23]. This encompasses a large class of models with different fluctuations in the couplings and, consequently, different physical behaviour. For the sake of space, we cannot go into detail here, but just briefly sketch the calculation.

The first step consists in mapping the Hamiltonian (12.31) onto a model of free fermions with creation and annihilation operators $\eta_j^\dagger$ and η_j,

$$H = \sum_{j=1}^{N} \Lambda_j\, \eta_j^\dagger \eta_j + C , \quad \{\eta_j, \eta_\ell\} = \{\eta_j^\dagger, \eta_\ell^\dagger\} = 0 , \quad \{\eta_j^\dagger, \eta_\ell\} = \delta_{j,\ell} , \tag{12.32}$$

by means of a Jordan–Wigner transformation [20]. In this way, the task of computing the spectrum of (12.31) is reduced from the diagonalization of a $2^N \times 2^N$ matrix to that of the $2N \times 2N$ eigenvalue problem

$$
\begin{pmatrix}
0 & h_1 & 0 & 0 & 0 & \cdots & 0 & \varepsilon_N \\
h_1 & 0 & \varepsilon_1 & 0 & 0 & \cdots & 0 & 0 \\
0 & \varepsilon_1 & 0 & h_2 & 0 & \cdots & 0 & 0 \\
0 & 0 & h_2 & 0 & \varepsilon_2 & \cdots & 0 & 0 \\
\vdots & \vdots & \ddots & \ddots & \ddots & \ddots & & \vdots \\
0 & 0 & \cdots & 0 & h_{N-1} & 0 & \varepsilon_{N-1} & 0 \\
0 & 0 & \cdots & 0 & 0 & \varepsilon_{N-1} & 0 & h_N \\
\varepsilon_N & 0 & \cdots & 0 & 0 & 0 & h_N & 0
\end{pmatrix}
\begin{pmatrix}
\Phi_1 \\ \Psi_1 \\ \Phi_2 \\ \Psi_2 \\ \vdots \\ \Psi_{N-1} \\ \Phi_N \\ \Psi_N
\end{pmatrix}
= \Lambda
\begin{pmatrix}
\Phi_1 \\ \Psi_1 \\ \Phi_2 \\ \Psi_2 \\ \vdots \\ \Psi_{N-1} \\ \Phi_N \\ \Psi_N
\end{pmatrix} , \tag{12.33}
$$

whose N positive solutions $0 \leq \Lambda_1 \leq \Lambda_2 \leq \ldots \leq \Lambda_N$ determine the entire spectrum of the Hamiltonian (12.31). The quantum phase transition appears when the lowest excitation energy tends to zero, $\Lambda_1 \to 0$, in the thermodynamic limit $N \to \infty$, so we can concentrate on the low-energy behaviour.

The renormalization equations can now be derived by exploiting the recursive structure of the sequence, effectively combining those sites that belong to a single letter at the previous stage of the substitution rule. This results in a mapping in parameter space, which can be expanded in powers of Λ at the fixed point $\Lambda = 0$. This corresponds to the critical point given by the condition

$$1 = \lim_{N \to \infty} \left| \frac{\varepsilon_1}{h_1} \frac{\varepsilon_2}{h_2} \cdots \frac{\varepsilon_N}{h_N} \right|^{1/N} = \left| \frac{\varepsilon_a}{h_a} \right|^{p_a} \left| \frac{\varepsilon_b}{h_b} \right|^{p_b} \tag{12.34}$$

where the last equality applies to the two-letter case discussed above. It can be parametrized as $|\varepsilon_a/h_a| = r^{-p_b}$ and $|\varepsilon_b/h_b| = r^{p_a}$ with $r \in \mathbb{R}^+$. The mapping determines the finite-size scaling of the smallest excitation energy Λ_1, and thus the critical behaviour. In complete accordance with the Harris–Luck criterion, we find that $\Lambda_1 \sim v(r)N^{-1}$ for irrelevant aperiodicity, as in the periodic case. For marginal aperiodicity, we observe $\Lambda_1 \sim N^{-z(r)}$ with a coupling-dependent non-universal exponent $z(r)$, corresponding to an anisotropic scaling of the correlation length with different exponents $\nu_\parallel$ and $\nu_\perp$. For relevant aperiodicity, we find that the lowest excitation energy vanishes exponentially, $\Lambda_1 \sim \exp[-\Delta(r)N^\omega]$, where ω is the fluctuation exponent (12.19). In fact, we can also calculate the non-universal coupling-dependent terms. For our examples, this yields $v(r) = 2\ln(r)/(r - 1/r)$ for the Fibonacci chain, the exponent $z(r) = \ln(r^{1/3} + r^{-1/3})/\ln(2)$ for the period-doubling sequence, and the coefficient $\Delta(r) = \sqrt{2\sqrt{13} - 7}|\ln(r)|$ for the binary non-Pisot sequence, i.e., for $k = 1, 2, 3$ in (12.26), respectively. This renormalization approach can also be applied to obtain information about the eigenvector in (12.33), thus determining the behaviour of the surface magnetization in an open quantum chain [51]. Furthermore, other free-fermion models such as the XY chain can also be treated [23,52].

12.6   Summary and Conclusions

The article gives a brief overview of recent results for aperiodic Ising models. According to the personal taste and scientific experience of the author, aperiodically ordered Ising models receive most attention, while random Ising models are only discussed from a qualitative perspective. It is shown that heuristic scaling arguments like the Harris–Luck criterion can provide a rather powerful tool which, in most cases, correctly predicts whether or not a certain disorder will affect the critical behaviour. Still, there are a number of tacit assumptions behind the scenes, and there do exist some exceptional systems which defy the relevance criterion, as shown by an exactly solvable example.

It is quite a difficult task to treat randomly disordered or aperiodically ordered planar Ising models analytically, and rather sophisticated techniques have been employed. However, it turns out that the Ising quantum chain with coupling constants modulated according to substitution sequences can be solved by an exact real-space renormalization approach, thus proving the validity of the Harris–Luck criterion for this entire class of models and, in addition, providing information about the non-universal coefficients that enter the scaling behaviour.

Acknowledgements. The author would like to thank Michael Baake, Anton Bovier, Joachim Hermisson, Wolfhard Janke, and Przemysław Repetowicz for helpful discussions.

References

1. R. Balian: *From Microphysics to Macrophysics*, Vol. I (Springer, Berlin 1991)
2. D. Frenkel: Physica A **263**, 26 (1999)
3. R.J. Baxter: *Exactly Solved Models in Statistical Mechanics* (Academic Press, London 1982)
4. W. Lenz: Physikalische Zeitschrift **21**, 613 (1920)
5. E. Ising: Zeitschrift für Physik **31**, 253 (1925)
6. L. Onsager: Phys. Rev. **65**, 117 (1944)
7. J.-B. Suck, M. Schreiber, P. Häussler (eds.): *Quasicrystals* (Springer, Berlin 2002) to appear
8. L.P. Kadanoff: 'Scaling, universality and operator algebras'. In: *Phase Transitions and Critical Phenomena* Vol. 5a, ed. by C. Domb, J.L. Lebowitz (Academic Press, London 1976) pp. 1–34
9. J.L. Cardy: *Scaling and Renormalization in Statistical Physics* (Cambridge University Press, Cambridge 1996)
10. P. DiFrancesco, P. Mathieu, D. Sénéchal: *Conformal Field Theory* (Springer, New York 1997)
11. B.M. McCoy, T.T. Wu: *The Two-Dimensional Ising Model* (Harvard University Press, Cambridge, Massachusetts 1973)
12. H.-O. Georgii: *Gibbs Measures and Phase Transitions* (de Gruyter, Berlin 1988)
13. S. Kobe: J. Stat. Phys. **88**, 991 (1997); Phys. Bl. **54**, 917 (1998)
14. U. Grimm, M. Baake: 'Aperiodic Ising models'. In: *The Mathematics of Long-Range Aperiodic Order*, ed. by R.V. Moody (Kluwer, Dordrecht 1997) pp. 199–237
15. R. Peierls: Proc. Cambridge Philos. Soc. **32**, 477 (1936)
16. R.S. Ellis: *Entropy, Large Deviations, and Statistical Mechanics* (Springer, New York 1985)
17. H.A. Kramers, G.H. Wannier: Phys. Rev. **60**, 252 (1941)
18. C.N. Yang: Phys. Rev. **85**, 808 (1952)
19. R.B. Stinchcombe: J. Phys. C **6**, 2459 (1973)
20. E. Lieb, T. Schultz, D. Mattis: Ann. Phys. (NY) **16**, 407 (1961)
21. J.B. Kogut: Rev. Mod. Phys. **51**, 659 (1979)
22. C. Chatelain, P.E. Berche, B. Berche: Eur. Phys. J. B **7**, 439 (1999)
23. J. Hermisson: *Aperiodische Ordnung und Magnetische Phasenübergänge* (Shaker, Aachen 1999)
24. A.B. Harris: J. Phys. C: Solid State Phys. **7**, 1671 (1974)
25. J.M. Luck: J. Stat. Phys. **72**, 417 (1993); Europhys. Lett. **24**, 359 (1993)

210   Uwe Grimm

26. R.B. Stinchcombe: 'Dilute magnetism'. In: *Phase Transitions and Critical Phenomena* Vol. 7, ed. by C. Domb, J.L. Lebowitz (Academic Press, London 1983) pp. 151–280
27. W. Selke: 'Monte Carlo simulations of dilute Ising models'. In: *Computer Simulation Studies in Condensed Matter Physics IV*, ed. by D.P. Landau, K.K. Mon, H.-B. Schüttler (Springer, Berlin 1993) pp. 18–27
28. W. Selke, L.N. Shchur, A.L. Talapov: 'Monte Carlo simulations of dilute Ising models'. In: *Annual Reviews of Computational Physics I*, ed. by D. Stauffer (World Scientific, Singapore 1994) pp. 17–54
29. Vik. S. Dotsenko, Vl. S. Dotsenko: Adv. Phys. **32**, 129 (1983)
30. B.N. Shalaev: Sov. Phys. Solid State **26**, 1811 (1984); Phys. Rep. **237**, 129 (194)
31. R. Shankar: Phys. Rev. Lett. **58**, 2466 (1987); **61**, 2390 (1988)
32. A.W.W. Ludwig: Phys. Rev. Lett. **61**, 2388 (1988); Nucl. Phys. B **330**, 639 (1990)
33. A. Roder, J. Adler, W. Janke: Phys. Rev. Lett. **80**, 4697 (1998)
34. W. Janke, R. Villanova, 'Ising spins on 3D random lattices'. In: *Computer Simulation Studies in Condensed Matter Physics XI*, ed. by D.P. Landau, H.-B. Schüttler (Springer, Berlin 1999) pp. 22–26
35. C.M. Newman: *Topics in Disordered Systems* (Birkhäuser, Basel 1997)
36. A. Bovier, P. Picco (eds.): *Mathematical Aspects of Spin Glasses and Neural Networks*, Progress in Probability 41 (Birkhäuser, Boston 1998)
37. H.-O. Georgii, O. Häggström, C. Maes: 'The random geometry of equilibrium phases'. In: *Phase Transitions and Critical Phenomena*, Vol. 18, ed. by C. Domb, J.L. Lebowitz (Academic Press, London 2001) pp. 1–142
38. H.-O. Georgii, Y. Higuchi: J. Math. Phys. **41**, 1153 (2000)
39. M. Baake: 'A guide to mathematical quasicrystals', math-ph/9901014, to appear in [7]
40. S. Wolfram: *Mathematica: A System for Doing Mathematics by Computer*, 2nd ed. (Addison–Wesley, Reading, Massachusetts 1991)
41. U. Grimm, M. Schreiber: 'Aperiodic tilings on the computer', cond-mat/9903010, to appear in [7]
42. R. Ammann, B. Grünbaum, G. Shephard: Discrete Comput. Geom. **8**, 1 (1992)
43. M. Baake, U. Grimm, R.J. Baxter: Int. J. Mod. Phys. B **8**, 3579 (1994)
44. C. Domb: 'Ising model' in: *Phase Transitions and Critical Phenomena* Vol. 3, ed. by C. Domb, M.S. Green (Academic Press, London 1974) pp. 357–458
45. P. Repetowicz, U. Grimm, M. Schreiber: J. Phys. A: Math. Gen. **32**, 4397 (1999)
46. P. Repetowicz, U. Grimm, M. Schreiber: Mat. Sci. Eng. A **294–296**, 638 (2000)
47. O. Redner, M. Baake: J. Phys. A: Math. Gen. **33**, 3097 (2000)
48. F. Iglói, L. Turban: Phys. Rev. Lett. **77**, 1206 (1996)
49. F. Iglói, L. Turban, D. Karevski, F. Szalma: Phys. Rev. B **56**, 11031 (1997)
50. J. Hermisson, U. Grimm, M. Baake: J. Phys. A: Math. Gen. **30**, 7315 (1997)
51. J. Hermisson, U. Grimm: Phys. Rev. B **57**, R673 (1998)
52. J. Hermisson: J. Phys. A: Math. Gen. **33**, 57 (2000)
53. See www.tu-chemnitz.de/physik/HERAUS/2000/Springer.html

13 Quantum Phase Transitions

Thomas Vojta

Summary. Quantum phase transitions occur at zero temperature when some non-thermal parameter like pressure, chemical composition or magnetic field is changed. They are caused by quantum fluctuations which are a consequence of Heisenberg's uncertainty principle. These lecture notes give a pedagogical introduction to quantum phase transitions. After collecting a few basic facts about phase transitions and critical behavior we discuss the importance of quantum mechanics and the relation between quantum and classical transitions as well as their experimental relevance. As a primary example we then consider the Ising model in a transverse field. We also briefly discuss quantum phase transitions in itinerant electron systems and their connection to non-Fermi liquid behavior.

13.1 Introduction: From the Melting of Ice to Quantum Criticality

When a piece of ice is first taken out of the freezer, its properties change only slowly with increasing temperature. At 0°C, however, a drastic change happens. The thermal motion of the water molecules becomes so strong that it destroys the crystal structure. The ice melts, and a new phase of water forms, the liquid phase. This process is an example of a phase transition. At the transition temperature of 0°C, the solid and liquid phases of water coexist. A finite amount of heat, the so-called latent heat, is needed to transform the ice into liquid water. Phase transitions which involve latent heat are usually called first-order transitions.

Another well known example of a phase transition is the magnetic transition of iron. At room temperature, iron is ferromagnetic, i.e., it possesses a spontaneous magnetization. With increasing temperature the magnetization decreases continuously. It vanishes at the Curie temperature of 770°C, and above this temperature iron is paramagnetic. This is a phase transition from a ferro- to a paramagnet. In contrast to the previous example, there is no phase coexistence at the transition temperature. Instead, the two phases become indistinguishable. Consequently, there is no latent heat. Phase transitions of this type are called continuous transitions or second-order transitions. They are the result of competition between order and thermal fluctuations. When approaching the transition point the typical length and time scales of the fluctuations diverge, leading to singularities in many physical quantities.

The diverging correlation length at a continuous phase transition was first observed in 1869 in a famous experiment by Andrews [1]. He discovered that the properties of the liquid and the vapor phases of carbon dioxide became indistinguishable at a temperature of about 31°C and pressure 73 atm. In the vicinity of this point the carbon dioxide became opaque, i.e., it strongly scattered visible light, indicating that the length scale of the density fluctuations had reached the wavelength of the light. Andrews called this special point in the phase diagram the critical point and the strong light scattering in its vicinity the critical opalescence.

More formally, one can define a phase transition as the occurrence of a singularity in the thermodynamic quantities as functions of the external parameters. Phase transitions have played, and continue to play, an essential role in shaping our world. The large-scale structure of the universe is the result of a sequence of phase transitions during the very early stages of its development. Even our everyday life is unimaginable without the never-ending transformations of water between ice, liquid and vapor. Understanding the properties of phase transitions, and in particular those of critical points, has been a great challenge for theoretical physics. More than a century went by between the first discoveries and the emergence of a consistent picture. However, the theoretical concepts established during this development, viz., scaling and the renormalization group [2], now belong to the central paradigms of modern physics.

Over the last decade, considerable attention has focused on a class of phase transitions which are qualitatively very different from the examples discussed above. These new transitions occur at zero temperature when a non-thermal parameter like pressure, chemical composition or magnetic field is changed. The fluctuations which destroy the long-range order in these transitions cannot be of thermal nature since thermal fluctuations do not exist at zero temperature. Instead, they are quantum fluctuations which are a consequence of Heisenberg's uncertainty principle. For this reason, phase transitions at zero temperature are called quantum phase transitions, in contrast to thermal or classical phase transitions at finite temperatures. (The justification for calling all thermal phase transitions classical will become clear in Sect. 13.3.)

As an illustration of classical and quantum phase transitions, we show the magnetic phase diagram of the transition metal compound MnSi in Fig. 13.1. At ambient pressure, MnSi is a paramagnetic metal for temperatures larger than $T_c = 30$ K. Below T_c it orders ferromagnetically[1] but remains metallic. This transition is a *thermal* continuous phase transition analogous to that in iron discussed above. Applying pressure reduces the transition temperature, and at about 14 kbar the magnetic phase vanishes. Thus, at 14 kbar MnSi undergoes a quantum phase transition from a ferro- to a paramagnet. While this is a very obvious quantum phase transition, its properties are rather complex due to the interplay between the magnetic degrees of freedom and other fermionic excitations (see [4] for details).

---

[1] Actually, the ordered state is a spin spiral in the (111) direction of the crystal. Its wavelength is very long (200 Å) so that the material behaves like a ferromagnet for most purposes.

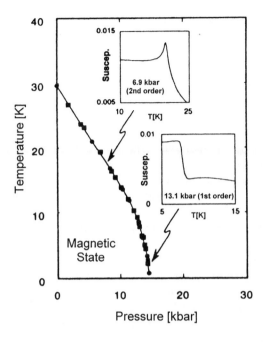

Fig. 13.1. Magnetic phase diagram of MnSi (after [3]). The peculiar behavior of the magnetic susceptibility shown in the *insets* is a consequence of the interplay between the magnetic and other fermionic degrees of freedom [4]

At first glance quantum phase transitions seem to be a purely academic problem since they occur at isolated parameter values and at zero temperature which is inaccessible in an experiment. However, within the last decade it has turned out that the opposite is true. Quantum phase transitions do have important, experimentally relevant consequences, and they are believed to provide keys to many new and exciting phenomena in condensed matter physics, such as the quantum Hall effects, the localization problem, non-Fermi liquid behavior in metals or high-T_c superconductivity.

These lecture notes are intended as a pedagogical introduction to quantum phase transitions. In Sect. 13.2, we first collect a few basic facts about phase transitions and critical behavior. In Sect. 13.3, we then investigate the importance of quantum mechanics for the physics of phase transitions and the relation between quantum and classical transitions. Section 13.4 is devoted to a detailed analysis of the physics in the vicinity of a quantum critical point. We then discuss two examples. In Sect. 13.5, we consider one of the paradigmatic models in this field, the Ising model in a transverse field, and in Sect. 13.6, we briefly discuss quantum phase transitions in itinerant electron systems and their connection to non-Fermi liquid behavior.

Those readers who wish to learn more details about quantum phase transitions and the methods used to study them, should consult one of the recent review articles, e.g., [5–7], or the excellent text book on quantum phase transitions by Sachdev [8]. Related topics are also discussed in Chaps. 14, 16 and 17 of this volume.

13.2   Basic Concepts of Phase Transitions and Critical Behavior

In this section we briefly collect the basic concepts of continuous phase transitions and critical behavior which are necessary for the later discussions. For a detailed presentation, the reader is referred to one of the textbooks on this subject, e.g., those by Ma [9] or Goldenfeld [10].

A continuous phase transition can usually be characterized by an order parameter, a concept first introduced by Landau [11]. An order parameter is a thermodynamic quantity that is zero in one phase (the disordered phase) and non-zero and non-unique in the other (the ordered phase). Very often the choice of an order parameter for a particular transition is obvious, as in the ferromagnetic transition, where the total magnetization is an order parameter. Sometimes, however, finding an appropriate order parameter is a complicated problem in itself, e.g., for the disorder-driven localization–delocalization transition of non-interacting electrons.

While the thermodynamic average of the order parameter is zero in the disordered phase, its fluctuations are non-zero. If the phase transition point, i.e., the critical point, is approached, the spatial correlations of the order parameter fluctuations become long-ranged. Close to the critical point, their typical length scale, the correlation length ξ, diverges as

$$\xi \propto |t|^{-\nu} , \tag{13.1}$$

where ν is the correlation length critical exponent and t is some dimensionless measure of the distance from the critical point. If the transition occurs at a non-zero temperature T_c, it can be defined as $t = (T - T_c)/T_c$. The divergence of the correlation length when approaching the transition is illustrated in Fig. 13.2, which shows computer simulation results for the phase transition in a two-dimensional Ising model.

In addition to the long-range correlations in space, there are analogous long-range correlations of the order parameter fluctuations in time. The typical time scale for a decay of the fluctuations is the correlation (or equilibration) time τ_c. As the critical point is approached, the correlation time diverges as

$$\tau_c \propto \xi^z \propto |t|^{-\nu z} , \tag{13.2}$$

where z is the dynamical critical exponent. Close to the critical point there is no characteristic length scale other than ξ and no characteristic time scale other than τ_c.[2] As already noted by Kadanoff [12], this is the physics behind Widom's scaling hypothesis [13], which we now discuss.

Let us consider a classical system, characterized by its Hamiltonian

$$H(p_i, q_i) = H_{\mathrm{kin}}(p_i) + H_{\mathrm{pot}}(q_i) , \tag{13.3}$$

---

[2] Note that a microscopic cutoff scale must be present to explain non-trivial critical behavior (see for example Goldenfeld [10] for details). In a solid, such a scale is the lattice spacing, for example.

Fig. 13.2. Snapshots of the spin configuration of a two-dimensional Ising model at different temperatures (a *black dot* corresponds to spin up, an *empty site* means spin down). From *left* to *right*: $T = 2T_c$, $T = 1.3T_c$, $T = T_c$. The correlation length increases with decreasing temperature and diverges at $T = T_c$. Here the system fluctuates at all length scales

where q_i and p_i are the generalized coordinates and momenta, and H_{kin} and H_{pot} are the kinetic and potential energies, respectively.[3] In such a system, statics and dynamics decouple, i.e., the momentum and position sums in the partition function

$$Z = \int \prod \mathrm{d}p_i \, e^{-H_{\text{kin}}/k_B T} \int \prod \mathrm{d}q_i \, e^{-H_{\text{pot}}/k_B T} = Z_{\text{kin}} Z_{\text{pot}} \tag{13.4}$$

are completely independent from one another. The kinetic contribution to the free energy density $f = -(k_B T/V) \log Z$ will not usually display any singularities, since it derives from the product of simple Gaussian integrals. Therefore one can study phase transitions and the critical behavior using effective time-independent theories like the classical Landau–Ginzburg–Wilson theory. In this type of theory, the free energy is expressed as a functional of the order parameter $M(r)$, which is time-independent but fluctuates in space. All other degrees of freedom have been integrated out in the derivation of the theory starting from a microscopic Hamiltonian. In its simplest form [2,11,14], valid for example for an Ising ferromagnet, the Landau–Ginzburg–Wilson functional $\Phi[M]$ reads

$$\Phi[M] = \int \mathrm{d}^d r \, M(r) \left[-\frac{\partial^2}{\partial r^2} + t \right] M(r) + u \int \mathrm{d}^d r \, M^4(r) - B \int \mathrm{d}^d r \, M(r) \,,$$

$$Z = \int D[M] e^{-\Phi[M]} \,, \tag{13.5}$$

where B is the field conjugate to the order parameter (the magnetic field, in the case of a ferromagnet).

Since close to the critical point the correlation length is the only relevant length scale, the physical properties must be unchanged if we rescale all lengths in the system by a common factor b, and at the same time adjust the external

---

[3] Velocity-dependent potentials, as in the case of charged particles in an electromagnetic field, are excluded.

Table 13.1. Definitions of the commonly used critical exponents in the language of magnetism, i.e., the order parameter is the magnetization $m = \langle M \rangle$, and the conjugate field is a magnetic field B. t denotes the distance from the critical point and d is the space dimensionality. The exponent y_B defined in (13.6) is related to δ by $y_B = d\delta/(1+\delta)$

	Exponent	Definition	Conditions
Specific heat	α	$c \propto |t|^{-\alpha}$	$t \to 0, B = 0$
Order parameter	β	$m \propto (-t)^{\beta}$	$t \to 0^-, B = 0$
Susceptibility	γ	$\chi \propto |t|^{-\gamma}$	$t \to 0, B = 0$
Critical isotherm	δ	$B \propto |m|^{\delta}\text{sign}(m)$	$B \to 0, t = 0$
Correlation length	ν	$\xi \propto |t|^{-\nu}$	$t \to 0, B = 0$
Correlation function	η	$G(r) \propto |r|^{-d+2-\eta}$	$t = 0, B = 0$
Dynamical exponent	z	$\tau_c \propto \xi^z$	$t \to 0, B = 0$

parameters in such a way that the correlation length retains its old value. This observation gives rise to the homogeneity relation for the free energy density,

$$f(t, B) = b^{-d}f(t\,b^{1/\nu}, B\,b^{y_B})\,. \tag{13.6}$$

Here y_B is another critical exponent and d is the space dimensionality. The scale factor b is an arbitrary positive number. Analogous homogeneity relations for other thermodynamic quantities can be obtained by differentiating the free energy. The homogeneity law (13.6) was first obtained phenomenologically by Widom [13]. It can be derived from first principles within the framework of renormalization group theory [2].

In addition to the critical exponents ν, y_B and z defined above, a number of other exponents are in common use. They describe the dependence of the order parameter and its correlations on the distance from the critical point and on the field conjugate to the order parameter. The definitions of the most commonly used critical exponents are summarized in Table 13.1.

Note that not all the exponents defined in Table 13.1 are independent from one another. The four thermodynamic exponents $\alpha, \beta, \gamma, \delta$ can all be obtained from the free energy (13.6), which contains only two independent exponents. They are therefore connected by the so-called scaling relations

$$2 - \alpha = 2\beta + \gamma\,, \qquad 2 - \alpha = \beta(\delta + 1)\,. \tag{13.7}$$

Analogously, the exponents of the correlation length and correlation function are connected by two so-called hyperscaling relations

$$2 - \alpha = d\nu\,, \qquad \gamma = (2 - \eta)\nu\,. \tag{13.8}$$

Since statics and dynamics decouple in classical statistics the dynamical exponent z is completely independent from all the others.

The set of critical exponents completely characterizes the critical behavior at a particular phase transition. One of the most remarkable features of continuous phase transitions is universality, i.e., the fact that the critical exponents are the same for entire classes of phase transitions which may occur in very different physical systems. These classes, the so-called universality classes, are determined only by the symmetries of the Hamiltonian and the spatial dimensionality of the system. This implies that the critical exponents of a phase transition occurring in nature can be determined exactly (at least in principle) by investigating any simplistic model system belonging to the same universality class, a fact that makes the field very attractive for theoretical physicists. The mechanism behind universality is again the divergence of the correlation length. Close to the critical point the system effectively averages over large volumes rendering the microscopic details of the Hamiltonian unimportant.

The critical behavior at a particular transition is crucially determined by the relevance or irrelevance of order parameter fluctuations. It turns out that fluctuations become increasingly important if the spatial dimensionality of the system is reduced. Above a certain dimension, called the upper critical dimension d_c^+, fluctuations are irrelevant, and the critical behavior is identical to that predicted by mean-field theory (for systems with short-range interactions and a scalar or vector order parameter $d_c^+ = 4$). Between d_c^+ and a second special dimension, called the lower critical dimension d_c^-, a phase transition still exists but the critical behavior is different from mean-field theory. Below d_c^- fluctuations become so strong that they completely suppress the ordered phase.

13.3   How Important is Quantum Mechanics?

The question to what extent quantum mechanics is important for understanding a continuous phase transition is a multi-layered question. One may ask, for example, whether quantum mechanics is necessary to explain the existence and properties of the ordered phase. This question can only be decided on a case-by-case basis, and very often quantum mechanics is essential, as for superconductors, for example. A different question to ask would be how important quantum mechanics is for the asymptotic behavior close to the critical point and thus for the determination of the universality class the transition belongs to.

It turns out that the latter question has a remarkably clear and simple answer: quantum mechanics does *not* play any role for the critical behavior if the transition occurs at a finite temperature. It does play a role, however, at zero temperature. In the following, we will first give a simple argument explaining these facts. To do so it is useful to distinguish fluctuations with predominantly thermal and quantum character depending on whether their thermal energy $k_B T$ is larger or smaller than the quantum energy scale $\hbar\omega_c$. We have seen in the preceeding section that the typical time scale τ_c of the fluctuations diverges as a continuous transition is approached. Correspondingly, the typical frequency scale ω_c goes to zero and with it the typical energy scale

$$\hbar\omega_c \propto |t|^{\nu z} . \tag{13.9}$$

Quantum fluctuations will be important as long as this typical energy scale is larger than the thermal energy $k_B T$. If the transition occurs at some finite temperature T_c, quantum mechanics will therefore become unimportant for $|t| < t_x$, with the crossover distance t_x given by

$$t_x \propto T_c^{1/\nu z} . \tag{13.10}$$

We thus find that critical behavior asymptotically close to the transition is entirely classical if the transition temperature T_c is non-zero. This justifies calling all finite-temperature phase transitions classical transitions, even if the properties of the ordered state are completely determined by quantum mechanics, as is the case, for example, for the superconducting phase transition of mercury at $T_c = 4.2$ K. In these cases, quantum fluctuations are obviously important on microscopic scales, while classical thermal fluctuations dominate on the macroscopic scales that control critical behavior. This also implies that only universal quantities like the critical exponents will be independent of quantum mechanics, while non-universal quantities like the critical temperature will generally depend on quantum mechanics.

If, however, the transition occurs at zero temperature as a function of a non-thermal parameter like the pressure p, the crossover distance t_x equals zero, since there are no thermal fluctuations. [Note that, at zero temperature, the distance t from the critical point cannot be defined via the reduced temperature. Instead, one can define $t = (p - p_c)/p_c$.] Thus, at zero temperature the condition $|t| < t_x$ is never fulfilled, and quantum mechanics will be important for the critical behavior. Consequently, transitions at zero temperature are called quantum phase transitions.

Let us now generalize the homogeneity law (13.6) to the case of a quantum phase transition. We consider a system characterized by a Hamiltonian H. In a quantum problem, the kinetic and potential part of H do not generally commute. In contrast to the classical partition function (13.4), the quantum mechanical partition function does *not* factorize, i.e., statics and dynamics are always coupled. The canonical density operator $e^{-H/k_B T}$ looks exactly like a time evolution operator in imaginary time τ if one identifies

$$1/k_B T = \tau = -i\Theta/\hbar , \tag{13.11}$$

where Θ denotes the real time. This naturally leads to the introduction of an imaginary time direction into the system. An order parameter field theory analogous to the classical Landau–Ginzburg–Wilson theory (13.5) therefore needs to be formulated in terms of space- and time-dependent fields. The simplest example of a quantum Landau–Ginzburg–Wilson functional, valid for an Ising model in a transverse field (see section 13.5), reads

$$\Phi[M] = \int_0^{1/k_B T} d\tau \int d^d r\, M(\mathbf{r}, \tau) \left[-\frac{\partial^2}{\partial r^2} - \frac{\partial^2}{\partial \tau^2} + t \right] M(\mathbf{r}, \tau) \tag{13.12}$$

$$+ u \int_0^{1/k_B T} d\tau \int d^d r\, M^4(\mathbf{r}, \tau) - B \int_0^{1/k_B T} d\tau \int d^d r\, M(\mathbf{r}, \tau) .$$

Let us note that the coupling of statics and dynamics in quantum statistical mechanics also leads to the fact that the universality classes for quantum phase transitions are smaller than those for classical transitions. Systems which belong to the same classical universality class may display different quantum critical behavior, if their dynamics differ.

The classical homogeneity law (13.6) for the free energy density can now easily be adapted to the case of a quantum phase transition. At zero temperature the imaginary time acts similarly to an additional spatial dimension since the extension of the system in this direction is infinite. According to (13.2), time scales like the zth power of a length. [In the simple example (13.12) space and time enter the theory symmetrically, leading to $z = 1$.] Therefore, the homogeneity law for the free energy density at zero temperature reads

$$f(t, B) = b^{-(d+z)} f(t\, b^{1/\nu}, B\, b^{y_B})\,. \tag{13.13}$$

Comparing this relation to the classical homogeneity law (13.6) directly shows that a quantum phase transition in d spatial dimensions is equivalent to a classical transition in $d + z$ spatial dimensions. Thus, for a quantum phase transition the upper critical dimension, above which mean-field critical behavior becomes exact, is reduced by z compared to the corresponding classical transition. Note, however, that the mapping of a quantum phase transition to the equivalent classical transition generally leads to unusual anisotropic classical systems. Furthermore, the mapping is valid for the thermodynamics only. Other properties like the real-time dynamics at finite temperatures require more careful considerations (see [8], for example).

Now the attentive reader may again ask: why are quantum phase transitions more than an academic problem? Any experiment is done at a non-zero temperature where, as we have explained above, the asymptotic critical behavior is classical. The answer is provided by the crossover condition (13.10). If the transition temperature T_c is very small, quantum fluctuations will remain important down to very small t, i.e., very close to the phase boundary. At a more technical level, the behavior at small but non-zero temperatures is determined by the crossover between two types of critical behavior, viz. quantum critical behavior at $T = 0$ and classical critical behavior at non-zero temperatures. Since the 'extension of the system in the imaginary time direction' is given by the inverse temperature $1/k_B T$ the corresponding crossover scaling is equivalent to finite size scaling in the imaginary time direction. The crossover from quantum to classical behavior will occur when the correlation time τ_c reaches $1/k_B T$ which is equivalent to the condition (13.10). By adding the temperature as an explicit parameter and taking into account that in the imaginary-time formalism it scales like an inverse time (13.11), we can generalize the quantum homogeneity law (13.13) to finite temperatures,

$$f(t, B, T) = b^{-(d+z)} f(t\, b^{1/\nu}, B\, b^{y_B}, T\, b^z)\,. \tag{13.14}$$

Once the critical exponents z, ν, and y_B and the scaling function f are known this relation completely determines the thermodynamic properties close to the quantum phase transition.

13.4   Quantum Critical Points

We now use the general scaling picture developed in the last section to discuss the physics in the vicinity of the quantum critical point. There are two qualitatively different types of phase diagram depending on the existence or nonexistence of long-range order at finite temperatures. These phase diagrams are shown schematically in Fig. 13.3. Here p stands for the (non-thermal) parameter which tunes the quantum phase transition. In addition to the phase boundary, the phase diagrams show a number of crossover lines where the properties of the system change smoothly. They separate regions in which fluctuations have different character.

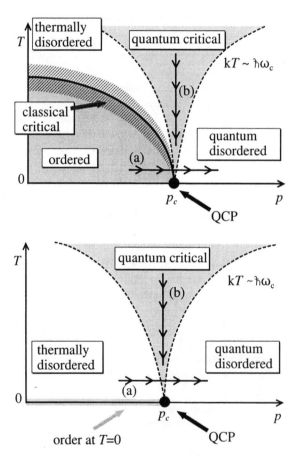

Fig. 13.3. Schematic phase diagrams in the vicinity of a quantum critical point (QCP) for situations where an ordered phase exists at finite temperatures (*top*) and situations where order exists at zero temperature only (*bottom*). The *solid line* marks the boundary between ordered and disordered phases. The different regions and paths (**a**) and (**b**) are discussed in the text

The first type of phase diagram describes situations where an ordered phase exists at finite temperatures. As discussed in the last section, classical fluctuations will dominate in the vicinity of the phase boundary (in the hatched region in Fig. 13.3). According to (13.10), this region becomes narrower with decreasing temperature. An experiment performed along path (a) will therefore observe a crossover from quantum critical behavior away from the transition to classical critical behavior asymptotically close to it. At very low temperatures, the classical region may become so narrow that it is actually unobservable in an experiment. In the quantum disordered region ($p > p_c$, small T), the physics is dominated by quantum fluctuations, and the system essentially looks as it does in its quantum disordered ground state at $p > p_c$. In contrast, in the thermally disordered region, the long-range order is mainly destroyed by thermal fluctuations.

Between the quantum disordered and the thermally disordered regions is the so-called quantum critical region [15], where both types of fluctuation are important. It is located at p_c but, somewhat counter-intuitively, at comparatively high temperatures. Its boundaries are also determined by (13.10) but generally with a different prefactor to that for the asymptotic classical region. The physics in the quantum critical region is controlled by the quantum critical point: the system 'looks critical' with respect to p (due to quantum fluctuations) but is driven away from criticality by thermal fluctuations (i.e., the critical singularities are exclusively cut off by the temperature T). In an experiment carried out along path (b), the physics will therefore be dominated by the critical fluctuations which diverge according to the temperature scaling at the quantum critical point.

The second type of phase diagram occurs if an ordered phase exists at zero temperature only (as is the case for two-dimensional quantum antiferromagnets). In this case there will be no true phase transition in any experiment. However, an experiment along path (a) will show a very sharp crossover which becomes more pronounced with decreasing temperature. Furthermore, the system will display quantum critical behavior in the above-mentioned quantum critical region close to p_c and at higher temperatures.

13.5   Example: Transverse Field Ising Model

In this section we want to illustrate the general ideas presented in Sects. 13.3 and 13.4 by discussing a paradigmatic example, viz. the Ising model in a transverse field. This model is also discussed, albeit in a different context, in Chap. 12 of this volume. An experimental realization of the model can be found in the low-temperature magnetic properties of $LiHoF_4$. This material is an ionic crystal, and at sufficiently low temperatures the only magnetic degrees of freedom are the spins of the holmium atoms. They have an easy axis, i.e., they prefer to point up or down with respect to a certain crystal axis. They can therefore be represented by Ising spin variables. Spins at different holmium atoms interact

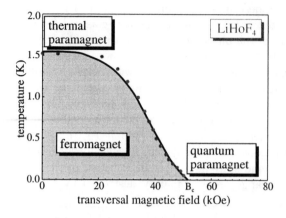

Fig. 13.4. Magnetic phase diagram of LiHoF$_4$ (after [16])

via a magnetic dipole–dipole interaction. Without any external magnetic field, the ground state is a fully polarized ferromagnet.[4]

In 1996 Bitko, Rosenbaum and Aeppli [16] measured the magnetic properties of LiHoF$_4$ as a function of temperature and a magnetic field which was applied perpendicularly to the preferred spin orientation. The resulting phase diagram is shown in Fig. 13.4. In order to understand this phase diagram, we now consider a minimal microscopic model for the relevant magnetic degrees of freedom in LiHoF$_4$, the Ising model in a transverse field. Choosing the z axis to be the Ising axis, its Hamiltonian is given by

$$H = -J \sum_{\langle ij \rangle} S_i^z S_j^z - h \sum_i S_i^x \,. \tag{13.15}$$

Here S_i^z and S_i^x are the z and x components of the holmium spin at lattice site i, respectively. The first term in the Hamiltonian describes the ferromagnetic interaction between the spins which we restrict to nearest neighbors for simplicity. The second term is the transverse magnetic field. For zero field $h = 0$, the model reduces to the well-known classical Ising model. At zero temperature, all spins are parallel. At a small but finite temperature, a few spins will flip into the opposite direction. With increasing temperature, the number and size of the flipped regions increases, reducing the total magnetization. At the critical temperature (about 1.5 K for LiHoF$_4$) the magnetization vanishes, and the system becomes paramagnetic. The resulting transition is a continuous classical phase transition caused by thermal fluctuations.

Let us now consider the influence of the transverse magnetic field. To do so, it is convenient to rewrite the field term as

$$-h \sum_i S_i^x = -h \sum_i (S_i^+ + S_i^-) \,, \tag{13.16}$$

---

[4] In the case of the dipole–dipole interaction, the ground state configuration depends on the geometry of the lattice. In LiHoF$_4$, it turns out to be ferromagnetic.

where S_i^+ and S_i^- are the spin flip operators at site i. From this representation it is easy to see that the transverse field will cause spin flips. These flips are the quantum fluctuations discussed in the preceeding sections. If the transverse field becomes larger than some critical field h_c (about 50 kOe in LiHoF$_4$), they will destroy the ferromagnetic long-range order in the system even at zero temperature. This transition is a quantum phase transition driven exclusively by quantum fluctuations.

For the transverse field Ising model, the quantum-to-classical mapping discussed in Sect. 13.3 can easily be demonstrated at a microscopic level. Consider a one-dimensional classical Ising chain with the Hamiltonian

$$H_{cl} = -J \sum_{i=1}^{N} S_i^z S_{i+1}^z \ . \tag{13.17}$$

Its partition function is given by

$$Z = \mathrm{Tr}\, \exp\left(-H_{cl}/T\right) = \mathrm{Tr}\, \exp\left(J \sum_{i=1}^{N} S_i^z S_{i+1}^z/T\right) = \mathrm{Tr}\, \prod_{i=1}^{N} M_{i,i+1} \ , \tag{13.18}$$

where $\mathbf{M}$ is the so called transfer matrix. It can be represented as

$$\mathbf{M} = \begin{pmatrix} e^{J/T} & e^{-J/T} \\ e^{-J/T} & e^{J/T} \end{pmatrix} = e^{J/T}(1 + e^{-2J/T} S^x) \approx \exp(J/T + e^{-2J/T} S^x) \ . \tag{13.19}$$

Except for a multiplicative constant the partition function of the classical Ising chain has the same form as that of a single quantum spin in a transverse field, $H_Q = -h\,S^x$, which can be written as

$$Z = \mathrm{Tr}\, e^{-H_Q/T_Q} = \mathrm{Tr}\, e^{hS^x/T_Q} = \mathrm{Tr}\, \prod_{i=1}^{N} e^{hS^x/(T_Q N)} \ . \tag{13.20}$$

Thus, a single quantum spin can be mapped onto a classical Ising chain. These considerations can easily be generalized to a d-dimensional transverse field (quantum) Ising model which can be mapped onto a $(d+1)$-dimensional classical Ising model. Consequently, the dynamical exponent z must be equal to unity for the quantum phase transitions of transverse field Ising models. Using a path integral approach analogous to Feynman's treatment of a single quantum particle, the quantum Landau–Ginzburg–Wilson functional (13.12) can be derived from (13.20) or its higher-dimensional analogs.

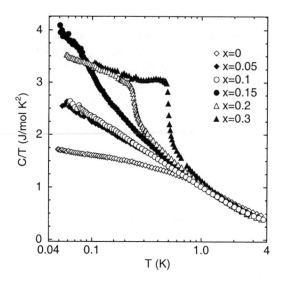

Fig. 13.5. Specific heat coefficient C/T of CeCu$_{6-x}$Au$_x$ as a function of temperature T for different gold concentrations x (after [20])

13.6   Quantum Phase Transitions and Non-Fermi Liquids

In this section we wish to discuss a particularly important consequence of quantum phase transitions, viz., non-Fermi liquid behavior in an itinerant electron system. In a normal metal the electrons form a Fermi liquid, a concept developed by Landau in the 1950s [17]. In this state the strongly interacting (via Coulomb potential) electrons behave essentially like almost non-interacting quasiparticles with renormalized parameters (like the effective mass). This permits very general and universal predictions for the low-temperature properties of metallic electrons: for sufficiently low temperatures, the specific heat is supposed to be linear in the temperature, the magnetic susceptibility approaches a constant, and the electric resistivity has the form $\rho(T) = \rho_0 + AT^2$ (where ρ_0 is the residual resistance caused by impurities).

The Fermi liquid concept is extremely successful, describing the vast majority of conducting materials. However, in the last few years there have been experimental observations that contradict the Fermi liquid picture, e.g., in the normal phase of the high-T_c superconducting materials [18] or in heavy fermion systems [19]. These are compounds of rare-earth elements or actinides where the quasiparticle effective mass is up to a few thousand times higher than the electron mass.

Figure 13.5 shows an example of such an observation, viz., the specific heat coefficient C/T of the heavy-fermion system CeCu$_{6-x}$Au$_x$ as a function of the temperature T. In a Fermi liquid, C/T should become constant for sufficiently low temperatures. Instead, in Fig. 13.5 C/T shows a pronounced temperature dependence. In particular, the sample with a gold concentration of $x = 0.1$

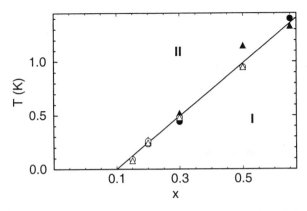

Fig. 13.6. Magnetic phase diagram of $CeCu_{6-x}Au_x$ (after [20]). I is the antiferromagnetic phase, II the paramagnetic phase. The Néel temperature was determined both for single crystals (*open symbols*) and polycrystals (*full symbols*) by measuring the specific heat (*triangles*) and the magnetic susceptibility (*circles*)

shows a logarithmic temperature dependence, $C/T \sim \log(1/T)$, over a wide temperature range.

In order to understand these deviations from the Fermi liquid and in particular the qualitative differences between the types of behavior at different x, it is helpful to relate the specific heat to the magnetic phase diagram of $CeCu_{6-x}Au_x$, shown in Fig. 13.6. Pure $CeCu_6$ is a paramagnet, but by alloying with gold it can become antiferromagnetic. The quantum phase transition is roughly at a critical gold concentration of $x_c = 0.1$ (and $T = 0$).

Let us first discuss the specific heat at the critical concentration $x_c = 0.1$. A comparison with the schematic phase diagram in Fig. 13.3 shows that at this concentration the entire experiment is done in the quantum critical region [path (b)]. When approaching the quantum critical point, i.e., with decreasing temperature, the antiferromagnetic fluctuations diverge. The electrons are scattered off these fluctuations and this hinders their movement, thereby increasing the effective mass of the quasiparticles. In the limit of zero temperature, the effective mass diverges and with it the specific heat coefficient C/T.

In an experiment at a gold concentration slightly above or below the critical concentration, the system will be in the quantum critical region at high temperatures. Here the specific heat agrees with that at the critical concentration. However, with decreasing temperature, the system leaves the quantum critical region, either towards the quantum disordered region (for $x < x_c$) or into the ordered phase (for $x > x_c$). In the first case there will be a crossover from the quantum critical behavior $C/T \sim \log(1/T)$ to conventional Fermi liquid behavior $C/T = $ const. This can be seen for the $x = 0$ data in Fig. 13.5. In the opposite case, $x > x_c$ the system undergoes an antiferromagnetic phase transition at some finite temperature, connected with a singularity in the specific heat. In Fig. 13.5, this singularity is manifested as a pronounced shoulder.

In conclusion, the non-Fermi liquid behavior of $CeCu_{6-x}Au_x$ can be completely explained (at least qualitatively) by the antiferromagnetic quantum phase transition at the critical gold concentration of $x_c = 0.1$. The deviations from the Fermi liquid occur in the quantum critical region where the electrons are scattered off the diverging magnetic fluctuations. Analogous considerations can be applied to other observables, e.g., the magnetic susceptibility or the electric resistivity.

13.7   Summary and Outlook

Quantum phase transitions are a fascinating subject in modern condensed matter physics. They open up new ways of looking at complex situations and materials for which conventional methods like perturbation theory fail. So far, only the simplest, and the most obvious cases have been studied in detail, which leaves a lot of interesting research for the future.

Acknowledgements. This work was supported by the DFG under grant Nos. Vo659/2 and SFB393/C2, and by the NSF under grant No. DMR–98–70597.

References

1. T. Andrews: Phil. Trans. R. Soc. **159**, 575 (1869)
2. K.G. Wilson: Phys. Rev. B **4**, 3174 (1971); ibid. 3184
3. C. Pfleiderer, G.J. McMullan, G.G. Lonzarich: Physica B **206** & **207**, 847 (1995)
4. T. Vojta, D. Belitz, R. Narayanan, T.R. Kirkpatrick: Z. Phys. B **103**, 451 (1997); D. Belitz, T.R. Kirkpatrick, T. Vojta: Phys. Rev. Lett. **82**, 4707 (1999)
5. S.L. Sondhi, S.M. Girvin, J.P. Carini, D. Shahar: Rev. Mod. Phys. **69**, 315 (1997)
6. T.R. Kirkpatrick, D. Belitz: 'Quantum phase transitions in electronic systems'. In: *Electron Correlations in the Solid State* ed. by N.H. March (Imperial College Press, London 1999)
7. T. Vojta: Ann. Phys. (Leipzig) **9**, 403 (2000)
8. S. Sachdev: *Quantum Phase Transitions* (Cambridge University Press, Cambridge 2000)
9. S.-K. Ma: *Modern Theory of Critical Phenomena* (Benjamin, Reading 1976)
10. N. Goldenfeld: *Lectures on Phase Transitions and the Renormalization Group* (Addison–Wesley, Reading 1992)
11. L.D. Landau: Phys. Z. Sowjetunion **11**, 26 (1937); ibid. 545; Zh. Eksp. Teor. Fiz **7**, 19 (1937); ibid. 627
12. L.P. Kadanoff: Physics **2**, 263 (1966)
13. B. Widom: J. Chem. Phys. **43**, 3892 (1965)
14. V.L. Ginzburg: Sov. Phys. Sol. State **2**, 1824 (1960)
15. S. Chakravarty, B.I. Halperin, D.R. Nelson: Phys. Rev. B **39**, 2344 (1989)
16. D. Bitko, T.F. Rosenbaum, G. Aeppli: Phys. Rev. Lett. **77**, 940 (1996)
17. L.D. Landau: Zh. Eksp. Teor. Fiz. **30**, 1058 (1956); ibid. **32**, 59 (1957) [Sov. Phys. JETP **3**, 920 (1956); ibid. **5**, 101 (1957)]
18. M.B. Maple: J. Magn. Magn. Mater. **177**, 18 (1998)
19. P. Coleman: Physica B **259–261**, 353 (1999)
20. H. von Löhneysen: J. Phys. Condens. Matter **8**, 9689 (1996)

14 Introduction to Energy Level Statistics

Bernhard Kramer

Summary. An introduction to the theory of random Hamiltonians is given with special emphasis on the statistics of energy level spacings. The physical origin of the repulsion of energy levels is discussed. The importance of the symmetries of the system for level repulsion is emphasized. Several recent physical examples of level statistics of disordered quantum coherent systems are discussed including localized and metallic systems, the universal behaviour of the level spacing distribution at the disorder-induced metal–insulator transition and the fluctuations in the Coulomb oscillations of the conductance of quantum dots.

14.1 Introduction

Physical theories are usually supposed to provide the tools for predicting the properties and the dynamical behaviour of systems. However, this aim cannot be achieved in the majority of the cases. For instance, in classical dynamics, it is well known that even a one-particle system can become chaotic, although the equations of motion are known and initial and boundary conditions can be precisely specified. Well known examples are chaotic billiards [1]. In many-particle physics, the situation becomes even more complicated since the equations of motion contain a macroscopically large number of interacting degrees of freedom ($\approx 10^{24}$) such that it is not even possible to specify initial and boundary conditions. Therefore, in the overwhelming majority of cases, even if described by conceptually simple mathematical rules, properties of physical systems show rather complex behaviour and almost always a certain degree of randomness.

Most striking (though not entirely physical) examples are the distribution of prime numbers [2] and the distribution of the zeroes of the Riemann ζ-function [3]. Well known early examples of many-particle quantum physics are the excitation spectra of heavy nuclei which are dominated by the strong interaction between the nucleons. Other physical examples are the hydrogen atom at strong magnetic fields, complex molecules, and the eigenmodes of classically chaotic microwave cavities. The statistics of the spectra of these systems are found to be surprisingly well described by the theory of Gaussian random matrices [3].

During recent years, due to the possibility of controlling electrons in nanostructures at millikelvin temperatures, quantum complex systems have become experimentally accessible. In addition, the low-temperature quantum transport properties of impure metal-like systems where localization is an essential ingredient have been found to be dominated by randomness as well as by interactions. These systems can also be described by random matrices. However, due to the presence of localization, strong deviations from the Gaussian behaviour are found.

In this contribution, a short introduction to random matrix theory for energy level statistics is given. As an application, we will discuss the distribution of the energy level spacings of an electron in a disordered potential, in particular when varying the disorder so that the system undergoes a transition from an insulating to a conducting state (Anderson transition) [4–10]. Additionally, we will briefly discuss fluctuations in the positions of the conductance peaks of a quantum dot in the regime of Coulomb blockade, where electronic transport is dominated by interactions as well as impurities [11,12].

14.2   Statistics of Energy Levels

A paradigmatic example of a Hamiltonian for a complex quantum system is

$$H = \sum_{j,k} H_{jk} |j\rangle\langle k| , \tag{14.1}$$

with the random matrix elements H_{jk} in the complete orthonormal set $\{|j\rangle\}$. In the simplest case, these are states associated with a regular lattice. In general, they may also be thought of as many-particle basis states. The set of $\{H_{jk}\}$ which characterizes the system, is called the configuration. Since H_{jk} are random, we replace the system in consideration by a statistical ensemble of macroscopically equivalent systems described by the normalized, multi-dimensional probability density distribution $P_H(\{H_{jk}\})$

$$\int \prod_{j\leq k} dH_{jk} P_H(\{H_{jk}\}) = 1 . \tag{14.2}$$

A one-particle example of such a system is the celebrated Anderson Hamiltonian defined on a lattice [13]

$$H = \sum_{j} \epsilon_j |j\rangle\langle j| + \sum_{[j,k]} |j\rangle\langle k| , \tag{14.3}$$

where the second sum runs over nearest neighbors only, and the lattice-site energies ϵ_j are distributed at random within the interval $[-W/2, W/2]$. The site states $\{|j\rangle\}$ are assumed to form a complete set, $\sum_j |j\rangle\langle j| = 1$. The first term in (14.3) corresponds to the (localizing) potential energy while the second (hopping) term represents the (delocalizing) kinetic energy.

Another example is

$$H = \sum_{[j\neq k]} |j\rangle\langle k| . \tag{14.4}$$

where j represent the sites of a continuous random network [14].

Other examples of random systems are single quantum particles confined within classically chaotic billiards like the 'Sinai billiard' or the 'stadium billiard', or interacting particles in a regularly shaped confining potential.

Physical systems that may be described by the above model Hamiltonians are random alloys, mixed crystals, amorphous semiconductors, (metallic) glasses, granular systems, large molecules, heavy nuclei, Rydberg atoms, and quantum dots with interacting electrons.

The random Schrödinger equation

$$H|\psi_\nu\rangle = E_\nu|\psi_\nu\rangle \tag{14.5}$$

has practically infinitely many random real eigenvalues E_ν and corresponding random eigenstates $|\psi_\nu\rangle$ which are characterized by random amplitudes and random phases.

In the following, we concentrate only on the statistics of the energy levels. Their probability density distribution, $P_E(\{E_\nu\})$, is assumed to be normalized

$$\int \prod_\nu dE_\nu P_E(\{E_\nu\}) = 1 \, . \tag{14.6}$$

It can in principle be calculated from P_H since $E_\nu = E_\nu(\{H_{jk}\})$.

For statistically independent levels, $P_E = \prod_\nu p(E_\nu)$. For instance, if N levels are distributed independently and completely at random within the interval W, $p(E) = W^{-1}$. In general, however, the levels will not be independent, but correlated. It then turns out to be useful to define reduced probability densities. For instance, the probability of finding a level within the interval dE_1 near E_1, independently of where all the other levels are, is

$$p_1(E_1)\,dE_1 = \int \prod_{\nu\neq 1} dE_\nu\, P_E(E_1, E_2, \dots)\,dE_1 \equiv \rho(E_1)\,dE_1 \, . \tag{14.7}$$

This represents the well-known 'density of states' $\rho(E) = N^{-1}\sum_\nu \delta(E - E_\nu)$. Correspondingly, 2-, 3-, and generally n-level distributions can be defined:

$$p_2(E_1, E_2) = \int \prod_{\nu\neq 1,2} dE_\nu\, P_E(E_1, E_2, E_3, \dots) \, ,$$

$$p_3(E_1, E_2, E_3) = \int \prod_{\nu\neq 1,2,3} dE_\nu\, P_E(E_1, E_2, E_3, E_4, \dots) \, ,$$

$$\vdots$$

$$p_n(E_1, \dots, E_n) = \int \prod_{\nu\neq 1,\dots,n} dE_\nu\, P_E(E_1, \dots, E_n, E_{n+1}, \dots) \, . \tag{14.8}$$

These contain the correlations between the energy levels. They may be further decomposed into products of conditional probability distributions

$$p_n(E_1, \dots, E_n) = c_1(E_1)\, c_2(E_1|E_2) \cdots c_n(E_1, \dots, E_{n-1}|E_n) \, , \tag{14.9}$$

where $c_n(E_1, \dots, E_{n-1}|E_n)$ is the probability density for finding the nth level near E_n provided that there are levels at $E_1, \dots, E_{n-1}$.

14.3   Level Spacing Distribution

For a complete statistical description of the spectrum one generally needs all the above correlation functions. However, one can obtain very important physical information about the spectrum by considering, in addition to the density of states, only the probability density distribution of the level spacings, $P(S)$, defined by

$$P(S)\, dS = (N-1)^{-1} \sum_\nu \delta(S - \Delta E_\nu)\, dS . \tag{14.10}$$

Here, ΔE_ν is the spacing between consecutive energy levels, $\Delta E_\nu \equiv E_{\nu+1} - E_\nu$, and $P(S)$ is the probability density for finding a level spacing near S in the interval dS. The spacing distribution contains information about all correlations between the levels, but $P(S) \neq p_2(E_1|E_2)$.

Using the above density distributions, averages are easily calculated. For instance, with the level spacing distribution, the mean energy level spacing

$$\langle \Delta \rangle = \frac{1}{N-1} \sum_j \Delta E_j = \frac{E_N - E_1}{N-1} \tag{14.11}$$

can be obtained as

$$\langle \Delta \rangle = \int SP(S)\, dS , \tag{14.12}$$

while the mean energy is related to the density of states by

$$\langle E \rangle = \int E\rho(E)\, dE . \tag{14.13}$$

We shall now consider two examples which illustrate the behaviour of $P(S)$ [3]. First we investigate the level spacing distribution of completely independent energy levels. The probability density for finding an energy level somewhere is then independent of E. Let us use the mean density of levels $N/(E_N - E_1) \equiv 1/\langle \Delta \rangle \equiv \rho$.

We divide the interval $(E, E+S)$ into m equal intervals of length S/m, with $m \to \infty$ eventually. Then, the probability that the interval $(E, E+S)$ does *not* contain any energy level is given by $(1 - \rho S/m)^m$ (since $\rho S/m$ is the probability of having a level within S/m). In the limit $m \to \infty$, this yields

$$P(S) = \rho e^{-\rho S} \tag{14.14}$$

for the normalized distribution of the spacings of uncorrelated levels; or, defining the dimensionless spacing $s \equiv \rho S = S/\langle \Delta \rangle$,

$$p(s) = e^{-s} . \tag{14.15}$$

This is nothing but the Poisson distribution, as expected.

In the second example we *assume* that there is a correlation between the energy levels such that large spacings are more probable than smaller ones:

$$p_0(s)\,ds = as\,ds\,,\tag{14.16}$$

where a is a constant. We shall discuss later how this assumption can be justified. Again, we divide the interval s into m subintervals and calculate the probability for finding the total interval empty of levels. For any finite m, this is given by

$$p_m(s)\,ds = as\,ds \prod_{\mu=0}^{m-1} \left(1 - \frac{\mu s}{m}\frac{as}{m}\right)\,.\tag{14.17}$$

Here, $\mu as^2/m^2$ is the probability of finding an energy level within the interval $(\mu s/m, \mu s/m + s/m)$, so that $1 - \mu as^2/m^2$ is the probability of finding *no* level in that interval. It is then easy to show that $\lim_{m\to\infty} p_m(s) = as \exp(-as^2/2)$. The normalized density distribution of level spacings is now

$$p_W(s) = \frac{\pi s}{2} e^{-\pi^2 s^2/4}\,.\tag{14.18}$$

This is the famous *Wigner surmise* for the distribution of the spacings of the levels in heavy nuclei [15,16]. For $s \to 0$, the probability vanishes. This indicates the repulsion of energy levels observed in many quantum systems.

14.4   Level Repulsion and Symmetry

Let us now investigate the conditions for level repulsion in more detail. We will see that this is generic for quantum systems and depends crucially on the symmetry [1].

We assume that the energy levels depend on some parameter α in such a way that they perform more or less random motions when changing α. Occasionally, some of the levels will come close to each other and tend to intersect. However, if this happens, there will generally be some coupling between the levels causing them eventually to repel each other. Close to the crossing points, the physics of the system will be dominated by this coupling so that one can describe the spectrum by a Hermitian two-level Hamiltonian

$$H = \begin{pmatrix} H_{11}(\alpha) & H_{12}(\alpha) \\ H_{21}(\alpha) & H_{22}(\alpha) \end{pmatrix}\,.\tag{14.19}$$

The solution of the corresponding eigenvalue problem is

$$E_\pm(\alpha) = \frac{H_{11} + H_{22}}{2} \pm \sqrt{\frac{(H_{11} - H_{22})^2}{4} + |H_{12}|^2}\,.\tag{14.20}$$

Whether or not the levels do cross will be determined by the discriminant

$$D = \frac{(H_{11} - H_{22})^2}{4} + (\mathrm{Re}H_{12})^2 + (\mathrm{Im}H_{12})^2\,.\tag{14.21}$$

In general, there will be no symmetry in a random system, apart possibly from time reversal invariance. In the following, we need therefore to know the basic facts about time reversal symmetry in quantum systems. They can be found in the standard books of quantum mechanics. For the sake of brevity, we will assume that the reader is familiar with them.

In order for the two levels to cross, several independent parameters, say β, have to become simultaneously zero when changing α. For systems with time reversal symmetry,

$$H = THT^{-1} , \tag{14.22}$$

where T is the time reversal operator, such that $T\psi(\tau) = \psi^*(\tau)$, τ denotes time, and $T^2 = 1$. The Hamiltonian can then be assumed real, $\mathrm{Im}H_{12} = 0$. There are two independent parameters, $\beta = 2$, namely $H_{11} - H_{22}$ and $\mathrm{Re}H_{12}$, which dictate whether or not the levels cross. For systems without time reversal symmetry, the Hamiltonian is Hermitian complex, $\mathrm{Im}H_{12} \neq 0$. In this case, we have three independent parameters, $\beta = 3$. The quantity β is often called the co-dimension of the level repulsion.

For a random system without spin, the above symmetry with respect to time reversal is the only one which is important. In the presence of spin, or more generally for half-integer angular momentum, we have $T^2 = -1$. Then, each eigenvalue is doubly degenerate (Kramers' degeneracy) and the above two-level Hamiltonian has to be represented by a 4×4-matrix of the form

$$H = \begin{pmatrix} a+b & 0 & c+if & -e-id \\ 0 & a+b & e-id & c+if \\ c+if & e+id & a-b & 0 \\ -e+id & c-if & 0 & a-b \end{pmatrix} , \tag{14.23}$$

which contains the matrix elements of the Hamiltonian in the states $|1\rangle$ and $|2\rangle$ of the two levels and their time reversals $T|1\rangle$ and $T|2\rangle$. The relations between the matrix elements can be obtained by using the properties of the time reversal operator $TT^\dagger = 1$, $T^2 = -1$ and the fact that H is Hermitian.

The resulting bi-quadratic secular equation has the solutions

$$E_\pm = a \pm \sqrt{b^2 + c^2 + d^2 + e^2 + f^2} , \tag{14.24}$$

such that the co-dimension of the level crossing is now $\beta = 5$.

We have thus found that the question as to whether or not two energy levels of a quantum system can be degenerate is generally dictated by β parameters. The co-dimension depends on the symmetry of the system,

$$\beta = \begin{cases} 2 & \text{orthogonal symmetry ,} \\ 3 & \text{unitary symmetry ,} \\ 5 & \text{symplectic symmetry .} \end{cases} \tag{14.25}$$

In the case of *orthogonal symmetry*, the system is time reversal invariant, $[H, T] = 0$ with $T^2 = 1$. The statistical ensemble is then invariant with respect to orthogonal transformations. The Hamiltonian can be chosen to be real.

For *unitary symmetry*, the system is not time reversal invariant, that is, $[H, T] \neq 0$, and the ensemble is invariant with respect to unitary transformations. The Hamiltonian is Hermitian complex.

In the case of *symplectic symmetry*, we have again $[H, T] = 0$ but $T^2 = -1$. This is the case for half-integer angular momentum. The ensemble is invariant under so-called symplectic transformations. The Hamiltonian can be represented by quaternions.

In a random system there are no other symmetries left. Accidental level crossing is the less probable the more vanishing parameters are necessary, in order to enforce degeneracy of the levels. This can be seen explicitly by the following calculation of the spacing distribution which holds asymptotically for small level spacings s.

We assume that the parameters of the level crossing are random variables distributed according to some continuous distribution function, $P_\beta(b_1, \dots, b_\beta)$, with the property $0 < P_\beta(0) < \infty$. Then, the level spacing distribution can be obtained from the well-known relation

$$p(s) = \langle \delta \left(s - S/\langle \Delta \rangle \right) \rangle \equiv \int \prod_{n=1}^{\beta} \mathrm{d}b_n \, P_\beta(b_1, \dots, b_\beta) \, \delta \left(s - b \right) , \qquad (14.26)$$

where $b = (\sum_n b_n^2)^{1/2}$ is the modulus of the vector of the parameters of the level crossing. With the substitution

$$b_n = s b_n' , \qquad n = 1, \dots, \beta , \qquad (14.27)$$

we arrive at

$$p(s) = s^{\beta-1} \int \prod_n \mathrm{d}b_n' \, P_\beta(s b_n') \, \delta(1 - b') . \qquad (14.28)$$

This gives the asymptotic form of the distribution

$$p(s) \propto s^{\beta-1} . \qquad (14.29)$$

This does indeed indicate universal level repulsion which is stronger if the co-dimension is larger and depends only on the symmetry of the system.

For large spacings, the levels become uncorrelated. The distribution can be shown to behave as a Gaussian [cf. (14.18)] in this region for certain model systems [3]. The above-described behaviour of the spacing distribution was first suggested by Wigner and is called the 'Wigner surmise' [15,16]. It is found for the spectra of nuclei as well as for classically chaotic quantum billiards [1].

14.5   Gaussian Ensembles

A class of systems which allow for the exact determination of the statistics of the energy levels are represented by the Gaussian ensembles. The form of the corresponding Hamiltonians is again determined by the symmetries. They can be defined as follows.

The *Gaussian orthogonal ensemble* is invariant under real orthogonal transformations R $(RR^\dagger = 1)$ such that with $H^R = R^\dagger H R$

$$P_H(H^R)\, \mathrm{d}H^R = P_H\, \mathrm{d}H \ . \tag{14.30}$$

Here, P_H denotes the distribution function of the matrix elements H_{jk} of the Hamiltonian and $\mathrm{d}H$ is the corresponding 'volume element', $\mathrm{d}H \equiv \prod_{j\le k} \mathrm{d}H_{jk}$. In addition, the matrix elements are statistically independent,

$$\langle H_{jk} H_{j'k'}\rangle \equiv \int \mathrm{d}H\, P_H(H) H_{jk} H_{j'k'} \propto \delta_{jj'}\delta_{kk'} \qquad (j \le k) \ . \tag{14.31}$$

The latter implies that the distribution can be written as a product of probability density distributions of the individual matrix elements,

$$P_H(H) \equiv \prod_{j\le k} p_{jk}(H_{jk}) \ . \tag{14.32}$$

The *Gaussian unitary ensemble* is not invariant under time reversal. The ensemble is invariant under unitary transformations U $(U^\dagger U = 1)$ such that, defining $H^U = U^\dagger H U$, one has

$$P_H(H^U)\, \mathrm{d}H^U = P_H\, \mathrm{d}H \ , \tag{14.33}$$

with the complex matrix elements $H_{jk} = H_{jk}^{(1)} + \mathrm{i}H_{jk}^{(2)}$ and the volume element $\mathrm{d}H \equiv \prod_{j\le k} \mathrm{d}H_{jk}^{(1)} \prod_{j<k} \mathrm{d}H_{jk}^{(2)}$. Statistical independence

$$\langle H_{jk}^m H_{j'k'}^{m'}\rangle \propto \delta_{jj'}\delta_{kk'}\delta_{mm'} \tag{14.34}$$

implies once again the form of a product distribution

$$P_H(H) = \prod_{j\le k} p_{jk}^{(1)}(H_{jk}^{(1)}) \prod_{j<k} p_{jk}^{(2)}(H_{jk}^{(2)}) \ . \tag{14.35}$$

For the *Gaussian symplectic ensemble*, the Hamiltonian can be represented by quaternions,

$$H_{jk} = \sum_{\lambda=0}^{3} H_{jk}^{(\lambda)} e_\lambda \ , \tag{14.36}$$

with the 2×2-matrices

$$e_0 = \begin{pmatrix} 1 & 0 \\ 0 & 1 \end{pmatrix}, \quad e_1 = \begin{pmatrix} \mathrm{i} & 0 \\ 0 & -\mathrm{i} \end{pmatrix}, \quad e_2 = \begin{pmatrix} 0 & -1 \\ 1 & 0 \end{pmatrix}, \quad e_3 = \begin{pmatrix} 0 & -\mathrm{i} \\ -\mathrm{i} & 0 \end{pmatrix} \ . \tag{14.37}$$

These fulfil the conditions

$$e_i e_j = -e_j e_i = e_k \quad (i,j,k \quad \text{cyclic}), \qquad e_j^2 = -1 \qquad (j = 1,2,3), \quad (14.38)$$

and they form a complete orthonormal set in the space of the 2×2-matrices.

The time reversal operator can be written in the form

$$T \equiv ZC \equiv \begin{pmatrix} e_2 & 0 & 0 & \cdots & 0 \\ 0 & e_2 & 0 & \cdots & 0 \\ \vdots & \vdots & \vdots & \vdots & \vdots \\ 0 & 0 & 0 & \cdots & e_2 \end{pmatrix} C \equiv e_2 \, C \cdot 1 . \qquad (14.39)$$

Here, Z is the so-called symplectic matrix and the operator C transforms a complex number into its conjugate. The matrices S ($S^\dagger S = 1$) that fulfil the condition $Z = SZS^t$ (S^t transpose of S) form the N-dimensional symplectic group.

The Gaussian symplectic ensemble then fulfils the following conditions. With the definition $H^S = SHS^{-1}$, invariance of the ensemble implies that

$$P_H(H^S) \, dH^S = P_H \, dH , \qquad (14.40)$$

where $dH \equiv \prod_{j \le k} dH_{jk}^{(0)} \prod_{j<k,\lambda=1}^{3} dH_{jk}^{(\lambda)}$ is the generalized volume element. The statistical independence of the matrix elements is again guaranteed by the product representation

$$P_H(H) = \prod_{j \le k} p_{jk}^{(0)}(H_{jk}^{(0)}) \prod_{j<k,\lambda=1}^{3} p_{jk}^{(\lambda)}(H_{jk}^{(\lambda)}) . \qquad (14.41)$$

With these definitions the probability density distribution of the Hamiltonian can be determined exactly. Since this is a lengthy but straightforward calculation, the reader is referred here to the literature [3]. Basically, the calculation consists of three steps

1. Rotation of the Hamiltonian with the operators $O = R, U, S$, depending on the symmetry of the system (rotation angle Θ),

$$H(\Theta) = O(\Theta) H O^{-1}(\Theta) . \qquad (14.42)$$

2. Invariance of the ensemble under these rotations implies a differential equation

$$\frac{dP_H}{d\Theta} = 0 . \qquad (14.43)$$

3. Solving this differential equation for an infinitesimal rotation yields, using the statistical independence of the Hamiltonian matrix elements,

$$P_H(H) = \exp\left(-a \mathrm{Tr} H^2 + b \mathrm{Tr} H + c\right) , \qquad (14.44)$$

where the constants a, b, c characterize the form of the distribution.

14.6   Level Spacing Distribution at the Anderson Transition

The Gaussian ensembles allow for the exact calculation of many properties of the system. This can be useful for quantities that are universal and thus independent of the microscopic details of the system considered. However, most physical Hamiltonians are not Gaussian, since the matrix elements are not statistically independent. Examples are Hamiltonians where exponential localization of quantum states is important (see above). In these cases, the results of the theory of random matrices cannot be applied without further ado. To illustrate this in more detail, we will now discuss the level spacing distribution at the Anderson transition.

Generally, for Hamiltonians which contain competing (delocalizing) kinetic energy terms as well as (localizing) random potential-energy terms, the eigenstates are localized in certain regions of the energy spectrum, such as the edges of the energy bands. This means that the probability amplitude is concentrated in a certain region of space, and decays exponentially on the average at infinity with a decay length ξ, the localization length [17]. Near the centers of the bands, the states will be extended as long as the disorder is not too large. Here disorder is understood to be represented by the width W of the random potential. In the energy regions of localized states, the energy levels will be completely independent. Thus, level spacings will be distributed according to a Poisson distribution (14.14). In the region of delocalized states, usually associated with metallic conduction, energy levels will be correlated and statistics will be characterized approximately by the Wigner surmise (14.18). The regions of localized and delocalized states are separated by critical energies, 'mobility edges' $E_{\mathrm{c}}(W)$. Here, the localization length diverges according to a power law $\xi \propto |E - E_{\mathrm{c}}|^{-\nu}$ with a critical exponent ν. The following question then arises: how does the level statistics change at such a critical point?

Critical points are defined as points where certain system properties, usually denoted as scaling variables, do not depend on the size of the system L. For the localization–delocalization transition, an example of such a scaling variable is the conductance. In the present problem, the conductance decays with increasing system size when the energy is located in the localized region. On the other hand, in the region of delocalized states, the conductance increases. At the transition point, the conductance is independent of the system size.

Such behaviour is also observed for the above statistics of the energy levels [4–9]. In the localized region, when the system size is finite, the states with localization lengths $\xi(E, W)$ larger than L will appear as delocalized. For a given disorder W, and for energies in the localized region near a critical point, the level statistics will be close to the Wigner distribution and show a crossover to Poisson behaviour only for sufficiently large L. Also in the delocalized region, Wigner behaviour of the level spacing will only be reached asymptotically for infinite L. Thus, the statistics of the energy levels will generally depend on the size of the system. Only at the critical points – the mobility edges – can the distribution be expected to be independent of L [18,19]. But what is its explicit form?

This question excited considerable interest amongst researchers working in the field of localization. Not only was this an opportunity to contribute to the field of statistical properties of random eigenvalues, but also, the prospects for applying the well-established tools of the theory of random matrices to the problem of the localization–delocalization transition appeared to be very promising.

For small s, the arguments given in Sects. 14.3 and 14.4 seem to be so general, basically employing only the level repulsion in a two-level system, that they should also be applicable at the transition point, where the quantum states are neither localized nor delocalized. Thus, when $s \to 0$ one can expect for the critical level spacing the same asymptotic behaviour as in the metallic region,

$$P_c(s) \propto s^{\beta-1} . \tag{14.45}$$

One question remains: what will happen at large spacings (relative to the mean!)?

This question was eventually answered by extensive numerical diagonalization of very large random matrices using the above Anderson model (14.3), after considerable debate in the scientific community [5,6]. Naively one would expect neighboring states with large energy separation to be decoupled, as in the region of localized states, leading to the Poisson behaviour

$$\ln P_c(s) \propto -s . \tag{14.46}$$

However, near the transition point it is well established that the wave functions are not really localized or delocalized: they are multifractal objects (see Chap. 16), and the above simple argument does not apply. In fact, it has been argued that the asymptotic form of P_c should be $\ln P_c(s) \propto -s^\alpha$ with $\alpha = 1 + (d\nu)^{-1}$, where d is the dimensionality of the system [20].

Numerical results show that the latter is not very probable. By analyzing the statistical properties of more than $20\,000\,000$ level spacings of the three-dimensional Anderson model close to the critical point, it has been found that the critical spacing distribution obeys an interpolation formula

$$I_c(s) = e^{\mu - \sqrt{\mu^2 + (A_c s)^2}} . \tag{14.47}$$

Here, $I(s)$ is defined as

$$I(s) = \int_s^\infty P(s')\,ds' , \tag{14.48}$$

which gives the probability of finding neighboring states with spacings larger than s. The parameters μ and A_c are found numerically to be ≈ 2.21 and 1.90, respectively. Equation (14.46) implies that for $s \to \infty$, the critical spacing distribution is Poissonian and displays level repulsion for small s.

14.7   Fluctuations of Energy Levels of Quantum Dots

Electrical transport through semiconductor quantum dots has been found to be dominated by the Coulomb interaction [21]. In the experiment, a small quantum dot is coupled by tunneling barriers to electron wires and the voltage at a third electrode, the gate voltage V_g, is used to vary the electrostatic potential. At sufficiently small temperatures, lower than the so-called charging energy, the structure is insulating since the charge is quantized and an electron cannot enter the dot due to Coulomb repulsion. However, when the potential is adjusted in such a way that the chemical potential μ of the leads equals the chemical potential μ_N of the electrons on the dot, electrons can enter and leave the dot. The quantity μ_N is nothing other than the difference between the ground state energies of $N+1$ and N electrons, E_{N+1} and E_N. As μ is approximately proportional to the voltage at the gate electrode, the conductance will show resonances at gate voltages where

$$\mu = \mu_N = E_{N+1} - E_N . \tag{14.49}$$

If the dot is represented by a small capacity C, the energy of N electrons is $(Ne)^2/2C$ and the positions of the conductance peaks correspond to

$$eV_g \propto \mu_N = \frac{Ne^2}{C} , \tag{14.50}$$

for $N \gg 1$. They are equidistant and separated by the charging energy e^2/C.

However, since the electrons in the dot are interacting, the ground state energies of N electrons corresponding to the peak number $N - 1$, $E_N^{(N-1)}$, and peak number N, $E_N^{(N)}$, will in general be different. In a mean field approach, we can write

$$E_N^{MF} = \frac{(Ne)^2}{2C} + \sum_{k=1}^{N} \epsilon_k , \tag{14.51}$$

where ϵ_k is the single-particle energy of the kth electron. Due to these interaction-induced fluctuations, the Coulomb-blockade peaks will not be equidistant but more or less randomly fluctuating. In addition, impurities in the vicinity of the quantum dot will also lead to random fluctuations in the peak positions,

$$\Delta^{(N)} \equiv \mu_N - \mu_{N-1} = \frac{e^2}{C} + \epsilon_{N+1} - \epsilon_N . \tag{14.52}$$

Thus the fluctuations of the charging energies in the Coulomb blockade will reflect the quantum fluctuations of the ground state energies of a quantum dot with different electron numbers. Will the fluctuations be described by random matrix theory?

Several experiments have been performed in order to investigate this question. However, the results obtained from GaAs as well as from In_2O_{3-x} dots do not seem to be consistent with the results of random matrix theory [11,12]. This is presently not very well understood and is a subject of continued experimental and theoretical work.

14.8   Conclusion

We have provided a short (but hopefully understandable) introduction to the random matrix theory which seems to account for certain aspects of several phenomena in different areas of quantum physics. As a challenge to those who want to enter this exciting field of modern mesoscopic quantum physics, we have also briefly discussed two examples where random matrix theory does not seem to provide the correct behaviour.

Acknowledgements. This work has been supported by the Deutsche Forschungsgemeinschaft within SFB 508 of the Universität Hamburg, the MURST via Cofinanziamento 98 and by the EU via TMR (FMRX-CT98-0180 and FMRX-CT96-0042).

References

1. F. Haake: *Quantum Signatures of Chaos*, Springer Series in Synergetics, Vol. 54 (Springer, Berlin 1991); T. Dittrich, P. Hänggi, G.-L. Ingold, B. Kramer, G. Schön, W. Zwerger: *Quantum Transport and Dissipation* (Wiley-VCH, Weinheim 1997)
2. C.E. Porter (Ed.): *Statistical Theories of Spectra: Fluctuations* (Academic Press, New York 1965)
3. M.L. Mehta: *Random Matrices* (Academic Press, New York 1991)
4. E. Hofstetter, M. Schreiber: Phys. Rev. Lett. **73**, 3137 (1994); Phys. Rev. Lett. **74**, 1874 (1995) erratum
5. I.Kh. Zharekeshev, B. Kramer: Phys. Rev. B **51**, R17239 (1995)
6. I.Kh. Zharekeshev, B. Kramer: Phys. Rev. Lett. **79**, 717 (1997)
7. J.X. Zhong, U. Grimm, R.A. Römer, M. Schreiber: Phys. Rev. Lett. **80**, 3996 (1998)
8. F. Milde, R.A. Römer, M. Schreiber: Phys. Rev. B **61**, 6028 (2000)
9. I.Kh. Zharekeshev, B. Kramer: In *Statistical and Dynamical Aspects of Mesoscopic Systems*, Lecture Notes in Physics 547, ed. by D. Reguerra, G. Platero, L.L. Bonilla, J.M. Rubi (Springer, Berlin 2000), p. 237
10. B. Kramer, O. Halfpap, I.Kh. Zharekeshev: In *Adv. Sol. St. Phys.*, ed. by B. Kramer, Vol. 39 (Vieweg, Braunschweig 1999) p. 253
11. U. Sivan, R. Berkovits, Y. Aloni, O. Prus, A. Auerbach, G. Ben-Yoseph: Phys. Rev. Lett. **77**, 1123 (1996)
12. F. Simmel, T. Heinzel, D.A. Wharam: Europhys. Lett. **38**, 123 (1997)
13. P.W. Anderson: Phys. Rev. **109**, 1492 (1958)
14. D. Weaire, M.F. Thorpe: Phys. Rev. B **4**, 2508 (1971)
15. E.P. Wigner: Ann. Math. **62**, 548 (1955); **65**, 203 (1957)
16. F.J. Dyson: J. Math. Phys. **3**, 140 (1962); **3**, 1199 (1962)
17. B. Kramer, A. MacKinnon: Rep. Progr. Phys. **56**, 1469 (1993)
18. B.I. Shklosvkii, B. Shapiro, B.R. Sears, P. Lambrianides, H.B. Shore: Phys. Rev. B **47**, 11487 (1993)
19. E. Hofstetter, M. Schreiber: Phys. Rev. B **48**, 16979 (1993); **49**, 14726 (1994)
20. A.G. Aronov, V.E. Kravtsov, I.V. Lerner: Phys. Rev. Lett. **74**, 1174 (1995)
21. M.A. Kastner: Rev. Mod. Phys. **64**, 849 (1992)

15 Randomness in Optical Spectra of Semiconductor Nanostructures

Erich Runge

Summary. Optical properties of semiconductor nanostructures are discussed, emphasizing the interplay of quantum mechanics and disorder, e.g., due to rough interfaces. Near the fundamental band gap, semiconductor optics is dominated by the excitonic response. For absorption, luminescence, and photoluminescence excitation spectra as well as for the time-dependent Rayleigh signal, the system can be described by a single-particle Schrödinger equation with an effective random potential for the center-of-mass motion of the exciton. Most results are illustrated using the one-dimensional Anderson Hamiltonian as model system. Based on simulation data, various optical experiments are suggested which exhibit typical random-matrix features such as level repulsion and Porter–Thomas-distributed matrix elements.

15.1 Spectra of Semiconductor Nanostructures

Light-emitting devices with active regions of nanometer size are crucial for the multi-billion dollar industry based on optical fiber communication or compact-disk players. At present, the central part of most of these devices is a quantum well (QW), a thin layer of one semiconducting material sandwiched between another semiconductor with larger bandgap (barriers, see Fig. 15.1). These semiconductor nanostructures are never perfect. Interface and surface roughness increase the electrical resistance and lead to a broadening of the optical spectra. Not surprisingly, enormous amounts of manpower and money have been spent in recent years to obtain a detailed understanding of these systems and their imperfections.

The present work addresses the interplay between disorder, such as rough interfaces, and quantum physics and the way this manifests itself in optical spectra. Special attention is paid to the numerical aspects and to the relationship with the Anderson disorder model [1]. The latter is a paradigmatic example of quantum mechanical metal–insulator transitions [2].

Due to Coulomb attraction, electrically injected or optically generated electrons and holes in a QW are highly correlated. At low carrier density, they form bound states, so-called *excitons*, which roughly correspond to hydrogen atoms, but with a modified binding energy $E_B = (\mu/m_0\,\varepsilon^2)$ Ry having a typical value of 1–10 meV due to a large dielectric constant $\varepsilon \geq 10$ and a small reduced effective mass $\mu^{-1} = m_e^{-1} + m_h^{-1}$. Further 'vertical' confinement of the exciton by the barriers leads to reduction of the 'lateral' (in-plane) extension and increases the binding energy to values up to 40 meV. In all technically important cases, the

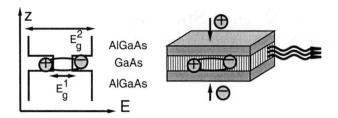

Fig. 15.1. Schematic illustration: formation of a QW and its potential use as a light emitter. Depending on the band offsets, layered deposition of semiconductor materials with different band gaps $E_g^1 < E_g^2$ leads to confinement of one or both constituents of the exciton

binding energy is much larger than the disorder-related local energy fluctuations. The exciton is thus expected to move as an entity through the QW. This argument is depicted in Fig. 15.2. It can be formalized in terms of a two-dimensional center-of-mass (COM) wavefunction which depends on the COM position $\boldsymbol{R}$ and where the COM states ψ_α obey a single-particle Schrödinger equation [3,4]

$$\varepsilon_\alpha \psi_\alpha(\boldsymbol{R}) = \left(-\frac{\hbar^2 \Delta_{\boldsymbol{R}}}{2M_{\mathrm{X}}} + v_{\mathrm{eff}}(\boldsymbol{R}) \right) \psi_\alpha(\boldsymbol{R}) , \tag{15.1}$$

with an exciton mass $M_{\mathrm{X}} = m_{\mathrm{e}} + m_{\mathrm{h}}$ and an effective potential

$$v_{\mathrm{eff}}(\boldsymbol{R}) = \int \mathrm{d}^2\varrho_a \sum_{a=\mathrm{e},\mathrm{h}} \frac{\mu^2}{m_a^2} \varphi_0^2 \left(\frac{\varrho_a - \boldsymbol{R}}{\alpha_a} \right) v_a(\varrho_a) , \tag{15.2}$$

which is obtained from the relative wavefunction $\varphi_0(\varrho_{\mathrm{e}} - \varrho_{\mathrm{h}})$ and the local single-particle energies $v_a(\varrho_a)$. Henceforth, eigenenergies ε_α will be given relative to the energy value of the disorder-free optically active exciton, i.e., the fundamental band gap minus the exciton binding energy.

The random potential v_{eff} is correlated over distances of at least the exciton Bohr radius, even if the underlying disorder v_a is short-range correlated. In this case, by virtue of the central limit theorem, v_{eff} is Gaussian distributed, and the statistical properties of the effective disorder can be completely described by the disorder strength $\sigma^2 = \langle v_{\mathrm{eff}}^2 \rangle$ and by φ_0 (determining the spatial correlation). As an aside, we notice that (15.2) has the form of a convolution and is therefore preferably evaluated numerically in momentum (Fourier) space.

Within certain approximations [4], the transition rate r_α between an exciton and a photon, i.e., the inverse of the exciton's radiative lifetime, is proportional to the absolute square of the optical matrix element

$$M_\alpha = \int \mathrm{d}^2R \, \psi_\alpha(\boldsymbol{R}) , \tag{15.3a}$$

$$r_\alpha = \mathcal{C} \, |M_\alpha|^2 . \tag{15.3b}$$

Fig. 15.2. Schematic illustration showing how the problem of an extended exciton in a QW with rough interfaces (characterized by islands of size a_{isl}) and alloy disorder can be mapped onto a single-particle problem with smooth effective potential $v_{eff}(R)$ for the center of mass (COM) motion. Potential variations occur on a length scale l_c (correlation length) which is of the order of the exciton size a_B

The absorption spectrum is simply

$$I_{abs}(\hbar\omega) = \sum_\alpha r_\alpha\,\delta(\hbar\omega - \varepsilon_\alpha)\,, \tag{15.4}$$

whereas the emitted photoluminescence (PL)

$$I_{PL}(\hbar\omega) = \sum_\alpha N_\alpha\,r_\alpha\,\delta(\hbar\omega - \varepsilon_\alpha) \tag{15.5}$$

involves the occupation N_α of the individual exciton states, which is not generally given by a thermal equilibrium distribution [4–6]. Thus, the luminescence (15.5) is strongly dependent on the excitation condition.

Recent techniques such as near-field scanning optical microscopy (NSOM) [7] or micro-PL (µPL) [5] allow excitation and/or detection of a few eigenstates at a time. In that case, the signals (15.4) and (15.5) consist of many sharp lines of spectral width typically less than 100 µeV. The weight and position of the individual peaks allow us to determine the *distributions* and *correlations* of eigenenergies ε_α and optical matrix elements M_α. The actual values ε_α and M_α are of little interest, since they show strong variations in going from sample to sample and from position to position on the same sample.

For comparison and later reference, we remark that the (local) density of states (DOS) can be expressed in a similar form to (15.4) and (15.5) as

$$\varrho_{DOS}(\hbar\omega) = \sum_\alpha \delta(\hbar\omega - \varepsilon_\alpha)\,, \tag{15.6}$$

and that PL excitation (PLE) spectroscopy records the light emitted at fixed detection frequency as a function of the excitation frequency.

15.2   Anderson Model, Wavefunction Localization, and Mobility Edge

A numerical solution of the continuous Schrödinger equation (15.1) typically involves a discretization on a real-space grid (spacing Δ)

$$\sum_{\langle ij\rangle} t\,\psi_{\alpha,i} + v_{eff,j}\,\psi_{\alpha,j} = \varepsilon_\alpha\,\psi_{\alpha,j}\,, \tag{15.7}$$

with $\langle ij \rangle$ marking nearest-neighbor sites and with the hopping matrix element $t = \hbar^2/2M_X\Delta^2$ as our energy unit. Henceforth, we shall take $t \equiv 1$. Equation (15.7) is often referred to as the *tight-binding model* with *random on-site energy*, by reference to concepts from quantum chemistry and electronic structure theory. If the emphasis is on the randomness, (15.7) is called the *Anderson model (of localization)*, honoring the seminal work of P.W. Anderson on disorder-driven metal–insulator transitions [8]. Here we focus on optical properties, while in Chaps. 14 and 16 the energy level statistics at the metal–insulator transition is discussed in more detail.

In the present context, the most important result of this model is that, in two dimensions (2D), all its solutions are exponentially localized [1,9–11]. Thus, considered as quantum eigenstates, all QW excitons are localized. However, as we will see below, for practical purposes, a distinction of 'free' and of 'localized' excitons separated by an 'effective mobility edge' can be useful. This scenario has been derived experimentally, and can also be justified by theory. In 3D, in contrast to the 2D case, a strict mobility edge exists separating exponentially localized states from states spanning the (infinite) system.

In our review and discussion of physical effects relevant for a 2D QW, we will illustrate these with 1D examples, which the reader can easily reproduce, using a few lines of MATHEMATICA code, for example, or slightly more lines using a higher programming language in conjunction with an eigensystem routine. Figures 15.3 and 15.4 show examples of eigenfunctions at all energies for different disorder strength in real space and in momentum space, respectively.

Gathering energies and optical matrix elements from many computer experiments (disorder realizations), predictions are obtained for the macroscopic DOS (not directly measurable) and the macroscopic absorption or optical density (see Fig. 15.5). This would also be the case in a (pointless) experiment where NSOM or μPL results are cumulated from different positions on the sample.

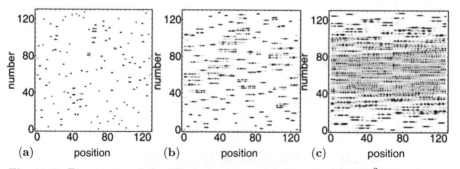

Fig. 15.3. Eigenstates of the 1D Anderson model. Density $|\psi_\alpha(R)|^2$ in real space. Wavefunctions have been numbered with increasing eigenenergy. Decreasing the disorder strength σ increases delocalization behavior: (**a**) $\sigma = 9.0$, (**b**) $\sigma = 1.5$, and (**c**) $\sigma = 0.45$ (system size $N_S = 128$). Uncorrelated Gaussian distributed disorder is used for all presented 1D calculations [12]. The strongest localization is found in the tails at very low and very high energies where the group velocity of the disorder-free exciton is small

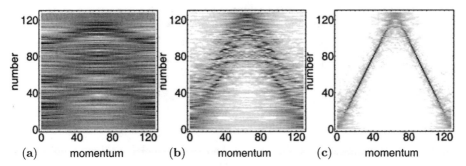

Fig. 15.4a–c. Eigenstates of the 1D Anderson model. Density in momentum space $|\psi_{\alpha q}|^2$ for the examples from Fig. 15.3. Note that momentum becomes a 'better quantum number' with decreasing disorder and closer distance to the band center [12]

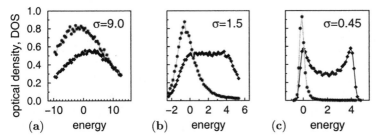

Fig. 15.5. Optical density (*) and density of states (♦) for the 1D Anderson model for (from *left* to *right*) $\sigma = 9.0$, 1.5, and 0.45 (data from 500 samples with $N_S = 128$). At unrealistically strong disorder (a), both DOS and optical density reflect the Gaussian distribution used for the on-site energy. Intermediate disorder (b) leads to an asymmetric optical density, shifted to slightly negative energies and with a strong tail towards high energies. This is in qualitative agreement with realistic 2D results (see Figs. 15.8 and 15.10 below). In the limit of weak disorder (c), the width of the inhomogeneously broadened line vanishes and a δ function develops at the energy of the zero-momentum state. Note the broadened van Hove singularities characteristic for the 1D DOS and small σ [12]

The same model Hamiltonian as in (15.7), albeit derived differently, has been used to determine the temperature dependence of absorption spectra and DOS in ordered systems [14].

Unfortunately, some features in Figs. 15.3c and 15.4c have to be regarded as artefacts. Obviously, inspection of the real-space wavefunctions in Fig. 15.3c or conversion of the k-space width in Fig. 15.4c into a characteristic length shows that many wavefunctions span the system from one end to the other. It is highly non-trivial to perform reliable calculations in this regime. Careful finite-size scaling to the infinite limit is usually necessary (see Chaps. 14 and 16 and references therein, and [1,13]). A convenient parameter for characterizing the spatial extension of a d-dimensional wavefunction is the participation

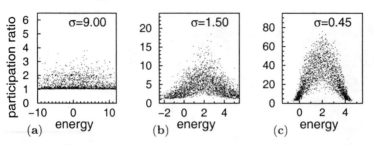

Fig. 15.6. Participation ratios for the 1D Anderson model [12]. Ten samples with $N_S = 128$ for **(a)** $\sigma = 9.0$, **(b)** $\sigma = 1.5$, and **(c)** $\sigma = 0.45$

ratio [1,10]

$$\ell_p^d = 1 \Big/ \int |\psi(\boldsymbol{R})|^4 \, d^d R \ . \tag{15.8}$$

1D results are presented in Fig. 15.6. For the weakest disorder in Fig. 15.6c, it is clear that only the wavefunctions in the tails are significantly smaller than the system size (see [13]).

Fortunately, we notice that the optical density and, more generally, all optical properties discussed in the present work are dominated by the best-localized wavefunctions! Consequently, finite-size effects are only of minor relevance in the calculation of excitonic spectra for semiconductor nanostructures.

We remark that in the case of weak disorder, the character of the wavefunctions changes with energy [10]. In the small or large energy tails, well-localized states are found which are located in regions of particularly low or high average potential values, respectively. In Fig. 15.3c, states can even be identified which have the character of excited states in the same local minimum as a corresponding state of lower energy. Closer towards the center of the band, states are much more extended. If one tries to describe these states in terms of local minima, they have to be regarded as resonances of excited states of *many* local minima. Alternatively, they can be considered as interference patterns of *many* plane-wave states with approximately the same energy (see Fig. 15.4). However, neither of these pictures suggests a simple explanation for why these states are strictly localized.

Even though calculations of optically active COM states are not much troubled by finite-size effects, they are sufficiently difficult due to the high dimensionality of the physical problem. It is well known that the computational effort of a full matrix diagonalization grows as N_S^3 with system size N_S, and hence with the sixth power of the linear extension of 2D systems. Fortunately, many powerful approaches have been developed that scale only slightly worse than linearly with N_S and allow in many cases, including the Anderson model, efficient determination of *some* eigenstates, e.g., those of lowest energy (see Chaps. 14 and 16 and references therein, and [10,15]). As we have seen, this is sufficient for the description of optical spectra. Examples of wavefunctions for $GaAs/Al_x Ga_{1-x} As$

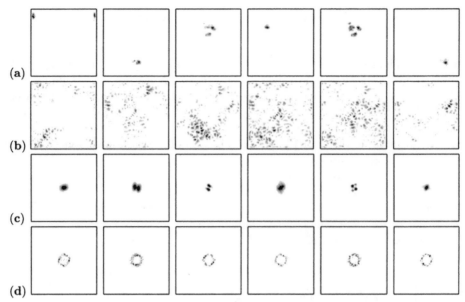

Fig. 15.7. Real space and Fourier space representations of the squared COM wavefunction for a QW with weak disorder on an area $\approx 1\ \mu m \times 1\ \mu m$ (GaAs/Al$_{0.3}$Ga$_{0.7}$As parameters: 5 nm width, $\sigma = 1.5$ meV, periodic boundary conditions). **(a)** Wavefunctions 19–24 (numbered with increasing energy) are well localized in real space and look approximately Gaussian. Some of them have ground-state character, some look like low-energy excited states. **(b)** Fourier transform of wavefunctions 19–24. The local ground state or excited state character is evident. **(c)** Wavefunctions 383–388 in the high-energy tail of the absorption line. **(d)** Fourier transform of wavefunctions 383–388. The concentration on a ring of given amplitude k and the randomness within this ring is visible. Note that all these states are in the energy region of the inhomogeneously broadened exciton line. The latter covers only the very-low energy part of the spectrum in the present case, which roughly corresponds to Fig. 15.5c

QW are given in Fig. 15.7. Calculated spectra can be found in [4] and references therein.

15.3   Remarks on Macroscopic Spectra

Before we return to the microscopic picture of disorder-localized wavefunctions and their statistical properties, we will briefly summarize the phenomenology of macroscopic spectra and its interpretations. Figure 15.8 summarizes theoretical results for absorption, PL, and PLE of a strained InGaAs/GaAs sample [6]. PL is redshifted relative to absorption. This reflects the simple effect that, after spectrally uniform excitation, excitons typically relax emitting phonons and later recombine in a state of lower energy. This so-called *Stokes shift* remains finite even at low temperatures, because the exciton lifetime is limited and because excitons trapped in local energy minima cannot reach other deeper minima.

Note that a thermal population of, e.g., a Gaussian-shaped tail shows an infinite Stokes shift for $T \to 0$.

Conceptually more interesting is the temperature-dependent PLE, shown in Fig. 15.8. The PLE line is highly asymmetric. Its high-energy tail coincides with absorption. On the low-energy side, a quite sharp cutoff behavior is found which slowly disappears with increasing temperature. The PLE peak is blueshifted relative to absorption and even more relative to PL. This can be understood in terms of an effective mobility edge [5] and the change of wavefunction character described in the context of Fig. 15.3c, which is much more pronounced in 2D [4]. At low temperatures and for fixed detection energy, i.e., selecting a few low-lying states, it is virtually impossible for an exciton excited at slightly higher energy, i.e., in a different local minimum, to be transfered to a minimum at the detection energy within its limited lifetime. Excitons generated in the fairly extended states above the effective mobility edge can relax into many different local minima, some of them probably being within the energy window of the detector.

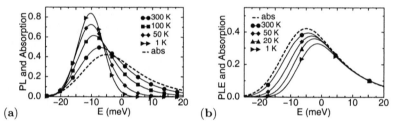

Fig. 15.8. Stokes shift of PL (**a**) and blueshift of PLE (**b**) relative to absorption for a strained InGaAs/GaAs QW sample with strong disorder at various temperatures (after [6])

15.4   Repulsion of Energy Levels

Individual energy values in a quantum system with randomness are of little interest. One approach is to consider ensemble averages, as has been demonstrated in Sect. 15.3. Of less practical importance, but more interesting, is the study of level correlations and related features (see Chaps. 14 and 16 and references therein). Indeed, many quantum systems show universal features which depend on qualitative properties such as time-reversal invariance, but not on quantitative details. These classes of Hamilton operators are often modeled by matrix ensembles, which are studied analytically in the so-called *random matrix theory* [16,17]. An early famous prediction is the Wigner–Dyson distribution (Wigner surmise). In our case, this states for the *Gaussian orthogonal ensemble* (GOE) of real symmetric Hamiltonian matrices that the neighbor level spacing distribution

$P(s)$ is

$$P(s) \equiv \frac{1}{\Delta N_S} \sum_{\nu} \delta(s - \varepsilon_{\alpha_{\nu+1}} + \varepsilon_{\alpha_\nu}) \approx a_1 \left| \frac{s}{\Delta} \right| e^{-b_1(s/\Delta)^2} , \qquad (15.9)$$

with average level distance $\Delta \sim 1/N_S$ and $a_1 = \pi/2$, $b_1 = \pi/4$. The characteristic level-repulsion dip $P(s) \sim |s|$ at zero spacing, $s \approx 0$, is *easily* found numerically for the system (15.7), as can be seen in Fig. 15.9, which shows the distribution $C(s)$ of *all* spacings. [Of course, at small level distance s, $C(s)$ and $P(s)$ coincide.] For a possible experiment, nearest-neighbor correlations P have the disadvantage that individual states have to be resolved and it is crucial to decide whether a peak contains one or two states. In contrast, the correlation C can be derived from a simple convolution. Furthermore, C can often be conveniently calculated from a time-dependent Schrödinger equation [3]. However, a level-repulsion dip should not be present, if the simulation were done properly with large system size [11]. As the localized wavefunctions in different parts of the (numerical) sample are statistically independent, the level-repulsion dip should disappear with increasing system size N_S as $P(s) \approx (1 - c/N_S) + (c/N_S)|s/\Delta|$ [17].

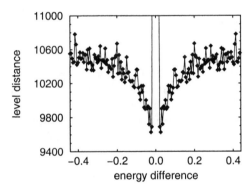

Fig. 15.9. Level correlation in a finite-size 1D Anderson model. Data for 1000 samples with $N_S = 128$ and $\sigma = 1.5$. Note the large vertical offset which strongly depends on N_S and localization length. The level self-correlation yields a δ function at $E_\alpha - E_\beta = 0$ [12]

In fact, the survival or disappearance of the level-repulsion dip with increasing system size has proven to be a powerful tool for studying the localization–delocalization transition (see Chaps. 14 and 16). The experimental equivalent of a small-system-size computer simulation is spatially-resolved spectroscopy with NSOM or µPL. We predicted a level-repulsion dip even for luminescence, which is easier to measure than absorption (see Fig. 15.10) [3]. Furthermore, from its relative weight, experimental estimates of the ratio of wavefunction extension and excitation spot size can be obtained. First experiments [18] seem to confirm this prediction and a statistical analysis of spatially resolved spectra promises to become a powerful tool in optical spectroscopy [19].

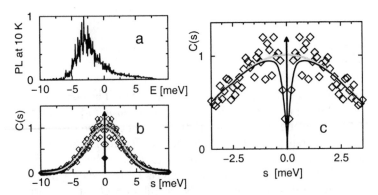

Fig. 15.10. (a) Simulated PL spectrum as a sum of 250 NSOM spectra for a 5 nm wide GaAs/Al$_{0.3}$Ga$_{0.7}$As QW ($\sigma = 8$ meV), each NSOM spectrum corresponding to an area 120 nm × 120 nm. (b) Spectral autocorrelation for the spectrum of (a) (*gray line*) and the sum of the 250 individual autocorrelations (*symbols*). Only a limited spectral region (approximately −6 meV to −1 meV) is used for the autocorrelations. (c) Enlargement of (b) with GOE fit for $C(s)$ to the data (from [3]). A level-repulsion dip is clearly visible (compare with Fig. 15.9)

15.5   Microscopic Relaxation Kinetics

We emphasize that the level-repulsion dip seen in Fig. 15.10b and c is only visible because a *small* excitation spot was simulated. The next three sections will cover predictions for physical experiments where the physics of level repulsion plays a more interesting role, since it is also visible for the infinite system [20].

Level repulsion is present even for infinite systems with localized states, if quantities are considered which involve not only energy differences but also matrix elements, e.g., of the type

$$C_{\hat{A}}(s) = \frac{1}{\Delta N_{\mathrm{S}}} \sum_{\alpha\beta} |\langle\alpha|\hat{A}|\beta\rangle|^2 \, \delta(s - \varepsilon_\beta + \varepsilon_\alpha) \,. \tag{15.10}$$

Well known examples are response functions. In the context of optical spectra for disordered systems, the phonon scattering rate is of the form (15.10), where $\hat{A}$ is the deformation potential: escape from a state ψ_α in a local minimum into a state ψ_β by absorption of a phonon of small thermal energy is only possible if ψ_β is simultaneously close to ψ_α with respect to both energy separation and distance in real space, i.e., if $\varepsilon_\beta - \varepsilon_\alpha \approx k_{\mathrm{B}}T$ and $|\langle\alpha|\hat{A}|\beta\rangle|^2 > 0$. Such transition rates are discussed in [6,21]. An example is reproduced in Fig. 15.11. A level-repulsion dip is present in the 2D case with realistic phonon scattering and might influence the low-temperature behavior of the anomalous Stokes shift [6]. From its construction, this quantity will show level repulsion even for large system size [9,11].

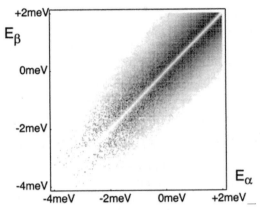

Fig. 15.11. Phonon scattering rates between states at energies E_α and E_β for a 4 nm wide GaAs/Al$_{0.3}$Ga$_{0.7}$As QW sample with $\sigma = 4$ meV. The level-repulsion dip along the diagonal is overemphasized due to the momentum dependence of deformation-potential scattering of acoustic phonons [21]

15.6   Radiative-Lifetime Distribution

A second well-known feature of complex and chaotic systems which can be described by random matrix theory is the Porter–Thomas distribution for the wavefunction values [13,16,17,22]

$$\mathcal{P}(x) = \langle \delta(x - N_S |\psi|^2) \rangle = \frac{e^{-x/2}}{\sqrt{2\pi\, x}} . \tag{15.11}$$

A simple heuristic argument extending the above picture of 'many interfering plane waves' is that, according to the central limit theorem, the wavefunction values can be considered to have a Gaussian distribution (Berry's conjecture). Equation (15.11) then follows from simple algebra. Again, these arguments are not valid for the localized states of the Anderson model. In particular, wavefunction values of localized states are not proportional to $\sqrt{N_S}$, as (15.11) would imply. However, it has recently been demonstrated numerically that the Fourier transform of the QW eigenstates shows a transition from a distribution derivable within the so-called *optimum fluctuation theory* [9] in the low-energy tail of the exciton peak to a Porter–Thomas distribution in its high-energy part [23].

   In view of the heuristics given above, this is not surprising and can be considered as a transition from well localized states in deep potential minima (Lifshitz tails [9]) to considerably more extended states which are localized only by quantum mechanical interference effects. The $k = 0$ element of the Fourier-transformed eigenstate ψ_α is the optical matrix element (15.3a). Its square is the radiative lifetime (15.3b). Thus, this transition and the Porter–Thomas distribution should be observable in spatially resolved spectroscopy, if good energy resolution is obtained simultaneously. A detailed discussion for realistic QW systems can be found in [23].

Again, the qualitative physics can be seen with little numerical effort in the 1D tight-binding model. The distribution of optical matrix elements is visualized in Fig. 15.12. Histograms taken at fixed energy (Fig. 15.13) display the described behavior. Again, the 1D results are in qualitative agreement with the predictions for realistic 2D QW systems of [23], reproduced as Fig. 15.14.

15.7   Rayleigh Scattering, Speckles, and Enhanced Back-Scattering

Rayleigh scattering of light is responsible for the blueness of the sky and the whiteness of a wall. It can be considered to be instantaneous, as long as the reflectivity varies weakly and smoothly with frequency. This is not the case for light resonant with the sharply peaked exciton line [24,25]. The exciton states ψ_α are excited resonantly; each state has an oscillating dipole moment radiating at frequency $\omega_\alpha = \varepsilon_\alpha/\hbar$ up to a few hundred ps after the excitation is switched off [26,27]. The quantum mechanical expectation value (thermal average or history average) for the electric field $\widehat{\mathcal{E}}$ emitted in direction q_{out} after excitation with q_{in} is (suppressing material-specific prefactors and describing directions by the 2D in-plane momentum of first exciton, then photon)

$$\widehat{\mathcal{E}}_{q_{\mathrm{out}}}(t) = \mathrm{e}^{-\mathrm{i}\omega_\alpha t}\, M^*_{\alpha,q_{\mathrm{out}}}\, M_{\alpha,q_{\mathrm{in}}} \,, \tag{15.12}$$

with the more general optical matrix element [see (15.3a)],

$$M_{\alpha,q} = \int \mathrm{d}^2R \; \psi_\alpha(\boldsymbol{R})\, \mathrm{e}^{\mathrm{i}\boldsymbol{q}\cdot\boldsymbol{R}} \,. \tag{15.13}$$

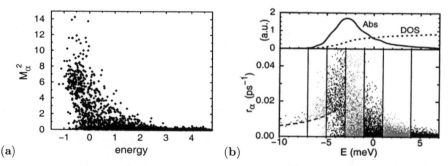

(a)   energy   (b)   E (meV)

Fig. 15.12. (a) Square of the optical matrix elements M_α^2 for the 1D Anderson model. Twenty samples with $N_{\mathrm{S}} = 256$ and $\sigma = 1.5$. A transition can clearly be seen within the inhomogeneously broadened line from a narrow distribution centered at a finite value (low-energy region, $E \leq -1$) to a broader distribution peaked at zero (high-energy region, $E \geq 0.3$) with a rich, non-universal behavior in the transition region, $-1 \leq E \leq +0.3$ [12]. (b) Realistic 2D results for radiative lifetime (*lower*) from [23] for comparison. Again, rather different distributions are found in different energy ranges. Within the exciton optical density (Abs) at the lower end of the disorder-broadened DOS (*upper*) the same transition appears from Lifshitz-tail states to Anderson-localized states

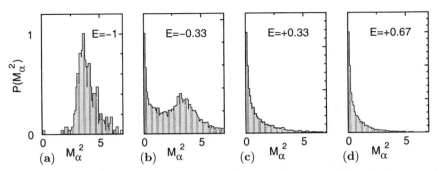

Fig. 15.13a–d. Distribution (non-normalized) of squared optical matrix elements for $\mathcal{P}(M_\alpha^2)$. 1000 samples with $N_S = 64$ and $\sigma = 1.5$ (see Fig. 15.12). Histograms present data near the energy values given in the panel headings. The high-energy distributions (**c**) and (**d**) follow a Porter–Thomas distribution (15.11), whereas the low-energy distribution peaks at the optimum fluctuation value [23]

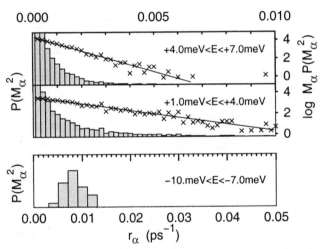

Fig. 15.14. Distribution $\mathcal{P}(r_\alpha)$, $r_\alpha \sim M_\alpha^2$ (histograms, linear scale) for 2D QW results from Fig. 15.12b and for different energy ranges. For the *upper two panels*, $\mathcal{P}(r_\alpha)\sqrt{r_\alpha}$ values are included (symbols, logarithmic scale on *right hand side*). For a Porter–Thomas distribution, a *straight line* results (from [23])

Obviously, the ensemble averaged electric field (determined by the excitonic polarization) [28]

$$E_q(t) = \left\langle \hat{\mathcal{E}}_q \right\rangle_t \tag{15.14}$$

vanishes for times $t \gg \hbar/\sigma$, because the eigenfrequencies are randomly spread with variance of order σ (see Fig. 15.5). At short times $t \approx 0$, a strong signal is found in the forward direction, $q_{\text{out}} \approx q_{\text{in}}$. This is the specularly reflected signal (the transmitted light is usually absorbed in the sample). The fact that the average field vanishes in other directions is a bit more subtle [26,28].

The sky is blue and a wall is white – this is due to the simple fact that vanishing expectation values for the field are compatible with a finite intensity

$$I_q(t) = \left\langle \widehat{\mathcal{E}}_q^+ \, \widehat{\mathcal{E}}_q \right\rangle = \left\langle \sum_{\alpha,\beta} e^{-i\omega_\alpha t} \, M^*_{\alpha,q_{\text{out}}} \, M_{\alpha,q_{\text{in}}} \, e^{i\omega_\beta t} \, M_{\beta,q_{\text{out}}} \, M^*_{\beta,q_{\text{in}}} \right\rangle .$$

(15.15)

Again, the sums include all states in the excitation area. One might ask why the scattering intensity obtained from this double sum only grows with the first power of the excitation area, i.e., the number of contributing states [26,28]. In the ensemble average, almost every term in (15.15) cancels. The only terms that contribute are those for which α and β are 'somehow correlated'. These will be pairs of localized states close to each other in real space (compare the discussion in Sect. 15.5). This leads to characteristic oscillations in the time-dependent Rayleigh signal [26], which may have been seen in recent experiments [29].

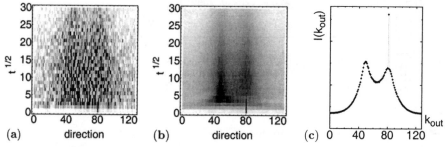

Fig. 15.15. Rayleigh scattering for the 1D Anderson model. (a) Time-resolved signal (15.15) of a single disorder realization with $\sigma = 1.5$ and system size $N_S = 128$, as a function of q_{out}. A decay $\exp(-\gamma t)$ with $\gamma = 0.01$ is included. The coherent initial signal in the specular direction q_{in} is seen as a *sharp dark line* at $t \approx 0$. Note the non-linear time scale showing both the short- and the long-term behavior. (b) Average of 500 samples. (c) Integral of the time-resolved signal (b), showing enhancements around the back-scattering and the specular direction, which has δ-function character [12]

Finally, we discuss another optical experiment suggested recently, which is determined by excitons and their localization properties. Again, it is demonstrated for the numerically easy 1D case. The resonantly scattered intensity fluctuates strongly as a function of both time and direction. These so-called *speckles* are shown in Fig. 15.15. Their detailed analysis yields information on the destruction of quantum coherence and the relaxation process [28,30].

In the ensemble average, the speckles disappear and a highly non-trivial dependence on time and direction evolves. The most noticeable features are the enhanced intensities in the back-scattering direction and near the specularly reflected beam (see Fig. 15.15b). These are even more clearly seen if data are integrated with respect to time. This is the same physics in the time domain as predicted in [27,31] for the frequency domain (see Fig. 15.16). Note that the frequency-resolved signal is not the Fourier transform of the intensity (15.15), but the squared Fourier transform of the field (15.14) [24].

Enhanced back-scattering is well-known from light and radar scattering of inhomogeneous media such as fog and from current transport in dirty metals [32]. Both are best understood in the formalism of weak localization, involving independent scatterers separated by more than the wavelength of the electromagnetic radiation or the electron. This assumption does not apply to QW excitons, a fact which leads to characteristic differences [27,31]. Note the narrowing in momentum space (direction), which reflects a wavefunction spread in real space at early times before the wavefunction 'realizes' that it is localized [33].

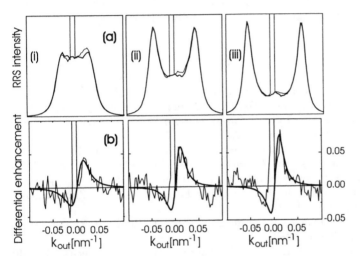

Fig. 15.16a,b. *Top row.* Angle-dependent, frequency-resolved Rayleigh intensity for various energies: (i) $\hbar\omega = 0$, (ii) 0.2, (iii) 0.4 meV above the center of the potential distribution. The incoming direction is $k_{in} = 0.005$ nm^{-1}. The signals in the plane of the incoming and reflected beam (*thick line*) and the perpendicular plane (*thin line*) are compared. *Bottom row.* Differential enhancement (*thin line*) and analytical 2-parameter fit (*thick line*) (after [31])

15.8   Summary

Much of the physics related to randomness and disorder in the spectra of 2D QW semiconductor nanostructures can be understood by means of the 1D Anderson model and in terms of features known from random matrix theory. The localization of excitons leads to a blueshift of PLE and a redshift of PL relative to the absorption line, which is asymmetrically broadened. The line shape can be interpreted in terms of a transition from Lifshitz-tail states to states localized by interference (Anderson localization). A Porter–Thomas distribution of radiative lifetimes is predicted for the latter. Quantum-mechanical level repulsion can be seen in spatially resolved spectroscopy, in low-temperature phonon-mediated relaxation, and in the time-dependent Rayleigh signal.

The reader is strongly encouraged to spend a few hours reproducing at least some of the 1D results presented in this discussion [12] and to experiment with correlated disorder!

Acknowledgments. Most results on realistic physical 2D QW systems were taken form a series of articles in collaboration with Roland Zimmermann and Vincenzo Savona (Berlin). I thank them as well as my experimental colleagues Uwe Jahn (PDI Berlin) and Wolfgang Langbein (University of Dortmund) for the close and pleasant collaboration. Useful remarks on the manuscript by Roland Hott (Forschungszentrum Karlsruhe) are gratefully acknowledged.

References

1. B. Kramer, A. Mac Kinnon: Rep. Prog. Phys. **56**, 1469 (1993)
2. M. Schreiber (Ed.): *Localization 1999: Disorder and Interaction in Transport Phenomena*, Special issue Ann. Phys. (Leipzig) **8** (1999) pp. 531–798 and SI-1–SI-300
3. E. Runge, R. Zimmermann: Phys. Stat. Solidi (b) **206**, 167 (1998)
4. E. Runge, R. Zimmermann: 'Optical properties of localized excitons in nanostructures: Theoretical aspects'. In: *Advances in Solid State Physics* Vol. 38, ed. by B. Kramer (Vieweg, Braunschweig 1998) pp. 251–263
5. U. Jahn, M. Ramsteiner, R. Hey, H.T. Grahn, E. Runge, R. Zimmermann: Phys. Rev. B **56**, R4387 (1997)
6. M. Grassi Alessi, F. Fragano, A. Patane, M. Capizzi, E. Runge, R. Zimmermann: Phys. Rev. B **61**, 10 985 (2000)
7. H.F. Hess, E. Betzig, T.D. Harris, L.N. Pfeiffer, K.W. West: Science **264**, 1740 (1994)
8. E. Abrahams, P.W. Anderson, D.C. Licciardello, T.V. Ramakrishnan: Phys. Rev. Lett. **42**, 673 (1979)
9. I.M. Lifshits, S.A. Gredeskul, L.A. Pastur: *Introduction to the Theory of Disordered Systems* (Wiley, New York 1988)
10. M. Schreiber: 'Multifractal Characteristics of Electronic Wave Functions in Disordered Systems'. In: *Computational Physics*, ed. by K.H. Hoffmann and M. Schreiber (Springer, Berlin 1996) pp. 147–165
11. A.D. Mirlin: Phys. Rep. **326**, 259 (1998)
12. MATHEMATICA code used for the 1D results can be obtained via the web page for this book, viz.,
    http://www.tu-chemnitz.de/physik/HERAEUS/2000/Springer.html
    or directly from:
    http://www-semic.physik.hu-berlin.de/runge/Anderson1D.
    The graphics output was edited for this publication.
13. K. Müller, B. Mehlig, F. Milde, M. Schreiber: Phys. Rev. Lett. **78**, 215 (1997)
14. Y. Toyozawa, M. Schreiber: J. Phys. Soc. Japan **51**, 1528 (1982); ibid. 1537; ibid. 1544

15. U. Elsner, V. Mehrmann, F. Milde, R.A. Römer, M. Schreiber: SIAM J. Sci. Comp. **20**, 2089 (1999)
16. M.L. Mehta: *Random Matrices*, 2nd edn. (Academic Press, San Diego 1990)
17. T. Guhr, A. Müller-Groeling, H. Weidenmüller: Phys. Rep. **299**, 190 (1998)
18. J.R. Guest, T.H. Stievater, D.G. Steel, D. Gammon, D.S. Katzer, D. Park: Presented at *Quantum Electronics and Laser Science Conference*, San Francisco, USA, May 7–12, 2000
19. G. von Freymann, E. Kurtz, C. Klingshirn, M. Wegener: Appl. Phys. Lett. **77**, 394 (2000)
20. E. Runge, R. Zimmermann: 'Level repulsion and wavefunction statistics in semiconductor quantum wells with localized excitons'. In: [2], pp. SI-229–SI-232
21. E. Runge, R. Zimmermann: Ann. Phys. (Leipzig) **7**, 417 (1998)
22. C.E. Porter (Ed.): *Statistical Theories of Spectra: Fluctuations* (Academic Press, New York 1965)
23. E. Runge, R. Zimmermann: Phys. Stat. Sol. (b) **221**, 269 (2000)
24. H. Stolz: *Time-Resolved Light Scattering from Excitons* (Springer, Berlin 1994)
25. J. Shah: *Ultrafast Spectroscopy of Semiconductors and Semiconductor Nanostructures* (Springer, Berlin 1996)
26. V. Savona, R. Zimmermann: Phys. Rev. B **60**, 4928 (1999)
27. V. Savona, E. Runge, R. Zimmermann, S. Haacke, B. Deveaud: Phys. Stat. Sol. (b) **221**, 365 (2000)
28. E. Runge, R. Zimmermann, 'Coherence properties of resonant secondary emission'. In: *Advances in Solid State Physics* Vol. 39, ed. by B. Kramer (Vieweg, Braunschweig 1999) pp. 423–432
29. V. Savona, S. Haacke, B. Deveaud: Phys. Rev. Lett. **84**, 183 (2000)
30. E. Runge, R. Zimmermann: Phys. Rev. B **61**, 4786 (2000)
31. V. Savona, E. Runge, R. Zimmermann: Phys. Rev. B **62**, R4805 (2000); and unpublished work
32. P. Sheng (Ed.): *Scattering and Localization of Classical Waves in Random Media* (World Scientific, Singapore 1990)
33. V.N. Prigodin, B.L. Altshuler, K.B. Efetov, S. Iida: Phys. Rev. Lett. **72**, 546 (1994)

16 Characterization of the Metal–Insulator Transition in the Anderson Model of Localization

Michael Schreiber and Frank Milde

Summary. In this chapter we discuss three different methods in statistical physics which have been successfully implemented to determine the metal–insulator transition and to characterize the electronic states in disordered systems described by the Anderson model of localization. First, we study the spatial decay of electronic states of the Anderson Hamiltonian along quasi-one-dimensional bars and use finite-size scaling to analyze the data and the transition in infinite three-dimensional samples. Second, we calculate the eigenfunctions and describe their spatial distribution by means of multifractal analysis. Third, we compute the eigenvalue spectrum and study the energy level statistics to determine the transition. Emphasis is laid on programming tricks to save computer time or to increase accuracy. As an example, some results of large-scale numerical investigations for anisotropic materials are presented, demonstrating that the three methods yield coinciding results. Several related topics of current research on the electronic properties of disordered materials are mentioned in which these statistical methods have been successfully applied.

16.1 The Metal–Insulator Transition in Disordered Systems

For a long time perfectly ordered crystals have been the subject of theoretical investigations in solid state physics. Due to their translational invariance many calculations can be performed analytically. Real solids, however, always contain at least some disorder, which might be caused by impurities in doped semiconductors, by defects in otherwise perfectly grown crystals, or by the random arrangement of the constituents in alloys. Even more obvious is the topological disorder in amorphous materials and glasses. While single impurities or defects can often be treated by perturbation theory, most analytical approaches are not sufficient to describe the physical properties of random materials in a realistic way. Therefore numerical methods have become more and more important for the modeling of such disordered systems. But even with today's advanced computers, it is often difficult to calculate material properties ab initio. Consequently, simple models are still of considerable significance.

One way of studying random systems which is widely used in computational solid state physics is the statistical analysis of eigenstates and eigenlevels of random operators. In this chapter we shall describe three different approaches of this kind. All three methods will be applied to the simple Anderson model of localization which has become a paradigm for the analysis of electronic states of disordered materials. It describes non-interacting electrons moving (or not) in a random potential V. Anderson suggested as early as 1958 [1] that particles moving in a potential which is in some sense random may show no diffusion at all. As we have learned since then, this localization may be induced by strong impurities in such a way that particles are bound at or near those impurities. The localization may also be caused by interference effects due to multiple scattering of the wave functions. Such localization by interference is known not only for electronic states, but also macroscopically for water waves [2] and for light waves [3].

In the spirit of Anderson's original paper we model the disordered systems keeping the lattice structure and restricting the randomness to the potential energies. In principle, one would have to solve the stationary Schrödinger equation

$$-\frac{1}{2m}\nabla^2\psi(x) + V(x)\,\psi(x) = E\,\psi(x) \tag{16.1}$$

for the wave functions $\psi(x)$ at position x, where m is the mass of the particle and E the energy. However, in the lattice model we are not interested in the behaviour of the wave function on length scales smaller than a lattice constant a. Then it is convenient to study the Schrödinger equation in site representation, which yields for a one-dimensional (1D) system [15]

$$t\,\psi_{n+1} + t\,\psi_{n-1} + \varepsilon_n\,\psi_n = E\,\psi_n . \tag{16.2}$$

Here ψ_n is the wave function at the position $x = na$, i.e., at the nth site. The transfer matrix elements t reflect the transfer of an electron from the nth site to neighbouring sites and are obtained by expressing the second derivative in the kinetic energy in (16.1) by finite differences. Up to a constant term, the parameters ε_n equal the potential energies V_n, which are chosen independently from a box distribution

$$P(\varepsilon_n) = W^{-1}\,\Theta(W/2 - |\varepsilon_n|) . \tag{16.3}$$

Here Θ denotes the step function.

The width W of the distribution (16.3) quantifies the strength of the randomness. For vanishing width, the ordered lattice is recovered. In this case the energy eigenvalues form energy bands of finite width and the allowed states corresponding to the energies in these bands are Bloch functions which are spread throughout the entire system. Electrons in such extended states will be highly mobile and contribute to charge transport. On the other hand, for large W, strong fluctuations of the potential energies, corresponding to sufficiently deep impurities, yield localized wave functions which decay exponentially in space. Electrons are bound in these localized states at the impurities and cannot contribute to charge transport. The theory of localization is concerned with the questions under which conditions localized or extended states occur in random systems, how these states can be characterized, how the delocalization–localization transition occurs between the two extreme situations of vanishing and strong disorder described above, and how the physical quantities behave near this metal–insulator transition (MIT). A more general discussion of these quantum phase transitions is presented in Chap. 13.

Starting from the perfect crystal and introducing some small potential fluctuations, the energy bands are broadened and localized states occur near the band edges. These are separated by the so-called mobility edges from the extended states in the band centre. The localization in the band tails can be understood as a spatial confinement of the electrons in potential wells [4], equivalent to localization at impurity sites. If one adds more or deeper impurities to the crystal, or if one considers energies not in the far tails but closer to the mobility edges, the electronic wave functions bound at the impurities or in different potential wells overlap, tunneling is enabled, and the electrons become less localized. Correspondingly, energetically close impurity levels hybridize. If more impurities or potential wells contribute to the eigenstates, the resulting fluctuations of the wave functions may be so strong that the exponential decay is completely masked.

In the band centre, the extended wave functions are of course also disturbed by the random potential. This can be understood as the effect of multiple scattering of an electronic wave at the impurities, leading to interference. For not too large disorder, the wave functions nevertheless remain extended but the resulting fluctuations of the wave functions may completely mask their extended character. With increasing disorder the energetic range in which extended states exist around the band centre becomes smaller. Increasing interference effects finally lead to localization of the last remaining extended states in the band centre at the critical disorder W_c. This value defines the MIT.

Experimentally this disorder-driven phase transition can be observed by changing the concentration of dopants or by applying pressure, for example. In this way the distance between the impurities is changed, and this corresponds to a change in the ratio of W and t in the Anderson model. Of course, the simplified model cannot be expected to allow a quantitative comparison between experiment and theory. However, the characteristic properties of the phase transition, in particular the critical exponent ν with which the localization length or the inverse conductivity diverge at W_c should not depend on the specific details of the Hamilton operator but only on the general symmetries. Therefore one of the goals of the MIT analysis must be the determination of ν. Some results will be presented below. We concentrate here on our numerical data for anisotropic systems. Such systems have been discussed previously [5], because they may be more realistically compared to particular experimental situations [5]. However, for the simple model that we use, we have found that the characteristic properties of the phase transition, like the critical exponent, are not changed by the anisotropy, i.e., the expected universality is present. Even experimentally, the value of the critical exponent has been under discussion for a long time and it is not clear whether a description in terms of non-interacting electrons is sufficient or whether electron–electron interactions have to be taken into account to solve the 'exponent puzzle' [6].

16.2   The Transfer Matrix Method

The Schrödinger equation (16.2) can be rewritten as a recursive equation

$$\psi_{n+1} = t^{-1}(E - \varepsilon_n)\psi_n - \psi_{n-1} , \qquad (16.4)$$

and this can be expressed in matrix notation together with the identity $\psi_n = \psi_n$ as

$$\begin{pmatrix} \psi_{n+1} \\ \psi_n \end{pmatrix} = \begin{pmatrix} (E-\varepsilon_n)/t & -1 \\ 1 & 0 \end{pmatrix} \begin{pmatrix} \psi_n \\ \psi_{n-1} \end{pmatrix} \equiv \mathbf{T}_n \begin{pmatrix} \psi_n \\ \psi_{n-1} \end{pmatrix} . \qquad (16.5)$$

Starting with arbitrary initial conditions ψ_1 and ψ_0, one obtains the amplitudes at sites $N+1$ and N by repeated multiplication of the random transfer matrices $\mathbf{T}_n$ defined in (16.5). For the product matrix

$$\mathcal{T}_N = \prod_{n=1}^{N} \mathbf{T}_n , \qquad (16.6)$$

the theorem of Oseledec [7] applies, guaranteeing the existence of

$$\Gamma \equiv \lim_{N \to \infty} \left(\mathcal{T}_N \mathcal{T}_N^\dagger \right)^{1/2N} \qquad (16.7)$$

in the thermodynamic limit, with eigenvalues e^{γ_m}. The Lyapunov exponents γ_m reflect the exponential increase or decrease of the wave functions for large distances from the origin. Due to the symplecticity of the product matrix (16.6), the two Lyapunov exponents γ_0 and γ_1 are not independent, but related by $\gamma_0 = -\gamma_1$. The localization length is thus given by

$$\lambda \equiv -\gamma_0^{-1} \equiv \gamma_1^{-1} . \qquad (16.8)$$

The above considerations can be easily generalized to quasi-1D systems with strip (or bar) geometry of $M \times N$ (or $M^2 \times N$) sites with $N \gg M$. In this case vectors ψ_n of length $M(M^2)$ comprise the amplitudes of the wave functions on the M (M^2) sites of the nth cross section of the strip (bar). There are $M(M^2)$ linearly independent ways to distribute the amplitudes of the initial state on the first cross section. These possibilities are simultaneously treated by assembling the respective vectors of expansion coefficients into an amplitude matrix ψ of size $M \times M$ ($M^2 \times M^2$). The transfer matrices $\mathbf{T}_n$ become $2M \times 2M$ ($2M^2 \times 2M^2$) random matrices and describe the transfer from cross section to cross section. This generalization is straightforward, but the random potential energies in (16.5) have to be replaced by matrices comprising not only these site energies but also the matrix elements t for the transfer within the cross section.

Details of this method have been described elsewhere [8]. Here we only repeat the most important points and add some new aspects to the previous discussion.

As in the 1D case, the product matrix (16.6) of the transfer matrices satisfies Oseledec's theorem. The Lyapunov exponents occur in $M(M^2)$ pairs γ_m

and $-\gamma_m$. They characterize how the amplitudes of the $M(M^2)$ initial states drift apart exponentially along the strip (bar). Their inverse values reflect the different characteristic length scales. The largest length which is given by the inverse of the smallest positive Lyapunov exponent determines the localization length λ_M.

In practice the eigenvalues of the product matrix are determined by the repeated multiplication of the transfer matrices on a set of $M(M^2)$ orthogonal initial amplitude vectors. Using only one arbitrary initial vector this Krylov series would yield the largest eigenvalue of the product matrix reflecting the fastest exponential increase. To determine the slowest exponential increase, i.e., the localization length, we have to determine all Lyapunov exponents. This can be achieved by starting from an orthogonal set of initial vectors and maintaining the orthogonality of the amplitude vectors during the iteration. However, the ratio of the smallest and the largest eigenvalue of the product matrix becomes comparable with the machine accuracy after very few multiplications so that the smallest eigenvalue would very soon be lost. This convergence problem can be circumvented by normalization of the amplitude vectors. To be specific, we orthogonalize each of the columns of the amplitude matrix onto the previous columns using the standard Gram–Schmidt orthogonalization procedure. The normalization of the column vectors yields the respective eigenvalues asymptotically. As a rule of thumb, this orthonormalization should be performed after every ten iteration steps. The mth normalization constants of all these Gram–Schmidt procedures are multiplied to determine the overall normalization of the mth vector and thus the mth eigenvalue. In practice, the logarithms are summed yielding the mth Lyapunov exponent. This sum can also be interpreted as an average over the respective Lyapunov exponents of many short strips of length 10. Accordingly, the fluctuations of these data can be used to determine the statistical error of the result in a straightforward way by computing the variance. This is an important detail of the calculation because the *accuracy* can thus be controlled during the iteration and the recursion can then be stopped when the requested accuracy is achieved.

Due to the *self-averaging* nature of the transfer matrix procedure, it is even possible to replace the iteration along a long strip or bar by an average over an ensemble of shorter strips or bars. This is demonstrated in Fig. 16.1 in which the distribution of a large number of different realizations of the random potential is compared with the variance corresponding to the statistical error of the separate iterations. We note that for a variance of 1%, the difference between the distribution of the Lyapunov exponent and its inverse (the localization length) is insignificant.

The Fortran program loc2d1.for available from our web site [9] performs the recursion for a quasi-1D strip. Due to the very simple structure of the transfer matrices, the matrix–matrix multiplication in the generalized form of (16.5) can be programmed in a very efficient way, avoiding all multiplications with t (which can be set to unity) and all summations of 0. The program performs the recursion until the preset accuracy or a preset maximum number of iteration

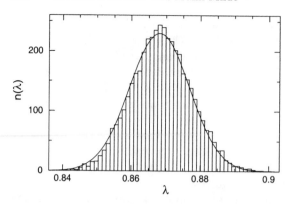

Fig. 16.1. Histogram of the localization lengths λ of 10.000 different bars with weakly coupled planes for $\gamma = 0.9, W = 8.5$, and $M = 5$. The statistical error of 1% of the resulting $\lambda = 0.8681$ is represented by the Gaussian (*solid line*) with width 0.008681. The anisotropy parameter γ is explained in the text

steps is reached. The results for the localization length and its variance are written into the file lambda_n.dat and can be visualized with the gnu-plot program lambda_n.gnu, depending on the length of the system N.

With a second version of the program, loc2d2.for, converged values of the localization length λ_M for different strip widths M are calculated and stored in lambda##.dat, where ## denotes a consecutive number for each combination of energy and disorder. With lambda_m.gnu, these data can be visualized through their dependence on M. Trying different parameter combinations, one should keep in mind that λ may become so large for small W that a reasonable accuracy is achieved only for extremely long strips. According to the Gersgorin circle theorem, the eigenvalues of the 2D version of (16.2) are restricted by $|E| \leq 4t+W/2$. Therefore reasonable values for E are given by this inequality. For small M, finite-size effects can be expected. Large M will exceed available computer resources, because the necessary computation time increases in proportion to M^5 in 2D systems and M^7 in 3D systems, and in proportion to the inverse square of the relative accuracy.

For large M the computation time is dominated by the orthonormalization. Consequently, the number of recursion steps performed between subsequent orthonormalizations should be as large as possible. However, if it is chosen too large, wrong results are calculated, a fact which is easily demonstrated by comparing various runs of loc2d1.for. In practice, it is therefore a big advantage to adjust the number of recursion steps automatically during the calculation by comparing the norm of the vector corresponding to the smallest positive Lyapunov exponent with the machine precision. If the contribution is relatively large, the number of recursion steps is reduced. If it is smaller than a preset value, the number of recursions is increased. We have observed that, for most parameter combinations, the number converges fast to a value in the range of 5 to

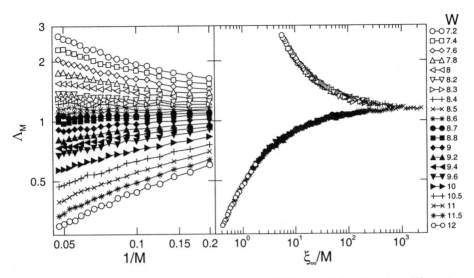

Fig. 16.2. Scaled localization lengths $\Lambda_M = \lambda_M/M$ for $\gamma = 0.9$ and various W as given on the *right edge* of the plot. In the *left part* of the plot, the data are presented versus $1/M$, whilst in the *right part* of the plot, the data are scaled by $\xi_\infty(W)$ for each W so as to minimize the deviations in the shifted data sets from the common scaling curve

30 and fluctuates only slightly afterwards. With this *adaptive orthogonalization*, a considerable amount of computation time can be saved.

Nevertheless, the calculation of a sufficiently large data set of localization lengths for different parameter values in 3D systems is too time-consuming for simple exercises. We therefore provide respective data sets `raw##.dat` at the web site [9], which can be plotted with `raw.gnu`. These results have been calculated for an anisotropic system with $t' = 1-\gamma = 0.4$ along the bar and $t = 1$ within the cross section for cross section sizes $M^2 = 6^2$ to 13^2. A typical example of data for even larger M^2 is shown in Fig. 16.2 for an even stronger anisotropy γ. The behaviour of $\Lambda_M = \lambda_M/M$ is plotted in this figure. These raw data can already be used to distinguish the regimes of extended and localized states: if Λ_M increases with increasing M, then the localization length grows faster than the extension of the system. Consequently, the localization length will become infinite in an infinitely wide system, corresponding to extended states and metallic behaviour. If, on the contrary, λ_M/M decreases with increasing M, then the electronic states will eventually fit completely into the bar for sufficiently large M. These states will be localized. The *fixed point* $\Lambda = \text{const.}$ determines the MIT. From Fig. 16.2 one can estimate the fixed point to occur at the critical disorder $W_c = 8.5$.

These raw data can be more quantitatively evaluated by *finite-size scaling* [8,10], if there exist properties which scale with the system size. We assume that Λ_M is such a suitable scaling variable so that it can be expressed as a function f of the system size M scaled by a parameter ξ which does not depend on M but only on W (and in general also on E, which is kept constant in our examples).

This one-parameter scaling relation

$$\Lambda = f(\xi/M) \tag{16.9}$$

is an ansatz, and the raw data must be checked to see whether or not they fulfill it. This can be done by a mean least squares fit of the data onto a common curve f, which requires a suitable adjustment of the scale M by means of the scaling parameter $\xi(W)$ for each disorder W. We have performed such a fit and supply the results for $\xi(W)$ in `xi.dat`. Multiplying the raw data by the corresponding scaling parameters with the program `scaling.for` yields the scaling curve which can be plotted by `scacurve.gnu`. For the raw data in Fig. 16.2, the respective scaling curve is also presented in that figure. It is obvious that the fit has been successful, yielding a common functional relationship f for all raw data. Thus the one-parameter scaling ansatz has been numerically verified.

It is important to note that this finite-size scaling procedure is much more than a simple extrapolation to large system sizes. The complete scaling curve is constructed from different pieces which have been calculated for the various parameter combinations. Thus the complete curve has been computed and, if the scaling ansatz is valid, the behaviour for large M for a certain parameter combination can be directly determined from the scaling curve without extrapolation.

The two branches of the scaling curve in Fig. 16.2 correspond to two qualitatively different types of raw data behaviour, namely the increase and decrease with increasing M. Thus the upper branch of the scaling curve corresponds to the metallic regime, the lower branch to the insulating regime. The branches touch at the MIT. Here the scaling parameter should diverge, which cannot be expected in the numerical scaling procedure due to rounding errors. Several sets of raw data overlap in this area due to the 1% error. As the mean least squares fit procedure minimizes the vertical deviations of the data from the scaling curve, a large horizontal shift of a data set in this regime would produce only a small difference in the vertical agreement of the data set with the scaling curve. This means that the fit cannot be very accurate in this regime. Such a rounding of divergencies is a common problem in the numerical analysis of phase transitions. It often appears possible to improve the fit by adjusting the scaling parameters by hand. For this purpose, we have supplied a second file of scaling parameters `xi.own` and encourage the reader to perform the entire scaling procedure by changing the data in this file, scaling the raw data again with `scaling.for` and plotting the resulting scaling curves with `scacurve.gnu`. Such a graphical construction often leads to a smoother scaling curve than the mean least squares fit [8].

In principle the *critical exponent* ν which describes the divergence of the scaling parameter $\xi(W)$ close to the critical disorder W_c can be determined from a non-linear fit to the data `xi.dat` or `xi.own`. Due to the rounded divergence in the calculated data, however, such a fit is not very accurate. It is much better to analyze the dependence of the raw data on W for various M, as shown in Fig. 16.3. At W_c, these curves should intersect in one point. In the vicinity of

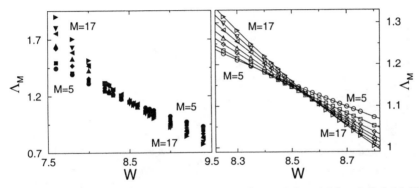

Fig. 16.3. Λ_M for weakly coupled planes with $\gamma = 0.9$ and $M = 5, 7, 9, 11, 13, 15, 17$. The accuracy of the data is 1% in the *left plot* and 0.07% in the *right plot*. The *solid lines* are cubic fits to the data

W_c they should behave as

$$\Lambda(W, M) = \Lambda(W_c) + \tilde{C}(W - W_c)M^{1/\nu} , \tag{16.10}$$

which can be obtained by linearizing the scaling relation (16.9) around the transition [11]. In this way ν can be determined directly from the raw data without the need to construct a scaling curve. A similar expression which is equivalent in the vicinity of the MIT can be derived for $\ln \Lambda$ [11]. The analysis can also be performed for the dependence of the scaling parameter on energy close to the MIT, as presented in Chap. 14. Although the calculation of the critical exponent from (16.10) is much more accurate than the determination of ν from the scaling curve, its accuracy is still limited by the required linearity of the data with respect to W, which restricts the fitting range; compare Fig. 16.3 which includes cubic fits to the data. Consequently, *nonlinear fitting procedures* have been used recently [12,13]. Moreover, a close inspection of Fig. 16.3 shows that the data for different M do not cross exactly in one point, but rather show a systematic shift especially for small M. Assuming that small deviations from the one-parameter scaling theory might be caused by an *irrelevant scaling variable*, respective corrections to the scaling function can be systematically taken into account [13]. Although the number of fitting parameters is considerably increased in this approach, a much better accuracy in the determination of ν becomes possible, provided that the number of raw data is sufficiently large [12].

A particularly clear example is shown in Fig. 16.4, where all the raw data for small M indicate localized behaviour. Only for larger M can the MIT at $W_c = 8.5$ be detected. Using the discussed corrections to scaling, namely nonlinear terms as well as an irrelevant scaling variable, a satisfactory fit can be achieved, like the one presented in Fig. 16.4.

It may be surprising at first sight that Figs. 16.3 and 16.4 have been calculated for the same parameters. The difference is the orientation of the transfer matrix bar with respect to the anisotropy γ of the transfer matrix elements t. In Fig. 16.4 the bar is oriented perpendicular to the weakly coupled planes, which

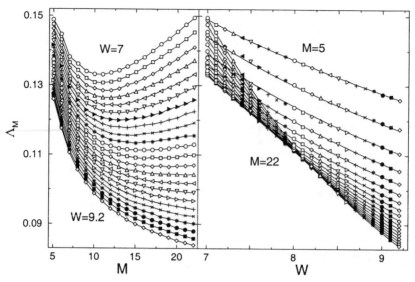

Fig. 16.4. λ_M for weakly coupled planes with $\gamma = 0.9$ in perpendicular orientation to the transfer matrix bar. The same data are plotted on the *left* and *right*. The *solid lines* are nonlinear fits including an irrelevant scaling variable

reduces the localization length in the direction of the bar by a factor of about $1 - \gamma$ and leads to a correspondingly small computing time. This advantage is lost, however, due to the *finite-size effects* which show up in Fig. 16.4, so that the calculation for much larger M becomes necessary.

For small W and large γ the finite-size effects for the perpendicular orientation become even more prominent. In this case the electronic states are more or less confined to the weakly coupled planes. The respective density of states is similar to the density for 2D systems and shows strong structures for small M which can be explained as broadened δ functions [12,14]. As a consequence, the density of states has a minimum for odd M and a maximum for even M in the band centre. This is reflected in an oscillating behaviour of Λ_M which seriously hinders the above-mentioned fit procedures [14]. One way to circumvent the problem is to consider only even or only odd system sizes.

16.3   Multifractal Analysis

As described in the introduction, disorder leads to strong spatial fluctuations of the wave functions, in particular near the MIT. In this section we present a method for analyzing these fluctuations quantitatively. For this purpose we determine the electronic wave functions of the Schrödinger equation (16.2), solving the corresponding eigenvalue problem by direct diagonalization. A detailed description has been given elsewhere [15]. Again we repeat here only the most important points and add some new aspects to the previous discussion.

It is important to note that the Hamiltonian matrix corresponding to (16.2) is very sparse and contains only 3 non-zero matrix elements per row and column for 1D, and 5 (7) non-zero matrix elements per row and column for the 2D (3D) analogue. It is therefore ideally suited for the application of the Lanczos algorithm, an iterative procedure requiring the repeated multiplication of the Hamiltonian matrix with some initial vectors. As in the previous section, this multiplication can be coded most efficiently by avoiding all multiplications with $t = 1$ and all summations of 0. In this way relatively large systems have been treated, with up to $M^3 = 111^3$ sites and periodic boundary conditions [12]. This is probably the largest system ever diagonalized for the Anderson problem of localization. A colour plot of the probability density of an electronic eigenstate of this system at the MIT ($W_c = 16.5t$) for $E = 0$ is included at the web site [9].

Of course, a determination of all eigenfunctions for such a system would produce a huge amount of data. This is not the purpose of the present analysis. Instead, we determine a few eigenstates in the band center $E = 0$. The Lanczos algorithm is most effective, however, at the band edges. We have nevertheless used the well-known implementation of Cullum and Willoughby [16] to determine a few eigenstates in the band center. It turned out that the Lanczos algorithm was the most effective one even in this case, compared with a variety of other modern algorithms [17], including the implicitly restarted Arnoldi method, a hybrid tridiagonalization algorithm for symmetric sparse matrices, testing different polynomial convergence accelerators as well as a shift-and-invert approach with direct solvers and with iterative solvers for the resulting system of linear equations. All considered modern methods turned out to be inapplicable for the largest system sizes, because the computation times and/or the memory requirements became much too large. We attribute the difficulty of these methods to the specific structure of the Anderson Hamiltonian. The periodic boundary conditions lead to non-zero matrix elements far away from the diagonal of the secular matrix, which cannot be transformed into a band matrix of small band width. We emphasize that the applied implementation of the Lanczos algorithm is a reliable tool, although numerical inaccuracies due to the finite-precision arithmetic lead to orthogonality problems similar to those for the transfer matrix method. As a consequence, so-called spurious or ghost eigenvalues appear in the spectrum which have to be identified. Due to the large matrix size, a repeated orthogonalization is not feasible, because it would require the storage of a very large number of vectors. The solution of this problem in the present implementation [16] is the comparison with a second diagonalization of a similar matrix which yields the same ghost eigenvalues, but not the same true eigenvalues. Another disadvantage of the Lanczos algorithm is the fact that it cannot detect degenerate eigenvalues. But for our disordered systems, this presents no problem, because no exact degeneracies occur for finite disorder.

We note that, due to the large matrix size, the algorithm can be most effectively run on vectorizing supercomputers. This is particularly true if the periodic boundary conditions are slightly altered and helical boundary conditions are used [15]. This yields do loops of length M^3. Physically such a change of boundary conditions is of no relevance, because it only means replacing the cubic unit cell by an affinely (slightly) distorted cell.

We have also implemented the algorithm on parallel computers. A reasonable efficiency has been achieved for a small number of processors [18], but for massively parallel systems the communication requirements between different processors turned out to be too large, unless the system size M was at least an order of magnitude larger than the number of processors.

Viewing the probability distribution $|\psi_n|^2$ of wave functions at or near the MIT, a fragmentation into high probability speckles of all sizes is the most prominent feature. For small disorder the probability distribution is more homogeneous, whilst for large disorder it is concentrated in certain areas. Mandelbrot [19] has called this behaviour curdling. A strong curdling of the electronic state occurs when an electron with sufficiently high or low energy is put into a system with too much disorder just in the same way as old milk curdles when it is poured into coffee which is too hot. At the web site [9], a number of 2D wave functions for various parameter combinations are included, together with a plot program to visualize the curdling.

The observed behaviour can be quantitatively described by a multifractal analysis. This concept originates from the scale invariance of eigenstates at the MIT. Approaching the MIT from the insulating side, the localization length diverges. On the metallic side the corresponding characteristic length, the coherence length, increases with increasing interference effects and also diverges at the MIT. At the MIT there is thus no relevant length scale in the system. This concept prompted the idea [20] that the wave functions at the MIT are scale invariant and display self-similarity. Such fractal behaviour is an appealing concept, because it allows us to accommodate two seemingly contradictory notions: on the metallic side of the transition the extended states are expected to be spread over the whole system, while on the localized side, a state should occupy only an infinitely small amount of space. A curdled wave function with speckles of all sizes can represent such a state, which extends all over the system but does not fill any finite fraction of the volume. Moreover, this concept enables us to assume a continuous transition of the characteristics of the wave function at the MIT. To determine these characteristics we consider the so-called box probability for finding an electron in a box of linear size L, viz.,

$$\mu_k(L) = \sum_{n(k)} |\psi_{in}|^2 , \quad k = 1, \dots, N_L . \tag{16.11}$$

The summation covers the L^2 or L^3 sites in the kth box. This measure and its qth moments show scaling in the limit of small box size $\delta = L/M$ with different powers of δ [15], thus yielding different fractal dimensions for the different moments. Accordingly, the wave function is a multifractal object. It is not really self-similar, but rather self-affine. Besides the (generalized) fractal dimensions, the singularity spectrum $f(\alpha)$ is commonly used for the characterization of multifractal entities. Here the Lipschitz–Hölder exponent α reflects the strength of the singularity of the box probability in the kth box:

$$\mu_k(\delta) \sim \delta^\alpha \ . \tag{16.12}$$

The set of all boxes with a particular value α is again a fractal, the number of boxes scales with the fractal dimension f:

$$\mathcal{N}(\alpha) \sim \delta^{-f(\alpha)} \ . \tag{16.13}$$

In principle the singularity spectrum can be obtained from the coarse-grained Lipschitz–Hölder exponent

$$\alpha_k = \log \mu_k(\delta) / \log \delta \tag{16.14}$$

for every box, counting the number $\mathcal{N}(\alpha)$ of boxes of a certain strength and evaluating the resulting histogram according to

$$f(\alpha) = - \log \mathcal{N}(\alpha) / \log \delta \tag{16.15}$$

in the limit $\delta \to 0$.

The box counting procedure and the histogram method are easily applied to the wave functions and can be programmed in a straightforward way. However, the convergence is not usually very good. One way to improve it is to use as many boxes as possible. For this purpose, one should average over different choices of the origin of the boxes, and one should not only use perfect subdivisions of the full system, i.e., not only integer values of M/L. In practice, this means considering all M^2 (M^3) possible non-equivalent positions for placing any L^2 (L^3) box in the periodically repeated system (which is straightforward due to the periodic boundary conditions). In this way the statistics can be significantly improved. Another improvement can be made by allowing slightly rectangular boxes to obtain more data points for the linear fits by which (16.14) and (16.15) are evaluated. The reader is encouraged to use these tricks in order to improve the statistics of the simple exercise programs, calculating the singularity spectrum as well as the multifractal dimensions.

For a more accurate computation of $f(\alpha)$, the parametric representation

$$\alpha(q) = \lim_{\delta \to 0} \sum_k \mu_k(q, L) \log \mu_k(1, L) / \log \delta \ , \tag{16.16}$$

$$f(q) = \lim_{\delta \to 0} \sum_k \mu_k(q, L) \log \mu_k(q, L) / \log \delta \tag{16.17}$$

should be determined, where $\mu_k(q, L)$ is the normalized qth moment of the measure (16.11). A standard linear regression is then used to evaluate α and f. Respective data are available at the web site [9], together with plot programs to display the data in different combinations. Comparing data for different system sizes and different values of the random potential energies, finite-size effects and

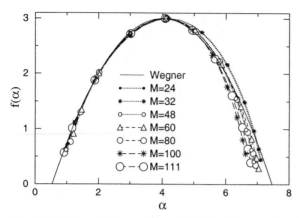

Fig. 16.5. Singularity spectrum $f(\alpha)$ for eigenstates of the isotropic Anderson model at the MIT for various system sizes M^3. *Symbols* indicate the values at the implicit parameter $q = 2, 1.5, 1, \ldots, -3$ (from *left* to *right*). The *solid line* reflects (16.18)

statistical fluctuations can be estimated. Results for the 3D Anderson model at the MIT are presented in Fig. 16.5. It is obvious that the result does not depend on the system size, corroborating the scale invariance of the wave functions at the MIT. We note that the singularity spectrum agrees around its maximum with the parabolic approximation

$$f(\alpha) = 3 - (4 - \alpha)^2/4 , \tag{16.18}$$

which was obtained analytically within the nonlinear σ model [21]. The independence of selected values of $\alpha(q)$ and $f(q)$ from the system size is explicitly demonstrated in Fig. 16.6. The error bars reflect the statistical accuracy due to averaging over 4–10 different random potential energy realizations. In fact, this independence can be used as a tool [22] for distinguishing localized and extended states. On the insulating as well as on the metallic side of the MIT, multifractal properties can also be observed up to a certain length scale. Close to the MIT, this scale is larger than the accessible system size, but even below this length scale the calculated singularity spectra already depend on the system size and approach the limiting nonfractal cases of homogeneously extended or exponentially localized states with increasing system size. They thus show a characteristic size dependence [23], in contrast to Fig. 16.6. The MIT in anisotropic systems has been determined in this way [24].

16.4   Energy Level Statistics

Besides the exponential decay of electronic wave functions and besides the spatial distribution of electronic eigenstates, the statistics of the eigenvalues can also be exploited to distinguish extended and localized states. If extended states are close in energy they hybridize, leading to energy level repulsion. On the other

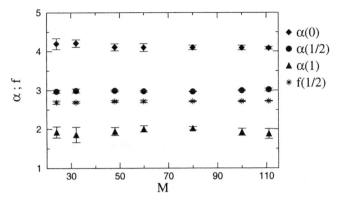

Fig. 16.6. Dependence of several multifractal characteristics $\alpha(q)$ and $f(q)$ from Fig. 16.5 on the system size. Note that $f(1) = \alpha(1)$

hand, electronic states which are localized in distant regions of space cannot overlap and therefore cannot hybridize. This leads to level clustering. A quantitative description of these characteristics is provided by random matrix theory, as introduced in Chap. 14. For this purpose, one studies the fluctuations of the energy spectrum, in particular the distribution $P(s)$ of the separation s between neighbouring energy levels. As mentioned in Chap. 14, the strongly localized states lead to a Poisson distribution, while for extended states we find the distribution of the Gaussian orthogonal ensemble (GOE). The Wigner surmise,

$$P_{\mathrm{GOE}}(s) \approx \frac{\pi}{2} s \, \exp\left(-\frac{\pi}{4} s^2\right) , \tag{16.19}$$

is derived for 2×2 Gaussian matrices, but turns out to be a good approximation for the level spacing distribution of larger systems as well. This holds true for the investigation of the Anderson Hamiltonian too, although in this case only the diagonal matrix elements are randomly distributed, while random matrix theory assumes all matrix elements to be random quantities. We note that the linear prefactor in (16.19) reflects the level repulsion.

For the determination of $P(s)$, we have diagonalized the Anderson Hamiltonian for systems up to $M^3 = 50^3$ sites, again using the Lanczos algorithm. A large number of eigenvalues, preferably all eigenvalues, must now be computed, in contrast to the calculation in the previous section, where only a few eigenstates were considered. On the other hand, we do not now need the eigenfunctions, and this reduces computational effort and makes parallelization somewhat more efficient. In practice one does not evaluate the complete spectrum, but only the central part, e.g., the central 50%. There are two reasons for this restriction: first, the behaviour of the states should not change significantly in the energy range under consideration, so that more or less localized and extended types of behaviour are not included in the same distribution. On the other hand, random matrix theory describes the fluctuations of the energy spectrum around the average density of states, which is not usually constant. It is therefore necessary, even for the central half of the spectrum, to apply an unfolding procedure to map the set of computed eigenvalues $\{E_i\}$ onto a new set $\{E_i'\}$ with constant average density,

$$E_i' = \bar{D}(E_i) = D(E_i) - D_{\rm f}(E_i) , \qquad (16.20)$$

after distinguishing the average $\bar{D}$ and the fluctuations $D_{\rm f}$ of the integrated density of states $D(E)$. This can be achieved by fitting $D(E_i)$ with cubic splines over a sufficiently large energy interval and taking the fitted curve as the average.

While the histogram of level spacings allows a quick distinction between the limiting Poisson and GOE cases, for the quantitative distinction, it is much better to determine the cumulative level-spacing distribution

$$I(s) = \int_s^\infty P(s') \, ds' , \qquad (16.21)$$

which yields a monotonically decreasing curve. This has been widely used to analyze the MIT in the Anderson model. In finite samples, one cannot of course expect to find the pure distributions of the limiting cases. But it turns out that computed distributions show a clear trend towards the limiting distributions with increasing system size. This can be exploited not only by plotting the distributions for different system sizes, but also in a quantitative way. For this purpose, one fits the distributions with an interpolating phenomenological formula [29]

$$P_{\rm phe}(s) = As^\beta (1 + C\beta s)^{f(\beta)} \exp\left[-\frac{\pi^2}{16}\beta s^2 - \frac{\pi}{2}\left(1 - \frac{\beta}{2}\right) s \right] , \qquad (16.22)$$

with $f(\beta) = 2^\beta (1 - \beta/2)/\beta - 0.16874$ and 3 parameters, two of which are determined by normalization. The finite-size dependence of the parameters A and C differs qualitatively in the metallic and in the insulating regime, thus allowing determination of the MIT. This method has also been used to analyze the localization of many interacting particles [30].

Besides $P(s)$, fluctuations in the energy spectrum can also be characterized by the variance $\Sigma_2(r)$ of the number of levels in an interval of width r [25], the distributions $P_n(s)$ of the separations of nth-nearest energy levels [26], or the deviation Δ_3 from a sequence of L uniformly spaced energy levels [27]. In particular the spectral rigidity Δ_3 shown in Fig. 16.7 has been widely used. In this figure the size dependence is obvious. Only at the MIT, where there are no characteristic length scales, are the statistical properties of the spectrum independent of the system size.

For a quantitative evaluation it has again turned out profitable to consider the integrated statistics

$$\alpha_M(W) = \int_0^{30} \Delta_3(L, W, M) \, dL , \qquad (16.23)$$

where the upper boundary of the integral has been chosen arbitrarily. It is obvious from Fig. 16.7 that α increases monotonically with increasing W between

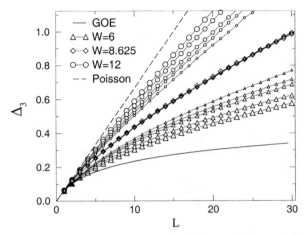

Fig. 16.7. Δ_3 for weakly coupled planes with $\gamma = 0.9$ and several system sizes $M = 13, 17, 21, 30, 40$, indicated by increasing *symbol size*

the limiting cases. In the localized region it increases with M, but for extended states it decreases with increasing M, while it remains constant at the MIT, as shown in Fig. 16.8. This behaviour is similar to the dependence of Λ_M in Fig. 16.2. This has led us to assume that α_M is also a suitable scaling variable for finite-size scaling. Indeed, a finite-size scaling procedure turns out to be successful, and the analysis of the raw data in analogy to the discussion of Fig. 16.3

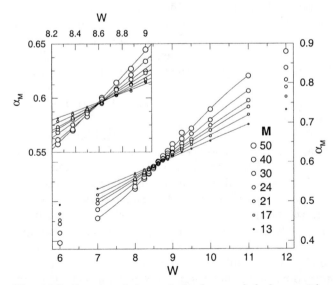

Fig. 16.8. Integrated Δ_3 statistics for coupled planes with $\gamma = 0.9$. *Lines* correspond to a non-linear fit. The *inset* shows an enlargement of the region around the MIT. Here, the lines are cubic fits

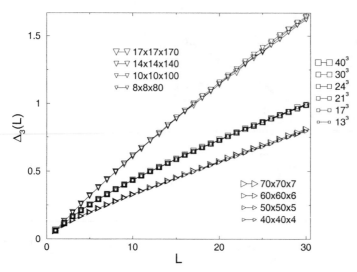

Fig. 16.9. Δ_3 statistics from cubic and non-cubic samples for coupled planes with $\gamma = 0.9$ and various system sizes at $W_c = 8.625$

was possible in a straightforward way [12], yielding the MIT at $W_c = 8.625$ for the data in Fig. 16.8 and the critical exponent $\nu \approx 1.45 \pm 0.2$.

It is interesting to note that W_c and ν do not depend on the specific details of the investigated system. As an example, we have considered non-cubic samples. In Fig. 16.9, the Δ_3 statistics is shown to be significantly different for long, for cubic, and for broad samples. However, in all cases, the size independence clearly reflects the fact that the system is at the MIT for the chosen parameter W_c. One should therefore use the finite-size (in)dependence, but not the similarity of a computed distribution with the Poisson or the GOE case as a criterion for characterizing the system.

The statistics of energy levels is also discussed in Chap. 15 with respect to its influence on absorption spectra.

It is also interesting to note that the GOE statistics has even been found in a purely deterministic system, namely in a quasicrystalline model. No random matrix elements occur in that case, and the non-diagonal hopping matrix elements only appear to be randomly scattered all over the matrix. In reality, they are determined by a deterministic construction scheme for the quasicrystals. Nevertheless, the GOE statistics was clearly obtained [27]. In contrast to the Anderson model, where randomness always requires an average over several disordered samples for a given parameter combination, there are no random fluctuations in the quasicrystalline model. As a result, the determination of the integrated level statistics (16.21) was so accurate that the usually insignificant deviation of the GOE distribution from the Wigner surmise could be distinguished in the data [27].

Finally we note that large-scale numerical investigations of the critical level statistics [28] have shown that it displays level repulsion for small s, but a Poisson tail for large s. This is also discussed in Chap. 14, where the dependence of the level spacing distribution on energy close to the MIT is investigated.

Acknowledgements. Helpful discussions with R.A. Römer and V. Uski are gratefully acknowledged. Some of the programs were provided by Th. Rieth. This work was supported by the DFG within SFB 393.

References

1. P.W. Anderson: Phys. Rev. **109**, 1492 (1958)
2. P.E. Lindelof, J. Nørregard, J. Hanberg: Phys. Scr. T **14**, 317 (1986)
3. P.-E. Wolf, G. Maret: Phys. Rev. Lett. **55**, 2696 (1985)
4. E.N. Economou, C.M. Soukoulis: Phys. Rev. B **28**, 1093 (1983)
5. W. Apel, T.M. Rice: J. Phys. C **16**, L1151 (1983); Q. Li, C.M. Soukoulis, E.N. Economou, G.S. Grest: Phys. Rev. B **40**, 2825 (1989); I. Zambetaki, Q. Li, E.N. Economou, C.M. Soukoulis: Phys. Rev. Lett. **76**, 3614 (1996); Q. Li, S. Katsoprinakis, E.N. Economou, C.M. Soukoulis: Phys. Rev. B **56**, R4297 (1997), cond-mat/9704104
6. H. Stupp, M. Hornung, M. Lakner, O. Madel, H.V. Löhneysen: Phys. Rev. Lett. **71**, 2634 (1993)
7. V.I. Oseledec: Trans. Moscow Math. Soc. **19**, 197 (1968)
8. B. Kramer, M. Schreiber: 'Transfer-Matrix Methods and Finite-Size Scaling for Disordered Systems'. In: *Computational Physics – Selected Methods, Simple Exercises, Serious Applications*, ed. by K.H. Hoffmann, M. Schreiber (Springer, Berlin, Heidelberg 1996) pp. 166–188
9. http://www.tu-chemnitz.de/physik/HERAEUS/2000/Springer.html
10. A. MacKinnon, B. Kramer: Z. Phys. B **53**, 1 (1983)
11. A. MacKinnon: J. Phys. Condens. Matter **6**, 2511 (1994)
12. F. Milde: Disorder-Induced Metal–Insulator Transition in Anisotropic Systems. Dissertation, Technische Universität Chemnitz (Chemnitz 2000)
13. K. Slevin, T. Ohtsuki: Phys. Rev. Lett. **82**, 382 (1999), cond-mat/9812065
14. F. Milde, R. A. Römer, M. Schreiber, V. Uski: Eur. Phys. J. B **15**, 685 (2000)
15. M. Schreiber: 'Multifractal Characteristics of Electronic Wave Functions in Disordered Systems'. In: *Computational Physics – Selected Methods, Simple Exercises, Serious Applications*, ed. by K.H. Hoffmann, M. Schreiber (Springer, Berlin, Heidelberg 1996) pp. 147–165
16. J. Cullum and R.A. Willoughby: *Lanczos Algorithms for Large Symmetric Eigenvalue Computations*, Vol. 1: *Theory*. (Birkhäuser, Boston 1985)
17. U. Elsner, V. Mehrmann, F. Milde, R.A. Römer, M. Schreiber: SIAM J. Sci. Comp. **20**, 2089 (1999)
18. M. Schreiber, F. Milde, R.A. Römer, U. Elsner, V. Mehrmann: Comp. Phys. Comm. **121–122**, 517 (1999)
19. B.B. Mandelbrot: *The Fractal Geometry of Nature* (Freemann, New York 1982)
20. H. Aoki: J. Phys. C: Solid State Phys. **16**, L205 (1983)
21. F. Wegner: Nucl. Phys. B **316**, 663 (1989)
22. M. Schreiber, H. Grussbach: J. Fractals **1**, 1037 (1993)

23. H. Grussbach, M. Schreiber: Phys. Rev. B **57**, 663 (1995)
24. F. Milde, R. A. Römer, M. Schreiber: Phys. Rev. B **55**, 9463 (1997)
25. M. Schreiber, U. Grimm, R.A. Römer, J.X. Zhong: Physica A **266**, 477 (1999)
26. U. Grimm, R.A. Römer, M. Schreiber, J.X. Zhong: Mat. Sci. and Eng. **294–296**, 564 (2001), cond-mat/9908063.
27. J.X. Zhong, U. Grimm, R.A. Römer, M. Schreiber: Phys. Rev. Lett. **80**, 3996 (1998)
28. I.K. Zharekeshev, B. Kramer: Phys. Rev. Lett. **79**, 717 (1997)
29. G. Casati, F. Izrailev, L. Molinari: J. Phys. A **24**, 4755 (1991)
30. F. Epperlein, M. Schreiber, T. Vojta: Phys. Rev. B **56**, 5890 (1997)

17 Percolation, Renormalization and Quantum Hall Transition

Rudolf A. Römer

Summary. In this article, I give a pedagogical introduction and overview of percolation theory. Special emphasis will be put on the review of some of the most prominent algorithms devised to study percolation numerically. The real-space renormalization group treatment of the percolation problem is then discussed. As a rather novel application of this approach to percolation, I will review recent results using similar real-space renormalization ideas that have been applied to the quantum Hall transition.

17.1 Introduction

Imagine a large chessboard, such as might be found in a park. It is fall, and all the master players have fled the cold a long time ago. You are taking a walk and enjoying the beautiful sunny afternoon, the many colors of the Indian summer in the trees and in the falling leaves around you. Looking at the chessboard, you see that some squares are already full of leaves, while others are still empty [1]. The pattern of the squares which are covered by leaves seems rather random. As you try to cross the chessboard, you see that there is a way to get from one side of the board to the opposite side by walking on leaf-covered squares only. This is percolation – or nearly.

Before continuing and explaining in detail what percolation is about, let me outline the content of this paper. In Sect. 17.2, I will review some of the most prominent and interesting results on classical percolation. Percolation theory is at the heart of many phenomena in statistical physics that are also topics in this book. Beyond the exact solutions of percolation in $d = 1$ and $d = \infty$ dimensions, further exact solutions in $2 \leq d < \infty$ only rarely exist. Thus computational methods, using high-performance computers and algorithms are needed for further progress and in Sects. 17.2.2–17.2.4, I explain in detail some of these algorithms. Section 17.3 is devoted to the real-space renormalization group (RG) approach to percolation. This provides an independent and very suggestive method for analytically computing results for the percolation problem, as well as a further numerical algorithm.

While many applications of percolation theory are mainly concerned with problems of classical statistical physics, I will show in Sect. 17.4 that the percolation approach can also give useful information at the quantum scale. In particular, I will show that aspects of the quantum Hall (QH) effect can be understood by a suitably generalized renormalization procedure of bond percolation in $d = 2$. This application allows the computation of critical exponents and conductance distributions at the QH transition and also opens the way for studies of scale-invariant, experimentally relevant macroscopic inhomogeneities. I summarize in Sect. 17.5.

17.2   Percolation

17.2.1   The Physics of Connectivity

From the chessboard example given above, we realize that the percolation problem deals with the spatial *connectivity* of occupied squares, rather than a simple counting to find out whether the number of such squares is in the majority. Then the obvious question to ask is: how many leaves are usually needed in order to allow passage across the board? Since leaves do not normally interact with each other, and friction-related forces can be assumed small compared to wind forces, we can model the situation by assuming that the leaves are *randomly* distributed on the board. Then we can define an occupation probability p as being the probability that a site is occupied (by at least one leaf). Thus our question can be rephrased in modern physics terminology as: is there a threshold value p_c at which there is a spanning cluster of occupied sites across an infinite lattice?

The first time this question was asked and the term *percolation* used was in the year 1957 in publications of Broadbent and Hammersley [2–4]. Since then a multitude of research articles, reviews and books have appeared on this subject. Certainly among the most readable such publications is the 1995 book by Stauffer and Aharony [5], where most of the relevant research articles have also been cited. Let me briefly summarize here some of the highlights that have been discovered in nearly 50 years of research on percolation.

The percolation problem in $d = 1$ can be solved exactly. Since the number of empty sites in a chain of length L is $(1-p)L$, there is always a finite probability for finding such an empty site in the infinite cluster at $L \to \infty$ and thus the percolation threshold is $p_c = 1$. Defining a correlation function $g(r) \propto \exp(-r/\xi)$ which measures the probability that a site at distance r from an occupied site at 0 belongs to the same cluster, we easily find $g(r) = p^r$ and thus $\xi = -1/\ln p \approx (p_c - p)^{-1}$. Therefore, close to the percolation threshold, the correlation length ξ diverges with an exponent $\nu = 1$.

In $d = 2$, the percolation problem provides perhaps the simplest example of a second-order phase transition. The order parameter of this transition is the probability $P(p)$ that an arbitrary site in the infinite lattice is part of an infinite cluster,[1] i.e.,

$$P(p) = \begin{cases} 0 , & p \leq p_c , \\ (p - p_c)^\beta , & p > p_c , \end{cases} \tag{17.1}$$

where β is a critical exponent similar to the exponent ν of the correlation length. The distribution of the sites in an infinite cluster at the percolation threshold

---

[1] For any finite lattice, 'infinite' means a cluster that reaches from top to bottom and/or left to right through the lattice.

can be described as a fractal (see Chap. 16), i.e., its average size M in boxes of length N increases as $\langle M(N)\rangle \propto N^D$, where D is the fractal dimension of the cluster. As in any second-order phase transition, much insight can be gained by a finite-size scaling analysis (see Chap. 16 and [6]). In particular, the exponents introduced above are related according to the scaling relation $\beta = (d - D)\nu$ [5]. Furthermore, it has been shown to an astonishing degree of accuracy, that the values of the exponents and the relations between them are independent of the type of lattice considered, i.e., square, triangular, honeycomb, etc., and also whether the percolation problem is defined for sites or bonds (see Fig. 17.1). This independence is called *universality*. In the following, we will see that universality does not apply for the percolation threshold p_c. Thus it is of importance to note that p_c is known exactly for site percolation on the triangular lattice and bond percolation on the square lattice: $p_c = 1/2$. In particular, the bond percolation problem has also received much attention from mathematicians [7].

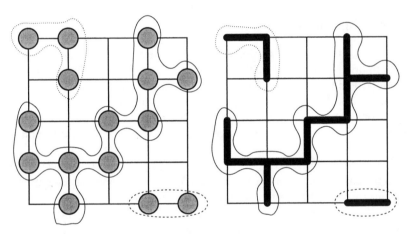

Fig. 17.1. Site (*left*) and bond (*right*) percolation on a square lattice. In site percolation, the sites of a lattice are occupied randomly and percolation is defined via, say, nearest-neighbor sites. In bond percolation, the bonds connecting the sites are used for percolation. The *thin outlines* define the 3 clusters in each panel. The *solid outline* indicates the percolating cluster

For higher dimensions, much of this picture remains unchanged, although the values of p_c *and* the critical exponents change. The upper critical dimension corresponds to $d = 6$, so that mean field theory is valid for $d \geq 6$ with exponents as given in Table 17.1.

Applications of percolation theory are numerous [8]. It is intimately connected to the theory of phase transitions as discussed in Chap. 13 and [6]. The connectivity problem is also relevant for diffusion in disordered media (see Chaps. 14 and 16) and networks (see Chap. 12). A simple model of forest fires is based on percolation ideas (see Chap. 8) and even models of stock market fluctuations [9] have been devised using ideas of percolation [10]. Percolation is

Table 17.1. Critical exponents ν and β and fractal dimension D for different spatial dimensions. For a more complete list see [5]

Exponent	$d = 2$	$d = 3$	$d = 4$	$d = 5$	$d = 6 - \epsilon$
ν	4/3	0.88	0.68	0.57	$1/2 + 5\epsilon/84$
β	5/36	0.41	0.64	0.84	$1 - \epsilon/7$
$D(p = p_c)$	91/48	2.53	3.06	3.54	$4 - 10\epsilon/21$
$D(p < p_c)$	1.56	2	12/5	2.8	–
$D(p > p_c)$	2	3	4	5	–

therefore a well-established field of statistical physics and it continues its vital progress with more than 230 publications in the cond-mat archives [11] alone.

17.2.2 The Coloring Algorithm

As stated above, there are only a few exact results available in percolation in $d \geq 2$. Thus in order to proceed further, one has to use computational methods.

The standard numerical algorithm of percolation theory is due to Hoshen and Kopelman [12]. The advantage with this algorithm is that it can analyze which site belongs to which cluster without having to store the complete lattice. Furthermore, this is done in one sweep across the lattice, thus reducing computer time.

At the heart of the Hoshen–Kopelman algorithm is a bookkeeping mechanism [5]. Look at the site percolation cluster in Fig. 17.1. Going from left to right and top to bottom through the cluster, we give each site a label (or color) as shown in Fig. 17.2. If its top or left neighbor is already occupied, then the new site belongs to the same cluster and gets the same label. Otherwise we choose a new label. In this way we can proceed through the cluster until we reach a problem in line 3 as shown in the left column of Fig. 17.2. According to our above rule the new site, indicated by the bold question mark, can be either 2 or 4. This indicates that all sites previously labelled by 2 and 4 actually belong to the same cluster. Thus we now introduce an index $\mathrm{Id}(\cdot)$ for each cluster label and define it in such a way that the index of a superfluous label, say 4, points to the right label, viz., $\mathrm{Id}(4) = 2$. Proceeding with our analysis into the 4th row of the lattice, we see in the center column of Fig. 17.2 that we again have to adjust our index at the position indicated in bold. Instead of labeling this site as 4, we choose $2 \equiv \mathrm{Id}(4)$, and consequently, $\mathrm{Id}(3) = 2$. In this way, we can easily check whether a cluster percolates from top to bottom of the lattice by simply checking whether the label of any occupied site in the bottom row of the lattice has an index equal to any of the labels in the top row. Furthermore, in addition to the top row, we only need to store the row presently under consideration and its predecessor. Thus the storage requirement is linear in lattice size L and not L^2 as it would be if we were to store the full lattice. Finally, the algorithm can

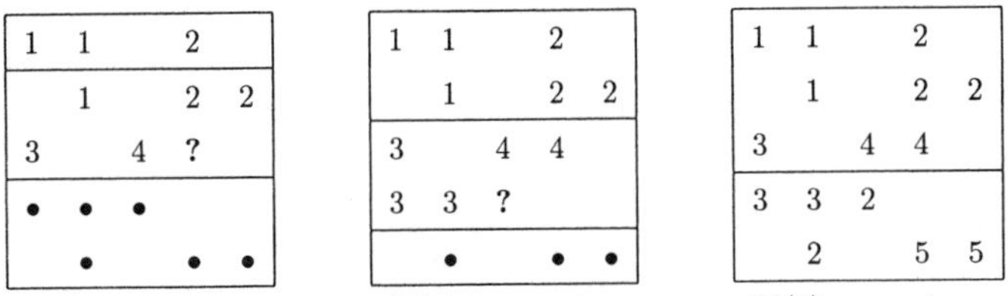

Id(1) = 1
Id(2) = 2
Id(3) = 3
Id(4) = 2

Id(1) = 1
Id(2) = 2
Id(3) = 2
Id(4) = 2

Id(1) = 1
Id(2) = 2
Id(3) = 2
Id(4) = 2
Id(5) = 5

Fig. 17.2. Schematic description of the Hoshen–Kopelman coloring algorithm for the site percolation problem on a square lattice as shown in Fig. 17.1. • denotes an occupied site, *numbers* denote cluster labels. The *horizontal lines* bracket the current and the previous row

also give information about all clusters, say, if needed for a fractal analysis of the non-percolating clusters. A Java implementation of such a coloring algorithm can be found in [13].

17.2.3 The Growth Algorithm

When we want to study primarily the geometrical properties of percolation clusters, another algorithm is more suitable, namely, one which generates the desired cluster structure directly. This algorithm is due to Leath [14] and works by a cluster-growth strategy. The idea of the algorithm is that we put an occupied site in the center of an otherwise empty lattice. Then we identify its nearest-neighbor sites as shown in Fig. 17.3. Next we occupy these sites according to the desired percolation probability p. We identify the new, yet undecided nearest-neighbor sites, occupy these again with probability p and repeat the procedure. The cluster continues to grow until either all sites at the boundary are unoccupied or the cluster has reached the boundary of the lattice. For $p < p_c$, the growth usually stops after a few iterations, while for $p > p_c$, percolating clusters are almost always generated. Therefore, besides giving information about the fractal structure of the percolating clusters, the Leath algorithm can also be used to estimate the value of p_c. A Java implementation of the Leath algorithm can be found in [15,16].

17.2.4 The Frontier Algorithm

As we have seen in the last section, the outer frontier of the percolation cluster is well defined. The fractal properties of this hull can be measured [17,18] and shown to yield $D_h = 1.74 \pm 0.02$. This suggests yet another algorithm for the determination of p_c [19,20]: generate a lattice with a constant gradient ∇p of the occupation probability p as shown in Fig. 17.4. By an algorithm similar to the one used in computing the hull of the percolating cluster, one traverses

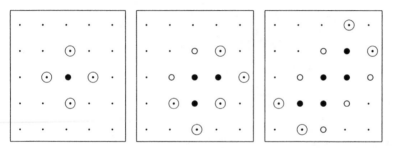

Fig. 17.3. Schematic description of the first three steps in the Leath growth algorithm for the site percolation cluster of Fig. 17.1. •, ○, and ⊙ denote occupied, empty, and undecided sites

the frontier of the occupied cluster and determines which sites belong to it and which belong to the empty cluster [18–20]. Then a very reliable estimate for the percolation threshold can be computed as

$$p_c = \frac{N_o p_{co} + N_e p_{ce}}{N_o + N_e} , \tag{17.2}$$

where N_o and N_e denote the number of sites in the occupied and empty frontiers and p_{co}, p_{ce} are the mean heights of the associated frontiers, respectively. Note that instead of actually generating the percolation lattice, the algorithm proceeds by just generating the sites needed for the construction of the frontier. Thus instead of dealing with, say, $\mathcal{O}(L^2)$ sites as in the Hoshen–Kopelman and Leath

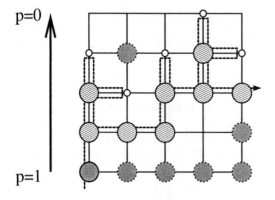

Fig. 17.4. Example of gradient percolation. All sites at $p = 1$ ($p = 0$) are occupied (empty). The 15 *large circles* correspond to occupied sites. The 8 *light shaded sites* belong to the frontier, whereas the 7 *dark shaded sites* are part of the interior of the cluster or belong to other clusters. The 5 *small circles* correspond to sites in the empty frontier. The *dashed line* indicates the frontier generating walk. Note that only the 14 sites with *solid circles* have actually been visited. The 6 other circles are shown here just for clarity and need not be generated. According to (17.2), we have $p_c = (8 \cdot 0.4375 + 5 \cdot 0.75)/13 = 0.55769$

Table 17.2. Various current estimates of p_c for site and bond percolation on different lattices in $d = 2$ [21–26]. The lattices are classified according to their number of first, second and higher nearest-neighbors. The upper index gives the corresponding number in the dual lattice

Lattice		Site p_c	Bond p_c
$3, 12^2$		$0.807\,904^a$	
$4, 6, 12$		$0.747\,806$	
$4, 8^2$		$0.729\,724$	
6^3	Honeycomb	$0.697\,043$	$0.652\,703^a$
$3, 6, 3, 6$	Kagomé	$0.652\,703^a$	$0.524\,405$
4^4	Dice	$0.584\,8$	$0.475\,595$
$3, 4, 6, 4$		$0.621\,819$	
4^4	Square	$0.592\,746\,0$	$0.500\,000^a$
$3^4, 6$		$0.579\,498$	
$3^2, 4, 3, 4$		$0.550\,806$	
$3^3, 4^2$		$0.550\,213$	
3^6	Triangular	$0.500\,000^a$	$0.347\,296^a$

^{a} Exactly known values.

algorithms for a lattice in $d = 2$, one only needs $\mathcal{O}(L)$ sites in the present algorithm.[2] This reduces computer time, and estimates for p_c can be obtained with high accuracy for a wide variety of lattices [21–26], as shown in Table 17.2.

17.3 Real-Space Renormalization

17.3.1 Making Use of Self-Similarity

As mentioned in Sect. 17.2, the transition at p_c corresponds to a second-order phase transition and the correlation length ξ is infinite. There is no particular length scale in the system and all clusters are statistically similar to each other. This *self-similarity* (see Chap. 16) underlies the renormalization description just as for the fractal analysis of percolation clusters.

We may use self-similarity in the following way. Let us replace a suitable collection of sites by *super*-sites and then study percolation of the super-lattice [27–31]. In general, the occupation probability p' of the super-lattice will be different from the original p. Furthermore, if the extent of the collection of sites

---

[2] Alas, this intuitively convincing argument is not strictly true. The percolation frontier is a fractal and as such scales $\propto L^{1.75}$ [17]. On the other hand, it is not the random number generation for L^2 sites in the Hoshen–Kopelman algorithm but rather the numerical determination of the percolating clusters which is numerically challenging.

in the lattice was b, then the super-lattice will have a lattice constant b. Thus $\xi = b\xi'$ and, with $\xi \propto |p - p_c|^{-\nu}$, we find

$$b|p' - p_c|^{-\nu} \equiv |p - p_c|^{-\nu} . \tag{17.3}$$

Hence,

$$\nu = \frac{\log b}{\log(dp'/dp)} . \tag{17.4}$$

As an example, let us consider the bond percolation problem on a square lattice [28–30]. Here we replace 5 bonds by a super-bond in the horizontal direction, as shown in Fig. 17.5. Summing all probabilities for a connected, horizontal super-bond as shown in Fig. 17.5, we find that

$$p' = p^5 + 5p^4(1 - p) + 8p^3(1 - p)^2 + 2p^2(1 - p)^3 . \tag{17.5}$$

At the transition, we have $p' = p$ and thus (17.5) has solutions $p = 0$, 0.5, and 1. The first and last solution correspond to a completely empty or occupied lattice and are trivial. The second solution reproduces the exact result of Table 17.2. From (17.4), we compute $\nu = 1.4274$, which is already within 8% of the exact result $4/3$. Thus the real-space RG scheme gives very good approximations to the known results. But beware, it may not always be that simple: the reader is encouraged to devise a similar RG scheme for site percolation on a square lattice.

17.3.2   Monte Carlo RG

The scheme of the last section is approximate since it cannot correctly handle situations like the one in Fig. 17.6. In order to improve on it, we can construct an

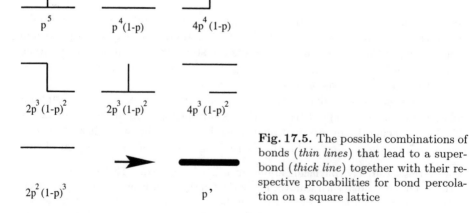

Fig. 17.5. The possible combinations of bonds (*thin lines*) that lead to a super-bond (*thick line*) together with their respective probabilities for bond percolation on a square lattice

RG scheme that uses a larger collection of bonds. The total number of (connected and non-connected) configurations in such a collection of n bonds is 2^n, putting severe bounds on the practicability of the approach for analytic calculations. However, the task is ideally suited for computers. On the web page for this book [33], I include a set of MATHEMATICA routines that compute the real-space RG for a $d = 2$ triangular lattice.

Fig. 17.6. Although the original bonds (*thin lines*) are not connected, the RG procedure outlined in the text nevertheless leads to two connected horizontal super-bonds (*thick lines*)

17.4 The Quantum Hall Effect

In 1980, von Klitzing et al. [32] found that the Hall resistance R_H of MOSFETs at strong magnetic field B exhibits a step-like behavior which is accompanied by a simultaneously vanishing longitudinal resistance R. This is in contrast to the classical Hall effect which gives a linear dependence of R_H on B. Even more surprisingly, the values of R_H at the transitions are given by universal constants, i.e., $(1/i)h/e^2$, where i is an integer.

Since its discovery this so-called integer quantum Hall effect (IQHE) has been extensively studied [34,35]. Besides semi-phenomenological models simply assuming a localization–delocalization transition, more general theories have been considered, e.g., gauge invariance [36], topological quantization [37], scattering [38] and field theoretical approaches [39].

17.4.1 Basics of the IQHE

A simple understanding of the IQHE can be gained by considering the Hamiltonian of a single electron in a magnetic field,

$$H_0 = \frac{1}{2m}\left(p + \frac{e}{c}A\right)^2 = \frac{\hbar\omega_c}{2l_B^2}\left(\xi^2 + \eta^2\right) , \tag{17.6}$$

where A denotes the vector potential and the Hamiltonian has been rewritten in guiding center coordinates $X = x - \zeta$, $Y = y - \eta$ and relative coordinates ζ, η [40]. Here, $\omega_c = eB/m$ is the frequency of the classical cyclotron motion and $l_B = (\hbar/eB)^{1/2}$ is the radius of the cyclotron motion. The spectrum of this Hamiltonian is simply the harmonic oscillator with $E_n = \left(n + \frac{1}{2}\right)\hbar\omega_c$, $n = 0$, 1, ... These Landau levels are infinitely degenerate since the Hamiltonian no longer contains X and Y. Thus the spectrum consists of δ-function peaks as indicated in Fig. 17.7. Introducing disorder into the model, by adding a smooth random potential $V(r)$ in (17.6), results in drift motion of the guiding center

$$\dot{X} = \frac{i}{\hbar}[H, X] = \frac{l_B^2}{\hbar}\frac{\partial V}{\partial y} , \quad \dot{Y} = \frac{i}{\hbar}[H, Y] = -\frac{l_B^2}{\hbar}\frac{\partial V}{\partial x} , \tag{17.7}$$

perpendicular to the gradient of $V(\mathbf{r})$ (see Fig. 17.8). Furthermore, the degeneracy of the Landau levels is lifted, and the δ-function density of states broadens [34], giving rise to a band-like structure as shown in Fig. 17.7. If the sample is penetrated by a *strong* magnetic field, the cyclotron motion is much smaller than the potential fluctuations. Consequently, the electron motion can be separated into cyclotron motion and motion of the guiding center along equipotential lines of the energy landscape.

The IQHE can then be understood as follows: assume that the center of the broadened Landau levels contain extended states that can support transport, whereas the other states are spatially localized and cannot. This is similar to the standard picture in the theory of Anderson localization (see Chaps. 14 and 16). Changing the Fermi energy E_F or the filling factor $\nu_f = 2\pi l_B^2 \rho_e = 2\pi\hbar\rho_e/eB \propto E_F$, where ρ_e denotes the electron density, we first have E_F in the region of localized states and both σ_{xx} and σ_{xy} are 0. When E_F reaches the region of extended states, there is transport, σ_{xx} is finite and $\sigma_{xy} = e^2/h$. Next, E_F again reaches a region of localized states and σ_{xx} drops back to 0 until we reach the extended states in the next Landau level.

This picture suggests the following effective *classical* high-field model [41] of the IQHE: neglecting the cyclotron motion (i.e., large B) and quantum effects (i.e., only one extended state), the classical electron transport through the sample with energy E_F only depends on the 'height' of the saddle points in the potential energy landscape $V(\mathbf{r})$. One obtains a classical bond-percolation problem [5], in which saddle points are mapped onto bonds. A bond is connecting only when the potential of the corresponding saddle point equals the energy of the electron E_F. From percolation theory, it follows [5] that an infinite system is conducting only when $E_F = \langle V \rangle$. Using this model one could already describe the localization–delocalization transition and thus the quantized plateaus in resistivity observed in IQHE [34]. But for bond percolation the correlation length diverges at the transition with an exponent of $\nu = 4/3$, in contrast to the value found in the QH experiments.

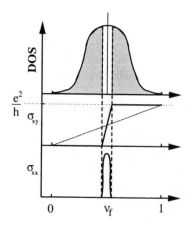

Fig. 17.7. Density of states (DOS), and transversal and longitudinal conductivity as a function of E_F or, equivalently, the filling factor ν_f or B^{-1} [35]. The peak in the *middle* of the band represents one δ-function peak of the clean Landau model. *Dark shaded regions* of the density of states correspond to localized states. The *thin dashed line* (with non-zero slope) for σ_{xy} indicates the classical Hall result

17.4.2   RG for the Chalker–Coddington Network Model

The Chalker–Coddington (CC) network model improved the high-field model by introducing quantum corrections [42], namely tunneling and interference. Tunneling occurs, in a semiclassical view, when electron orbits come close enough to each other for the electron cyclotron motions to overlap. This happens at the saddle points, which now act as quantum scatterers connecting two incoming with two outgoing channels by a scattering matrix, as shown in Fig. 17.8. Similar to bond percolation, a network can be constructed such that the saddle points are mapped onto bonds. While moving along an equipotential line, an electron accumulates a random phase which reflects the disorder of $V(r)$. Results for this quantum percolation also show one extended state in the middle of the Landau band. The critical properties at the transition, especially the value of the exponent $\nu \approx 2.4$ [43], agree with experiment [44,45].

As explained for the bond percolation problem, we now apply the RG method to the CC model. The RG structure which builds the new super-saddle points is displayed in Fig. 17.9. It consists of 5 saddle points drawn as bonds. The links (and phase factors) connecting the saddle points are indicated by arrows pointing in the direction of the electron motion due to the magnetic field. Each saddle point acts as a scatterer connecting the 2 incoming $I_{1,2}$ with the 2 outgoing channels $O_{1,2}$,

$$\begin{pmatrix} O_1 \\ O_2 \end{pmatrix} = \begin{pmatrix} t_i & r_i \\ r_i & -t_i \end{pmatrix} \begin{pmatrix} I_1 \\ I_2 \end{pmatrix} , \tag{17.8}$$

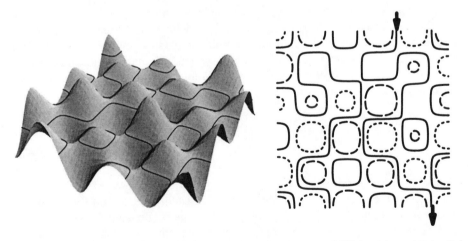

Fig. 17.8. *Left*: schematic plot of a smooth random potential $V(r)$ with equipotential lines at $E = \langle V \rangle$ indicated in *black*. *Right*: equipotential lines of the same potential for $E = \langle V \rangle - E_{\max}/2$, $\langle V \rangle$, and $\langle V \rangle + E_{\max}/2$ corresponding to *long dashed, solid* and *short dashed lines*. Note the *solid line* percolating the system from top to bottom as indicated by the *arrows*

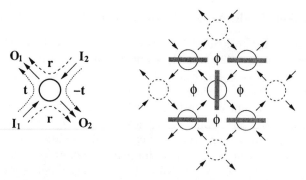

Fig. 17.9. *Left*: a single saddle point (*circle*) connected to incoming and outgoing currents I_i, O_i via transmission and reflection amplitudes t and r. *Right*: a network of 5 saddle points can be renormalized into a single super-saddle point by an RG approach very similar to the bond percolation problem of Sect. 17.3. Phases are denoted schematically by ϕs

with reflection coefficients r_i and transmission coefficients t_i, which are assumed to be real numbers. The complex phase factors enter later via the links between the saddle points. By this definition – including the minus sign – the unitarity constraint $t_i^2 + r_i^2 = 1$ is fulfilled a priori. The amplitude for transmission of the incoming electron to another equipotential line and the amplitude for reflection and thus staying on the same equipotential line add up to unity – electrons do not get lost.

In order to obtain the scattering equation of the super-saddle point, we now need to connect the 5 scattering equations according to Fig. 17.9. For each link the amplitude of the incoming channels is defined by the amplitude of the outgoing channel of the previous saddle point multiplied by the corresponding complex phase factor $e^{i\phi_k}$. This results in a system of 5 matrix equations, which has to be solved. One obtains an RG equation for the transmission coefficient t' of the super-saddle point [46], analogous to (17.5),

$$t' = \frac{t_{15}(r_{234}e^{i\phi_2} - 1) + t_{24}e^{i(\phi_3+\phi_4)}(r_{135}e^{-i\phi_1} - 1) + t_3(t_{25}e^{i\phi_3} + t_{14}e^{i\phi_4})}{(r_3 - r_{24}e^{i\phi_2})(r_3 - r_{15}e^{i\phi_1}) + (t_3 - t_{45}e^{i\phi_4})(t_3 - t_{12}e^{i\phi_3})},$$

$$(17.9)$$

depending on the products $t_{i...j} = t_i \cdot ... \cdot t_j$, $r_{i...j} = r_i \cdot ... \cdot r_j$ of transmission and reflection coefficients t_i and r_i of the $i = 1, \ldots, 5$ saddle points and the 4 random phases ϕ_k accumulated along equipotentials in the original lattice. For further algebraic simplification one can apply a useful transformation of the amplitudes $t_i = (e^{z_i} + 1)^{-1/2}$ and $r_i = (e^{-z_i} + 1)^{-1/2}$ to heights z_i relative to heights V_i of the saddle points. The conductance G is connected to the transmission coefficient t by $G = |t|^2 e^2/h$ [47].

17.4.3 Conductance Distributions at the QH Transition

For the numerical determination of the conductance distribution, we first choose an initial probability distribution P_0 of transmission coefficients t. The distribution is discretized in at least 1000 bins. Thus the bin width is typically $0.001e/\sqrt{h}$ for the interval $t \in [0, e/\sqrt{h}]$.

Using the initial distribution $P_0(t)$, we now randomly select many different transmission coefficients and insert them into the RG equation (17.9). Furthermore, the phases ϕ_j, $j = 1, \ldots, 4$ are also chosen randomly, but according to a uniform distribution $\phi_j \in [0, 2\pi]$. By this method, at least 10^7 super-transmission coefficients t' are calculated and their distribution $P_1(t')$ is stored. Next, P_1 is averaged using a Savitzky–Golay smoothing filter [48] in order to decrease statistical fluctuations. This process is then repeated using P_1 as the new initial distribution.

The iteration process is stopped when the distribution P_i is no longer distinguishable from its predecessor P_{i-1} and we have reached the desired fixed-point (FP) distribution $P_c(t)$. However, due to numerical instabilities, small deviations from symmetry add up in such a way that, typically after 15–20 iterations, the distributions become unstable and converge towards the classical FPs of no transmission or complete transmission, similarly to the classical percolation case. Figure 17.10 shows this behavior for one of the RG iterations. The FP distribution $P_c(G)$ shows a flat minimum around $G = 0.5e^2/h$ and sharp peaks at $G = 0$ and $G = e^2/h$. It is symmetric with $\langle G \rangle = 0.498e^2/h$. This is in agreement with

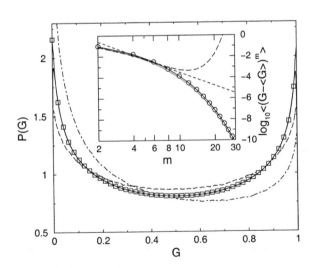

Fig. 17.10. Conductance distribution at a QH plateau-to-plateau transition. *Squares* correspond to the fixed-point distribution, *dashed* and *dot-dashed* lines to the initial distribution and an unstable distribution, respectively. The *solid line* indicates an analytical fit of the FP distribution $P_c(t)$. *Inset*: moments of the FP distribution $P_c(G)$. The *dashed lines* indicate various predictions based on extrapolations of results for small m [49]. The *dotted line* denotes the moments of a constant distribution

previous theoretical [50,51] and experimental [52] results, whereas our results contain far fewer statistical fluctuations. Furthermore we determine moments $\langle (G - \langle G \rangle)^m \rangle$ of the FP distribution $P_c(G)$. As shown in Fig. 17.10 for small moments up to $m = 6$, our results agree with the work of Wang et al. [49], who computed moments $m \leq 8.5$. But more interesting is the fact that the obtained moments of the FP distribution can hardly be distinguished from the moments of a simple constant distribution, thus indicating the influence of the broad flat minimum of the FP distribution around $G = 0.5e^2/h$.

To determine the critical exponent, we next perturb the FP distribution slightly, i.e., we construct a distribution with shifted average G_0. Then we perform an RG iteration and compute the new average G_1 of $P_1(G)$. Tracing the shift of the perturbed average G_n for several initial shifts G_0, we expect to find a linear dependence of G_n on G_0 for each iteration step n. The critical exponent is then related to the slope dG_n/dG_0 [53]. Figure 17.11 shows the resulting ν through its dependence on the iteration step and thus the system size. The curve converges close to $\nu \approx 2.4$, i.e., the value obtained by Lee et al. [43]. Note that the 'system size' is more properly called a system magnification, since we start the RG iteration with an FP distribution valid for an infinite system and then magnify the system in the course of the iteration by a factor 2^n.

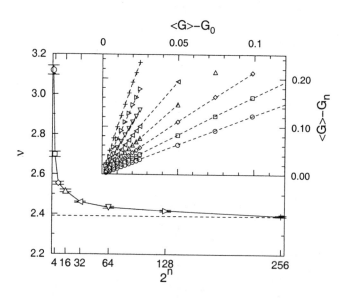

Fig. 17.11. Critical exponent ν as a function of magnification factor 2^n for RG step n. The *dashed line* shows the final result $\nu = 2.39$. *Inset:* the shift of the average G_n of $P(G)$ is linear in G_0. The *dashed lines* indicate linear fits to the data

17.5 Summary and Conclusions

The percolation model represents perhaps the simplest example of a system exhibiting complex behavior, although its constituents – the sites and bonds – are chosen completely uncorrelated. Of course, the complexity enters through the connectivity requirement for percolating clusters. I have reviewed several numerical algorithms for quantitatively measuring various aspects of the percolation problem. The specific choice reflects my purely personal preferences and I am happy to note that other algorithms such as breadth- and depth-first algorithms [54] have been introduced by P. Grassberger in Chap. 11.

The real-space RG provides an instructive use of the underlying self-similarity of the percolation model at the transition. Furthermore, it can be used to study very large effective system sizes. This is needed in many applications. As an example, I briefly reviewed and studied the QH transition and computed conductance distributions, moments and the critical exponent. These results can be compared with experimental measurements and shown to be in quite good agreement.

Acknowledgements. The author would like to thank Phillip Cain, Ralf Hambach, Mikhail E. Raikh, and Andreas Rösler for many helpful discussions. This work was supported by the NSF-DAAD collaborative research grant INT-9815194, the DFG within SFB 393 and the DFG-Schwerpunktprogramm 'Quanten-Hall-Systeme'.

References

1. We note that the WEH-Ferienkurs, following which this article was prepared, took place during *early* fall.
2. S.R. Broadbent, J.M. Hammersley: Proc. Camb. Philos. Soc. **53**, 629 (1957)
3. J.M. Hammersley: Proc. Camb. Philos. Soc. **53**, 642 (1957)
4. J.M. Hammersley: Ann. Math. Statist. **28**, 790 (1957)
5. D. Stauffer, A. Aharony: *Perkolationstheorie* (VCH, Weinheim 1995)
6. K. Binder: Rep. Prog. Phys. **60**, 487 (1997)
7. G. Grimmett: *Percolation* (Springer, Berlin 1989)
8. A. Bunde, S. Havlin (Eds.): *Percolation and Disordered Systems: Theory and Applications* (North-Holland, Amsterdam 1999)
9. J. Voit: *The Statistical Mechanics of Capital Markets* (Springer, Heidelberg 2001)
10. J. Goldenberg, B. Libai, S. Solomon, N. Jan, D. Stauffer: *Marketing Percolation* (2000). Cond-mat/9905426
11. http://de.arXiv.org/, 1993–2000
12. J. Hoshen, R. Kopelman: Phys. Rev. B **14**, 3438 (1976)
13. C. Adami: (1997), http://www.krl.caltech.edu/~adami/CD1/Percolation/percolation.html, likely to change without prior notice
14. P. Leath: Phys. Rev. B **14**, 5056 (1976)
15. W. Kinzel, G. Reents: *Physics by Computer* (Springer, Berlin 1998)

16. W. Kinzel, G. Reents: (1999),
    http://wptx15.physik.uni-wuerzburg.de/TP3/applet_java/percgr.html,
    likely to change without prior notice
17. R.F. Voss: J. Phys. A: Math. Gen. **17**(7), L373 (1984)
18. R.M. Ziff, P.T. Cummings, G. Stell: J. Phys. A Math. Gen. **17**, 3009 (1984)
19. M. Rosso, J.F. Gouyet, B. Sapoval: Phys. Rev. B **32**, 6053 (1985)
20. R.M. Ziff, B. Sapoval: J. Phys. A Math. Gen. **19**, L1169 (1986)
21. P.N. Suding, R.M. Ziff: Phys. Rev. E **60**, 275 (1999)
22. M.F. Sykes, J.W. Essam: Phys. Rev. Lett. **10**, 3 (1963)
23. R.M. Ziff, P.N. Suding: J. Phys. A Math. Gen. **30**, 5351 (1997)
24. C.D. Lorenz, R.M. Ziff: Phys. Rev. B **57**, 230 (1998)
25. P. Kleban, R.M. Ziff: Phys. Rev. B **57**, R8075 (1998)
26. M.E.J. Newman, R.M. Ziff: Phys. Rev. Lett. **85**, 4104 (2000). Cond-mat/0005264
27. A.B. Harris, T.C. Lubensky, W.K. Holcomb, C. Dasgupta: Phys. Rev. Lett. **35**, 327 (1975)
28. P.J. Reynolds, W. Klein, H.E. Stanley: J. Phys. C Solid State Phys. **10**, L167 (1977)
29. J. Bernasconi: Phys. Rev. B **18**, 2185 (1978)
30. P.J. Reynolds, H.E. Stanley, W. Klein: Phys. Rev. B **21**, 1223 (1980)
31. P.D. Eschbach, D. Stauffer, H. Herrmann: Phys. Rev. B **23**, 422 (1981)
32. K. von Klitzing, G. Dorda, M. Pepper: Phys. Rev. Lett. **45**, 494 (1980)
33. http://www.tu-chemnitz.de/physik/HERAEUS/2000/Springer.html
34. M. Janssen, O. Viewweger, U. Fastenrath, J. Hajdu: *Introduction to the Theory of the Integer Quantum Hall effect* (VCH, Weinheim 1994)
35. T. Chakraborty, P. Pietiläinen: *The Quantum Hall Effects* (Springer, Berlin 1995)
36. R.B. Laughlin: Phys. Rev. B **23**, 5632 (1981)
37. D.J. Thouless, M. Kohmoto, M.P. Nightingale, M. den Nijs: Phys. Rev. Lett. **49**, 405 (1982)
38. R.E. Prange: Phys. Rev. B **23**, 4802 (1981)
39. A.M.M. Pruisken: Nucl. Phys. B **235**, 277 (1984)
40. D.L. Landau: Z. Phys. **64**, 629 (1930)
41. S.V. Iordanskii: Solid State Commun. **43**, 1 (1982)
42. J.T. Chalker, P.D. Coddington: J. Phys. Condens. Matter **21**, 2665 (1988)
43. D.H. Lee, Z. Wang, S. Kivelson: Phys. Rev. Lett. **70**, 4130 (1993)
44. S. Koch, R.J. Haug, K. von Klitzing, K. Ploog: Phys. Rev. B **43**, 6828 (1991)
45. R.T.F. van Schaijk, A. de Visser, S.M. Olsthoorn, H.P. Wei, A.M.M. Pruisken: Phys. Rev. Lett. **84**, 1567 (2000)
46. A.G. Galstyan, M.E. Raikh: Phys. Rev. B **56**, 1422 (1997)
47. M. Büttiker, Y. Imry, R. Landauer, S. Pinhas: Phys. Rev. B **31**, 6207 (1985)
48. W.H. Press, B.P. Flannery, S.A. Teukolsky, W.T. Vetterling: *Numerical Recipes in FORTRAN*, 2nd edn. (Cambridge University Press, Cambridge 1992)
49. Z. Wang, B. Jovanovic, D.H. Lee: Phys. Rev. Lett. **77**, 4426 (1996)
50. A. Weymer, M. Janssen: Ann. Phys. (Leipzig) **7**, 159 (1998). Cond-mat/9805063
51. Y. Avishai, Y. Band, D. Brown: Phys. Rev. B **60**, 8992 (1999)
52. D.H. Cobden, E. Kogan: Phys. Rev. B **54**, R17 316 (1996)
53. P. Cain, R.A. Römer, M.E. Raikh, M. Schreiber: *Integer Quantum Hall Transition in the Presence of a Quenched Disorder*, submitted to Phys. Rev. B (2001). Cond-mat/0104045
54. A. Aho, J.E. Hopcroft, J.D. Ullman: *Data Structures and Algorithms* (Addison–Wesley, New York 1983)

Index

Printing: Mercedes-Druck, Berlin
Binding: Stürtz AG, Würzburg